BROKEN PROMISES

Popular Protest, Indian Nationalism and the Congress Party in Bihar, 1935–1946

BROKEN PROMISES

Popular Protest, Indian Nationalism and the Congress Party in Bihar, 1935–1946

VINITA DAMODARAN

DELHI
OXFORD UNIVERSITY PRESS
OXFORD NEW YORK
1992

Oxford University Press, Walton Street, Oxford OX2 6DP

New York Toronto
Delhi Bombay Calcutta Madras Karachi
Kuala Lumpur Singapore Hong Kong Tokyo
Nairobi Dar es Salaam
Melbourne Auckland
and associates in
Berlin Ibadan

Computerset by Rastrixi, New Delhi 110030
Printed in India at Rekha Printers Pvt. Ltd., New Delhi 110020
and published by S.K. Mookerjee, Oxford University Press
YMCA Library Building, Jai Singh Road, New Delhi 110001

To
my parents
Coroth Damodaran
and
Edavalath Seetha

Contents

LIST OF MAPS

LIST OF TABLES

LIST OF FIGURES

LIST OF ENCLOSURES

Preface

This study is concerned with popular protest, Congress strategies of action and the national struggle in one province of India, Bihar, from 1935 to 1946. The period 1937-9 was a very eventful phase in the history of Indian politics. A contemporary observer surveying the political scene in 1937 in India would have found a convergence of views between the imperial policy makers and the nationalist intelligentsia. Almost for the first time since the constitutional experiments began the Congress was poised to accept and work the 1935 Government of India Act along the lines visualized by the imperial policy makers. While there have been several studies on Indian nationalism that include the period there has been relatively little work carried out on the specifics of the Congress acceptance of office. In fact, there has been no substantive study of the Bihar ministries and it is this interpretative gap that my research work has attempted to fill.

The ministry phase was particularly interesting for, in the course of it, the Congress party changed from being a movement of opposition to being a party in power. Mass aspirations fuelled by the Congress acceptance of office resulted in a large scale *bakasht* agitation in several districts in rural Bihar. The implications of this large scale rural protest for Gandhian strategies of action and the overall nationalist programme are important for they allow us to understand the constitutive elements of the new national state that was constructed in Bihar in 1937 and 1946. The example of Bihar is also important in a comparative sense for it enables us to build up a much broader analytical picture of the relationship between colonial rule, peasant action, the erosion of common property resources and the nature of the Congress response to rural grievances and to low-level rebellion.

A section of chapter six of this book has appeared in a different form in *Studies in History* 6, 1, 1990, while a part of chapter five is due to appear in *Modern Asian Studies* in 1992.

In the course of completing this book I received help from several sources. My early interest in the social and political history of Bihar was nurtured by Professor Bipan Chandra, my tutor at Jawaharlal Nehru University. It was due to the generosity of the Commonwealth

Fund that I was able to pursue these studies further, while based at Cambridge. During my research at Cambridge I received considerable guidance from my supervisor, Professor Anthony Low, for which I am very grateful. I am also particularly indebted to Dr John Iliffe and Dr Peter Robb who were kind enough to read my manuscript and make suggestions. I also thank Dr David Arnold and Dr Chris Bayly who read parts of the manuscript. Over the years I have greatly benefited from discussions with friends and teachers at the Jawaharlal Nehru University. I would like to thank Dr S. Gopal, Dr Madhavan Palat, Dr Sabyasachi Bhattacharya, Dr Majid Siddiqui, Mridula Mukherjee, Aditya Mukherjee, Bhagwan Singh Josh, Vishalakshi Menon, Gurpreet Mahajan, Sucheta Mahajan, Anthony Thomas, Madhu Suri, Mannika Chopra and Sabita Singh. I have also been fortunate to receive assistance from Dr Lionel Carter, Dr Ejaz Huq, Mrs Shiela Sinha, Mr Uday Raj Sinha, Dr Maya Unnithan Kumar and Adrian Walford. Richard Grove, in particular, was a constant source of encouragement.

I would like to thank the staff of the following institutions in which I have consulted archival matter: the Cambridge University Library, the Centre for South Asian Studies at Cambridge, The India Office Library in London, Nehru Memorial Museum and Library, (New Delhi), The National Archives of India, (New Delhi), the State Central Record Office, Patna, and the Central Intelligence Department Library in Patna.

I also received financial support from the Smuts fund, the Charles Wallace India Trust, Sidney Sussex College and the Indian Council of Historical Research. This made possible two extended field trips to India.

The book was completed in the congenial environment of St Edmund's College. I would like to thank the master and fellows of the College for granting me a fellowship that enabled me to finish the work. I would also like to thank Oxford University Press for producing this book quicker than anticipated. Finally, I would like to gratefully acknowledge the support of my family, Sunil Damodar and Deepa Damodar and their children, Trishna and Rohit who provided a welcome distraction from the pressures of work on my field trips to India. My parents, Mr K.K. Damodaran and Mrs Seetha Damodaran, in particular, have consistently supported my academic interests. To them my debt is immeasurable.

St Edmund's College, Cambridge
December 1991

Abbreviations

AICC	All India Congress Committee
BSA	Bihar State Archives
CID	Criminal Investigation Department
CSP	Congress Socialist Party
CWMG	Collected Works of Mahatma Gandhi
DCC	District Congress Committee
FMBP	Freedom Movement in Bihar Papers.
FR	Fortnightly Report for Bihar
INA	Indian National Army
IOL	India Office Library
NAI	National Archives of India
PCC	Provincial Congress Committee
Rs	Rupees
SWJN	Selected Works of Jawaharlal Nehru
TCC	Town Congress Committee
TOP	Transfer of Power volumes
UP	United Provinces

Introduction

This book sets out to examine the nature of the nationalist movement in one major Indian province, Bihar, during the 1930s and 1940s. The choice of Bihar rests simply on the assumption that in many ways it was in the vanguard of popular nationalism. Concentrating on an eleven-year period, the work examines the changing relationship between the Congress and its mass base in the penultimate phase of the Raj. This phase of the national movement was particularly important as, in the course of it, the Congress underwent the transition from being a movement of opposition to being a party in power. Twice within the period, in 1937-9 and again in 1946, the Congress accepted office. This had far-reaching implications for the nature of the organization and its relationship with its mass constituency. Soon after the Congress assumed office in 1937 a large scale peasant agitation emerged which aimed to persuade the Congress to abide by its promises to the people and restore peasant rights eroded by landlord attacks and a growing resource crisis. The study examines the critical ecological transition that fuelled these agitations. Following the withdrawal from office in 1939, the pressures on the Congress shifted dramatically. This process was to culminate in the Quit India movement in 1942. Immediately after this mass upsurge, the Congress retreated underground and became a more amorphous organization allied to dacoit gangs in the countryside. The activities of these underground organizations, known as the Azad Dastas, have never previously been described and provide an instructive picture of the nature of popular nationalism. It was only in 1946 that the right wing in the party was able to regain the political initiative as the organization was restructured to resume power, this time in preparation for total independence. This in effect marked the victory of the right wing in the Bihar Congress and meant that the organization was purged of its more radical and populist elements. A new nation state was being formed.

A HISTORIOGRAPHICAL RÉSUMÉ

The present study has developed out of a dissatisfaction with the existing historical analyses of the Indian national movement. Indian historiography in the 1970s moved away from studies of national level politics that had hitherto dominated it to examine the political dynamics operative at the level of locality and province. The old Cambridge school was the first to challenge the notion of a unified national movement which had characterized earlier historical writing right through from R.C. Majumdar to R.P. Dutt.[1] It emphasized instead the areas of local power in which local social conflicts were played out under the banner of nationalism. However, a tendency also developed among historians of this school to concentrate on personalities and cliques.[2] Their argument was further developed and extended into an understanding of twentieth-century Congress politics by historians such as Judith Brown and B.R. Tomlinson.[3] The polarization of the Congress party along ideological lines was ascribed to factional in-fighting, while real issues of power and ideology were glossed over. At another level there was a complete neglect of the real nature of popular movements and the reasons for the participation of particular sections of the peasantry and urban poor in nationalist agitations. The patronage of influential men was seen as sufficient to draw large sections of people into the movement through no volition of their own, while the nature and importance of popular participation and consciousness were ignored. The interpretation offered by the Cambridge school was essentially a neo-Namierite explanation of Indian history. It made no rigorous attempt to study or understand nationalism on the ground or at the grassroots.

An alternative view has developed among a group of Marxist historians, in the tradition of Bipan Chandra and P.C. Joshi. This view recognizes a variety of levels of participation in the national movement. These historians have emphasized the 'bourgeois-dominated'

[1] R.C. Majumdar, *History of the Freedom Movement in India*, vol. 1, Calcutta, 1962; R.P. Dutt, *India Today*, London, 1940.

[2] Historians such as Anil Seal, John Gallagher and Gordon Johnson argued that local politics in nineteenth-century India were organized by factions into systems apparently insulated from each other. 'These factions', they asserted, 'were the stuff of most Indian politics . . . on the unsteady base of local squabbles for spoils rested the larger political systems of India.' A. Seal, 'Imperialism and Nationalism in India' in J. Gallagher, *et al.* (eds.), *Locality, Province and Nation, Essays on Indian Politics 1840-1940*, London, 1973, p.5.

[3] B.R. Tomlinson's *The Indian National Congress and the Raj, 1929-1942*, London, 1976, falls back on factional arguments and patron-client linkages to explain popular agitation. See also Judith Brown, *Gandhi's Rise to Power in Indian Politics 1915-1922*, London, 1972.

nature of the nationalist struggle and thus followed in the footsteps of R.P. Dutt. However, when it came to understanding the links between the people and the bourgeoisie they fell back on well-worn notions of the primacy of elite mobilization.[4] More recently Bipan Chandra, while continuing to recognize the middle-class nature of the Congress leadership, has shied away from critically assessing the nature of Congress policies in government. Thus his analysis of the national movement has been limited by his inability to view the Congress in this period as anything other than a truly progressive organization.[5] Another view, which also recognized the contribution of specific social groups to the national movement, was that put forward in a collection of articles edited by D.A. Low. These emphasized the role of 'dominant peasants' in nationalist agitations. Such groups were identified as closest to the Congress in the countryside and the main force behind rural agitations. However, it can be questioned whether or not these groups were actually the prime movers in specific agitations.[6]

In more recent times the 'Subaltern' project, pioneered by Ranajit Guha, has thrown a quite different light upon the history of nationalism. The *Subaltern Studies* volumes have been concerned with seeking a way out of what was characterized as the 'elitism of modern Indian historiography'. The subalternists argue, quite simply, that the politics of the people is left out of this historiography. Based on Gramscian notions of the 'subaltern', the term is used to denote an entire people as subordinate in terms of class, caste and gender. The elite/subaltern dichotomy does have certain analytical uses. It is not so much the categorical rigidity of these concepts that is important. Instead, one needs to judge their value by exploring the possible uses they can be put to in explanation. Many of the contributors to *Subaltern Studies*, notably Shahid Amin and Gyan Pandey, have perceptively explored popular consciousness and the nature of mass actions.[7]

[4] P.C. Joshi, 'Development Perspectives in India: Some Reflections on Nehru and Gandhi' in B.R. Nanda and V.C. Joshi (eds.), *Studies in Modern Indian History*, New Delhi, 1972, pp. 137-8; Bipan Chandra, *Nationalism and Colonialism in India*, Delhi, 1979.

[5] See Bipan Chandra, *et al.* (eds.), *India's Struggle for Independence*, Delhi, 1989.

[6] See D.A. Low, *The Congress and the Raj: Facets of the Indian Struggle 1917-47*, London, 1977.

[7] See Shahid Amin, 'Gandhi as Mahatma', *Subaltern Studies*, vol. 3 and Gyan

Nonetheless, their approach has several shortcomings. My main difficulty with the project is that it steers clear of areas of subaltern/elite interaction. Indeed, as has been recently argued, some of the contributors take the argument of subaltern autonomy quite literally and use it as a rigid category rather than conceptualizing the subaltern/elite dichotomy as a relationship of power. Thus the notion of subaltern autonomy is taken to extreme lengths by some of its contributors, notably by Henningham.[8] According to his account, the subaltern groups seem to generate their own traditions and pursue their political projects quite independently of the Indian National Congress. However, towards the end of his article he is forced to concede that what brought the 'elite' and 'subaltern' groupings together was the shared ideology of nationalism.

The subalternist preoccupation with an autonomous domain of low-level action has led it to neglect the nature of the state, forms of intervention or even to explore the Gramscian notion of hegemony in any meaningful way.[9] Instead, the nature of the state and its relations to the prevailing political and economic forces need, I feel, to be examined far more intensively to explain the background and timing of many popular actions. Indeed, in this context, it can be demonstrated that elite ideologies played a critical role in limiting many popular actions, especially in the 1930s and 40s. Any study which does not examine these problems will thus tend to be incom-

---

Pandey's 'Rallying around the Cow, Sectarian Unrest in Bhojpur', *Subaltern Studies*, vol. 2.

[8] Henningham, for example, argues that there were two streams in the Quit India agitation in Bihar, quite independent of each other: one of rich peasants and landlords and another of poor low-caste peasants. Certainly it is possible to distinguish different levels of participation and motivation in the movement. However, the social breadth of the insurrection, subsuming an ideology of nationalism as well as a popular consensus based on a shared assumption of the need to challenge the colonial state, makes Henningham's identification of two independent streams in the movement quite untenable. S. Henningham, 'The Quit India Agitation in Bihar and the Eastern United Provinces: The Dual Revolt' in *Subaltern Studies*, vol. 2, pp. 130-77.

[9] Gramsci is careful when he ascribes autonomy to subaltern activity, arguing that 'Subaltern groups are always subject to the activity of ruling groups even when they rebel and rise up : only permanent victory breaks their subordination, and that not immediately. In reality even when they appear triumphant the subaltern groups are merely anxious to defend themselves. . . Every trace of independent activity of the subaltern should be of incalculable value for the historian'. See A. Gramsci, *Prison Notebooks*, London, 1971, p. 55.

plete and will give one only a partial understanding of the nature of popular protest.

The emphasis in the subaltern project on the existence of a unified, unchanging peasantry also needs to be questioned. Guha claims to restore the peasant to his own history, arguing that the peasant does not exist as a subject in the discourse of radical historiography, having been replaced by an abstraction called the 'worker' or the 'peasant', and being thereby subsumed in a classificatory mode that conceals the real historical personality of the 'insurgent'.[10] This is a valid criticism. However, in the very course of his critique, Guha goes on to fall into the same terminological trap in his own characterization of the 'insurgent', a term which is again little more than an abstraction. Guha's idealism is most apparent in his analyses of the social and economic conditions which generate rebellion. According to him, 'factors of economic and political deprivation do not at all relate to the peasant's consciousness or do so only negatively'. Thus, in one article, Guha emphasizes the religious consciousness of the Santhal insurgent, arguing that, in the case of a Santhal insurrection, it is not possible to speak of insurgency except as a form of religious consciousness. However, this 'religious consciousness' of the peasantry is not subjected to any determination and he boldly assumes that the peasantry possesses an ideal form of pure 'peasant' consciousness.[11] This is an argument that is difficult to sustain because the religious consciousness of the Santhal insurgent in the 1850s, as of the Mundas in the 1890s, developed from a perception of their surbordination and oppression at the hands of an outsider Hindu landlord class. Indeed, the religious aspect of rebel consciousness may have been emphasized by British officials (upon whose discourses Guha bases his evidence), to the exclusion of the social, economic or political content of that

[10] This view has been echoed more recently by Hobsbawm and serves to make one aware of the pitfalls of economic determinism. See E.J. Hobsbawm, 'Some Reflections on History from Below' in F. Krantz (ed.), *History from Below, Studies in Popular Protest and Popular Ideology in Honour of George Rude*, Montreal, 1985, p. 65.

[11] However it should be noted that Guha is not prepared to extend this ideal consciousness to the later colonial period, by which time, as he points out, 'the politics of nationalism and socialism had penetrated the countryside on a large scale', preferring instead to concentrate on early peasant movements. Other historians of the Subaltern group, however, such as Henningham extend this autonomous world of the peasant to the 1920s and 1930s, apparently quite uncritically.

consciousness, to construct an image of the inherent irrationality of the peasantry.

The other subaltern historians have also tended to underplay the impact of differentiation among the peasantry, while the nature of tenurial relations figures only marginally in their analyses. Indeed, in a peasantry existing in a class-divided society it is difficult to believe that peasant consciousness exists entirely independently of existing social divisions and ideological interventions. This apparently uncritical acceptance by the subaltern historians of a 'true peasant consciousness' has led many of the subaltern contributors to fall short of a more thorough understanding of the working of different levels of popular consciousness and of the frequently 'contradictory' nature of such consciousness and action.[12] This is particularly true of their inability to explain the dramatic rise of sectarian or communal consciousness among the peasantry.[13]

The terminological problem remains, nevertheless, a recurrent one. In this respect one may do well to heed the warnings recently offered by Polly Hill. According to this line of thinking the careless use of broad generic terms, particularly in the substitution of the atavistic 'peasant' for any other term, may simply betray an unwillingness to be rigorously thorough, local or specific in anthropological or historical investigation. Where real economic decisions, (as distinct from academic theorizing), have been based on the loose use of terms as a substitute for social knowledge, the social consequences have often been, as Hill has shown, predictably disastrous.[14] The danger, then,

[12] Gramsci uses the term contradictory consciousness, one that has certain analytical uses to identify the nature of the perceptions of subaltern groups. 'His (the worker's) theoretical consciousness can indeed be historically in opposition to his activity. One might almost say that he has two theoretical consciousnesses (or one contradictory consciousness): one which is implicit in his activity and which in reality unites him with all his fellow workers in the practical transformation of the real world; and one superficially explicit or verbal, which he has inherited from the past and uncritically absorbed.' See Gramsci, *Prison Notebooks*, p. 333.

[13] Gyan Pandey is one of the few subaltern historians to actually grapple with this problem. However, even he is unable to explain the communal riots of the 1930s and 40s and ascribes them to the effects of elite propaganda at the national level in the period. In doing so he contradicts himself for he calls into question the notion of subaltern autonomy.

[14] P. Hill, *Development Economics on Trial: The Anthropological Case for a Prosecution*, Cambridge, 1987, pp. 8-15.

appears to lie both in simplistic terminology and in allowing the narratives and actions of pre-determined social groups to become academically sacrosanct. Ultimately, then, the greatest weakness of the subaltern project may be that it tends, as Bayly has recently suggested, 'to frustrate the writing of a rounded history, as did elitism'.[15]

THE PRESENT STUDY

The study has emerged from an understanding of the shortcomings within existing historical analyses of the interactions between 'elite' and 'popular' groups in the context of social change and rebellion. It aims, instead, to recreate a more 'rounded history' by reconstructing the world of elite politics and by examining the impact of the state, institutions and economic and social change on the peasantry and other subordinate rural classes. In this way one may begin to understand the historical processes that have formed and informed popular consciousness and actions. The work also seeks to examine the evolution of the economic and social conditions in which these actions were generated.[16]

In 1937 the Congress party had accepted office in seven out of eleven provinces. It thus gained control of state institutions, at least at the provincial level. This development had long-term consequences both for the colonial state, in weakening its power and authority in key areas and eroding its legitimacy, and also for the Congress organization which began to exhibit serious internal strains as a result of the transition to power. At the level of imperial authority the latter years of the 1930s saw a weakening of imperial power and a withdrawal of resources. At the level of the national struggle, the popular heady nationalism of the early decades gave way in 1937-9 to a more pragmatic approach which revealed the growing distance between the ideological content of the nationalist discourse and the real politics which nationalism actually conducted.

The 1930s was the period when the nationalist struggle was at its height. Colonial hegemony had been seriously eroded and the Indian

[15] C.A. Bayly, 'Rallying around the Subaltern', *Journal of Peasant Studies* vol. 16, no. 1, October 1988, p. 115.

[16] As Levi-Strauss has pointed out, popular consciousness is generally constituted as a result of the interaction of several institutions and structures, rather than of one. C. Levi-Strauss, *The Savage Mind*, London, 1966, p. 155.

National Congress had successfully built up its own mass base and legitimized its role as the foremost anti-imperialist organization in the country. Its own hegemonic ideology had elevated the issue of 'nationalist unity' into an icon. Pitting itself against the reality of a colonial rule which appeared before it as a semi-permanent historical truth, nationalism sought to open out the feasibility of an entirely new political dispensation.[17] These were political claims that were formulated, propagated and defended in the battlefield of politics, and were claims which the colonial state continuously denied. Subtly, as Bernard Cohn has shown, a new nationalist discourse, and a popular nationalist idiom, organized around *prabhat pheries*, Gandhi's *bhajans* and the *charkhas* (as a symbol of the new age), had asserted themselves through the spread of the nationalist message.[18] These were to provide the cultural basis of the idea of a unified nation state. This task of the Congress was facilitated by its political and ideological programme through which the largest possible alliance had been built up between different classes, and under which the lower and subordinate classes had been mobilized.

By the early 1930s the Congress meant many different things to many different people. Much work on the 'decoding' of the meaning and interpretation of the Gandhian message at the local level, and on the disparate movements that erupted under the umbrella of the national movement has been undertaken by the historians of the subaltern school.[19] As a movement of opposition, the Congress was able to incorporate a set of very disparate struggles into the national movement, and then appropriate a rich variety of motivations to add to the body of the same discourse.

However, this integrative characteristic of the Congress was to be eroded with its acceptance of office in 1937, as the political situation and the pressures on the Congress shifted dramatically. The populist discourse created through mass struggles of opposition against imperialism, which had helped to weld classes and regions together, proved inappropriate or inadequate once the Congress replaced the Raj at the provincial level. As a party in power, the Congress was

[17] On the invention of the nation state and the construction of a national identity, see B. Anderson, *Imagined Communities: Reflection on the Origin and Spread of Nationalism*, London, 1983.

[18] B. Cohn, 'Representing Victorian Authority' in T. Ranger and E.J. Hobsbawm (eds.), *The Invention of Tradition*, Cambridge, 1983, p. 209.

[19] The research carried out by Shahid Amin stands out in this respect.

forced to fall back on the institutions of its imperial predecessors and often forced to reproduce the structures of imperial authority. In fact, the institutions of the courts and the police under Congress government in Bihar worked almost entirely in favour of the rural and urban elites. This development was not necessarily inevitable. A period of transition was underway, with several ideologies and premises battling for supremacy.[20] The Congress itself, freshly emerged from the civil disobedience movement, incorporated a variety of different ideas and tendencies ranging right through from the far left to the far right. In the early years in office, however, the struggle to gain control of the Congress organization escalated and a new balance of power between the different tendencies asserted itself. The somewhat unequal struggle waged between the CSP/Kisan Sabha and the Congress in Bihar over the capturing of Congress committees epitomizes this. Once the Congress was in office the nationalist discourse was ruthlessly adapted, and even perfected, to a discourse of order and the rational organization of power.

The implications of these developments have not been sufficiently examined by historians. Many of the standard histories concentrate either on the period of the civil disobedience or on the climactic years of the war and its aftermath.[21] The critical period during which the Congress became a political party (and at two periods, temporarily, a government) rather than a movement, the implications of this change at the level of institutions, and the dynamics of the transition have been largely neglected. Certainly, Gyan Pandey's work on the civil disobedience movement in the United Provinces represents a sound empirical approach to the nature of popular nationalism in a province in the early 1930s and to the imperfect mobilization programme of the Congress. Pandey, however, ends his analysis in 1934, concluding

[20] As Brenner has pointed out, if in the mode of production, the economic is the determinant in the last instance, in the mode of transition, it is the political practice of struggle that is the determinant. See R. Brenner, 'Agrarian Class Structure and Economic Development in Pre-industrial Europe', *Past and Present*, 70, (February 1976), p. 52.

[21] Apart from B.R. Tomlinson's work which is on the political groupings and factionalism of Congress politics and Rani Shankardass' study on Bombay, no study exists which encompasses the problems cited above. See R. Shankardass, *The Congress Raj: Provincial Autonomy in Bombay*, Delhi, 1982, and B.R. Tomlinson, *The Indian National Congress and the Raj 1929-1942*. By concentrating on the politics of the national and provincial elites, Tomlinson fails completely to explore the terrain of mass politics and popular actions.

that 'as the organization and strength of the Congress increased and the scale of the campaigns widened, the social depth of the movement actually diminished'. I would tend to argue that this weakening of the social commitment of the Congress and the concomitant strains in its organization date from a considerably later time, during the period of office acceptance, and from a point when the Congress was forced to spell out its policy in the agrarian and labour fields, firmly identifying itself with specific interest-groups. These were the small *zamindar* and rich peasant conclave in the countryside, and the petty and big business communities in urban areas.

Popular aspirations generated by the advent of the Congress in office resulted, on frequent occasions, in large scale urban and rural rioting for better wages among the working classes and for the return of *bakasht* lands among the peasantry. In the rural areas a large scale peasant movement fuelled by the erosion of common property rights, growing ecological marginalization, and the effects of the Depression erupted in several districts in the province. Increasingly, the Congress dissociated itself from these popular movements, first by condemning the actions as resulting from the machinations of the Congress socialists, and then by actively suppressing them with all the instruments of state repression at its disposal. In power, and given the now strong possibility of replacing the Raj, the Congress no longer needed to incorporate all movements of protest into the nationalist struggle and was beginning to distinguish between what were 'permitted' struggles and those that were outside the ambit of the national movement. As a result, the hegemonic ideology of the Congress came to be increasingly challenged by contending ideologies, especially by those of the left. It may be argued that no hegemony can be so penetrative and pervasive as to negate resistance against it. In the case of the Congress ministry it can be noted that, once in power, the Congress ceased to wield the moral and hegemonic influence it had wielded as a movement encompassing all classes and successfully appropriating all (or most) struggles into the body of the same nationalist discourse.[22] The class alliances of the Congress had, to some extent, also crystallized clearly and it had become more closely linked to dominant groups in the society. In the context of increasing peasant grievances this

[22] An important exception relates to the largely untold, but frequently very violent, history of popular rebellion against forest laws and departments, where the Congress had relatively little influence.

development was not to go unchallenged. The political articulation of the Congress Socialist party and the Kisan Sabha gave force to these movements of protest in the countryside and directly attacked the legitimacy of the Congress as the foremost anti-imperialist and progressive organization. To counter this challenge the Congress party in power was forced to use considerable violence and coercion against the left-wing groups and peasant activists in the rural areas in Bihar. At another level, it attempted to regain legitimacy through legislative interventions which were aimed at a merely cosmetic redressal of grievances. Thus the agrarian legislation in Bihar fell far short of popular expectations. Briefly, it clearly favoured the small zamindar. Similarly, the Congress/Zamindar agreements of 1937 and 1938 easily secured the support of the wealthier social groups to the Congress, albeit at the cost of a few compromises to save face among the richer sections of the peasantry in terms of bakasht concessions. On the other hand, the interests of large sections of the poor peasantry and the landless agricultural labourers went by the board. The political picture in the agrarian arena was repeated in Congress activities in the labour field. Here the Congress favoured compromise and arbitration, making labour strikes as a means of protest difficult, and thereby easily gaining the support of the industrialists and capitalist classes.[23]

The nationalist discourse subtly changed as it became part of the justification for social order and political power. As the Congress set about curbing the Kisan Sabha agitation in Bihar, and expelling the socialists, a clear distancing emerged between the content of the nationalist discourse (which still operated along the lines of the Congress being a supra-party and the 'principal mediator between classes') and the kind of politics it was actually conducting. The speeches of Nehru, Prasad and Patel were degenerating into mere posturing and rhetoric.[24]

[23] See V. Damodaran, 'Some Aspects of the Congress Ministry in Bihar, 1937-39' (unpublished M. Phil. dissertation, J.N.U., 1985).

[24] Gramsci has noted in this context that, 'Although every party is the expression of a social group, and of one social group only, nevertheless in certain given conditions certain parties represent a single social group precisely in so far as they exercise a balancing and arbitrating function between the interests of their group and those of other groups, and succeed in securing the development of the group they represent with the consent and assistance of the allied group—if not out and out with that of groups which are definitely hostile'. The Congress may be said to have operated in such a manner once its class base became formally structured. See A. Gramsci, *Prison Notebooks*, p. 148.

At the level of popular perceptions the challenge to Congress legitimacy was short-lived. No alternative to the Congress programme for a political and social order was effectively asserted by the Left, however much it was discussed. Why not, one may ask? Some tentative answers can be suggested. The socialists attempted to lead and direct the movement of opposition against Congress activities in the legislatures and were partially successful, as can be seen from the scale and depth of the Kisan Sabha agitation in Bihar in 1937-9. However, they also continued to work within the Congress organization. Though there was a growing divide between the official Congress leadership and the CSP there was no rupture of relations in this period. Instead, the CSP propagandized and agitated among the people for the entry of the right kind of Congressmen into the Congress, and for those who would be sympathetic to the demands of the people, rather than attempting to attack the legitimacy of the Congress organization itself.[25] As such, in the popular mind, the Congress retained its image as an organization of the people, for in the popular perception the Kisan Sabha took on the mantle of the Congress.

The Congress was thus able to survive the first phase of office acceptance with its image virtually intact, and by 1942, the Quit India movement had whipped up popular passions to new heights. The imprisonment of the Congress leadership followed. This was the period when the official Congress institutions were peripheralized and mass actions were unconstrained by internal party politics. The nature of popular actions left its own mark on the lower level Congress organization. In 1943-4, at the local level, the Congress ceased to exercise any control and links were forged between dacoit groups in the countryside and the socialist-led Azad Dastas. A transformation

[25] In some ways, by the end of the first ministry phase, the CSP had been incorporated by the Congress which may be said to have been an expression of the continued intellectual, moral and political hegemony of the Congress. Gramsci's concept of 'transformismo' seems to be relevant in this regard. The term was used to describe the process of the extension of the Action party in Italy from 1848 onwards whereby the left and the right parties which emerged from the Risorgimento tended to converge in terms of a programme during the years that followed. He has noted that, 'The extension of the ruling class involved the gradual but continuous absorption, achieved by methods which varied in their effectiveness, of the active elements produced by allied groups and even of those which came from antagonistic groups and seemed irreconcilably hostile. In this sense political leadership became merely an aspect of the function of domination'. See A. Gramsci, *Prison Notebooks*, pp. 58-9.

of the party occurred at the local level as it became less formally structured and more loosely organized.

By 1946, however, the Congress leadership was beginning to reassert itself. This meant that, when the Congress came to form a second ministry, many of the tendencies of the first Congress ministry were further entrenched and structuralized. The Congress record in office for the period was a dismal one. A second peasant agitation over bakasht and other grievances quickly developed. However, independence was now near so that when the nation state was formed in 1947, it was able to incorporate the ideology and the symbols of the national struggle into the new nation-state. Thus Nehru was to argue, apparently convincingly, that the state would represent the balanced aggregate interest of the people as a whole. It would not be dominated by any particular group or class. It would stand above conflicts and provide an autonomous political will to direct the economy in the interest of the people as a whole. This was, of course, merely an idealistic notion. Moreover, the record of the second Congress ministry soon showed that, for large sections of the poorer peasantry and for the agricultural labourer, the 'nation' had yet to come into its own. Gandhi was greatly saddened by this course of events. He insisted on the renewal of the moral battle. It had by then become obvious that the main body of the Congress leadership was now fully engaged in the task of running a modern state machinery, using the full range of coercive instruments to do so. In a speech in April 1947 he felt compelled to ask, 'Where is the Congress today? It is disintegrating. I am crying in the wilderness.'

One area where the Congress suffered a serious loss of legitimacy and where an alternative was successfully asserted in the 1930s and 1940s was among the Muslim masses. The Congress acceptance of office had led the Muslim splinter parties to unite under the banner of the League. The constitutional experiment and the subsequent Congress sweep to power in the Muslim minority provinces had made it clear to the Muslim parties that any real devolution of power at the centre would succeed in empowering the Congress alone to the detriment of other parties, unless they regrouped and strengthened their resources. The Muslim League therefore emerged with renewed strength after 1939. In the intervening period it had successfully built up the image of Congress rule as a Hindu raj, resulting in a dramatic increase in communal tensions. Hindu communal parties also had a field day and the Congress could do little to assuage tensions. Its inability to attack Hindu communal forces and the inefficiency of the police in dealing with communal riots gave credence to Muslim fears that the Congress in Bihar was being progressively Hinduized. The propaganda of the communal parties also enhanced the fears and suspicions growing between the two communities at the popular level. Bihar saw serious communal outbreaks both in 1937-9 and 1946-7. By the end of 1939 the Congress in Bihar had lost what little support it had had among the Muslim masses. It would be a little over-simplistic to ascribe this loss of Muslim support only to the communalist interventions of the League. The evidence from Bihar suggests some rather more interesting conclusions. In particular, in many areas, a complex relationship developed between agrarian and communal rioting in the period 1937 to 1946. At the level of popular protest this raises important questions about the nature of popular consciousness. These form the basis of a discussion in the last chapter. However, they also indicate the need for considerable further research.

My study raises a number of general questions, as well as taking an analytical view of the development of the nationalist movement at a critical phase. Above all, it considers the dynamics of the relationship between a purely agitatory movement and the constraints imposed by the acceptance of a role in government. It examines the tensions this caused and the transformations wrought in the political standpoints, radicalism and personal commitments of politicians. In other words, it is looking specifically at the process undergone by a political party as it acquires real executive power and still has to function with some awkwardness as a mass movement in order to acquire further political concessions from the colonial government. It specifically examines, too, the relationship of the Congress with its mass base and the nature of popular action and protest in this period. How does a party lose touch with the masses? What were the political dynamics of compromise?

CHAPTER ONE

Working the Imperial Constitution : Office Acceptance, the Indian National Congress and Bihar, 1931-7

In 1937 the Indian National Congress seemed poised to take part in the latest imperial constitutional experiment, the Government of India Act of 1935. How had this remarkable change come about? The party had always opposed participation in the structures of government under colonialism and now it was willing to collaborate with it. Nationalist historiography has neglected to pay much real attention to this change in Congress tactics and policy. This indifference has been particularly marked in the case of leftist historiography which has tended to concentrate on phases of mass anti-imperialist action and to ignore the area of colonial constitutional politics. The current historiographical emphasis on 'subaltern' activity has also caused historians to neglect issues of elite policy. This constitutes a serious neglect since it is quite clear that overall Congress policy in the period was consistently and deeply involved with the patterns and progress of low-level and popular protest and action.[1] The decision to accept office and gain partial control of the state structure had dramatic and immediate effects on popular actions. At first it generated mass aspirations. Disillusion soon set in, however, to be swiftly followed by the rise of popular resistance to interventions of the new Congress proto-state. The internal dynamic of this social cycle of aspiration, disillusion and subsequent resistance is one of the core elements in this study.

[1] Such a preoccupation leads this group of historians to ignore the impact of the state and the political party or prevents them from examining the role of the intellectuals in any meaningful way.

The decision to accept office had far-reaching implications for the Congress relationship with its mass base. That the Congress was aware of the dangers of accepting office within the constraints imposed by an imperial constitution had been apparent earlier on in the debates over office acceptance. However, the partial control of the state structure afforded by the 1935 India Act meant that the Congress was forced to consider the political benefits of acceptance of office and the extent to which a real increment of power might be achieved through a careful working of the Act. In the counter-hegemonic struggle *vis-à-vis* the colonial state some sections of the Congress, notably the Gandhian group, believed that partial control of the state apparatus would tilt the real political balance in favour of the Congress. The views of this group of Gandhian rightists, and particularly men like Rajendra Prasad and Vallabhai Patel, differed markedly from those of Gandhi himself, not simply on this but on most matters pertaining to the nature of the state, modern civilization, bourgeois rationality and notions of progress.[2] However, they were also well aware that Gandhian ideology could be used to mobilize large sections of the Indian people into the national movement.

The Congress relationship with its popular base had evolved through the two mass movements based on non-cooperation and civil disobedience. It has recently been suggested, especially by Partha Chatterjee, that very specific notions of responsibility and control determined the attitudes of the Congress leadership in this period of mass struggle.[3] The necessity to incorporate the masses in a broad-based anti-imperialist struggle was seen as all-important.[4] The unpredictable nature of mass agitation, however, often worried the more rightist elements of the leadership. They, in essence, felt that the

[2] Gandhi was against power relations which became progressively elaborated, rationalized and centralized under the auspices of state institutions. He noted, 'The power to control national life through national representatives . . . will become unnecessary if the national life becomes so perfect as to be self-controlled. It will then be a state of enlightened anarchy in which each person will become his own ruler. He will conduct himself in such a way that his behaviour will not hamper the well-being of his neighbours. In an ideal state there will be no political institutions and therefore no political power'. See *Collected Works of Mahatma Gandhi*, (hereafter CWMG), vol. lxviii, p. 265.

[3] See Partha Chatterjee, *Nationalist Thought and the Colonial World*, Calcutta, 1984, p. 152.

[4] See D.A. Low, 'Introduction' in D.A. Low (ed.), *The Congress and the Raj*, p. 17.

peasants should be controlled and led by responsible leaders who would show them how they might fit, entirely in accordance with their 'true' or 'rational' interests, into the national movement. Gandhi's message was interpreted at the popular level in many different ways, enthusiastically drawing large sections of the people into the movement. In a very real sense, however, his deep personal charisma and rapport with the masses were skilfully appropriated by a rightist leadership. In this respect, Gandhi's message of class harmony became a highly convenient weapon in the hands of the rightists, and one that was co-opted in the attempt to create a bourgeois state very different in conception from that idealized by the leader himself.

Until the late 1930s, it should be remembered that the Congress was still a loosely structured organization encompassing several different trends and ideologies within an overall nationalist framework. The domination of the Gandhian group was not yet complete, but was instead placed constantly under challenge from an emerging core of left-wing politicians within the Congress whose perspective differed substantially from that of the Gandhian group. This latter group also saw the mass movement as one that had to be channelled from above, along the 'right' lines. However, their vision of the future development of India was conceived in essentially socialist terms. As such, their view of the state structure was framed in terms that would be democratic and, above all, responsive to popular demands. Such a state, it was felt, could be achieved only after independence. In this view, any compromise with the colonial government over a partial devolution of power might risk a dispersal of mass energies and end up in popular disillusion and 'bourgeois' reformism.[5] The decision to accept office was thus long debated within Congress circles and was not taken lightly or easily. It was brought about by a complex set of determining factors. These consisted broadly in the Congress party's perception of its role after civil disobedience (i.e. in its inability to launch another mass movement, in the given constitutional impasse) and the degree to which real power could be achieved by a careful working of the 1935 Act and the effective domination of the Gandhian group in the party organization.

Some scholars have argued that the decision to accept office was largely a result of the drive for personal power on the part of local and

[5] It might be argued, of course, that after independence this was an accurate prediction.

provincial elites.[6] These 'dominant provincial interests', as Tomlinson has called them, were men whose central concern was to maintain the position which they had long nurtured through the internal politics of the Provincial Congress Committees, and then to extend this position into a dominance of politics in their own provinces or districts. Such provincial elites, it was argued, were opposed to any radicalization of the Congress. In consequence, the Gandhian leadership increasingly drew on their opposition to defeat the 'socialists'. For Tomlinson the two major groups within the Congress, the Gandhians (attempting to play the political game by using provincial elites) and the socialists, were not ideological groupings. They were factions vying for power. In his view, factional conflict and identity constituted the main activity of Congressmen. To understand the actions of any group, it has been argued, it was necessary to understand their position in this faction system. The political game was played, it was asserted, on the basis of an astute manoeuvring for power through vertical alliances for mutual political benefit.

This factional theory has limitations. Above all the explanatory power of the term faction is clearly circumscribed. Tomlinson, in particular, makes no attempt to study the polarization of the Congress party on class or ideological lines. His study therefore degenerates into a survey of 'factional conflicts' and 'alliances' with no attempt being made to go beyond an examination of personal cliques and groups. In fact, Tomlinson's distinction between province and nation is also suspect when it comes to considering the ideological cleavages that ran through the various levels of the Congress party. Given the weak analytical power of faction-based studies of the Indian nationalist movement it does seem essential now to begin instead to try and explore more real issues of ideology and power in examining Congress strategies and tactics during this period.

A first step in this direction would involve a close inspection of the political scenario which developed after the civil disobedience movement of 1931.[7] At a surprisingly early stage the British had been

[6] B.R. Tomlinson, *The Indian National Congress and the Raj*, p. 44.

[7] The Congress civil disobedience campaign, started by Gandhi in March 1930, was a result of the failure of the colonial government to react satisfactorily to the Nehru Report and to the passage of the resolution on complete independence at Lahore in 1929. The resulting mass movement had exceeded even Congress expectations. As a result several disparate struggles broke out under the aegis of the Congress. These fissiparous tendencies worried the Gandhian

forced to parley with Indian leaders on relatively equal terms and to seek ways in which to divert the Congress from a policy of direct action.[8] An Act was proposed at the second Round Table Conference which sought to extend the political alliances of the Raj in Indian society by devolving some measure of constitutional power.[9] As Tomlinson has shown, these changes were designed to co-opt a wider popular sanction in order to lend support to a state whose essential function would lie in maintaining the pre-existing financial and military structure of India.[10]

The Act also envisaged federation as an essential part of the scheme, with safeguards in such matters as defence, external affairs, and the position of minorities. The franchise was to be increased, but only very marginally. In August 1932 the British introduced the first of a phase of constitutional concessions, involving a communal award. Separate communal groups were to vote for provincial assembly seats allotted to Muslim, European and Sikh constituencies. The untouchables were also to be given separate electorates under the

---

rightists who, while well aware of the need to engage the masses in political resistance, were also emphatic about the need to control resistance. Nevertheless, within the broad parameters of the Congress programme, the working committee did permit extensive provincial flexibility. In Bihar, for example, Rajendra Prasad decided on an attack on the chowkidari tax and this became a main theme of the Congress civil disobedience campaign in the province. This was a tax levied for local policing purposes. Above all, the movement demonstrated the ability of the Congress to unite the people around itself.

[8] Lord Willingdon to Lord Zetland, 13.7.1933, quoted in B.R. Tomlinson, *The Indian National Congress and the Raj*, p. 35. In 1931 Lord Irwin and Gandhi had negotiatied a settlement and civil disobedience had been suspended. However, at the Second Round Table Conference in that year the British continued to be inflexible and did not concede the basic nationalist demand for the immediate grant of dominion status. The failure of the talks led to the resumption of civil disobedience in 1932; but by 1933 the Congress plan of action was losing its appeal. It was petering out under severe repression while reserves of morale and finance began to run low. Indeed, British policy makers showed no signs of backing down from their earlier policy statement at the Round Table Conference, wherein the main points of the proposed Government of India Act were outlined.

[9] Details of the Second Round Table Conference, Indian Annual Register, vol. 1, 1932, pp. 5-6.

[10] See also Carl Bridge, *Holding India to the Empire*, Delhi, 1986, p. 158. Bridge argues that by devolving power at the provincial level the colonial state hoped to secure power in the centre.

award.[11] The proposals originating in the Round Table Conference and various committees at Westminster were then embodied in the Government of India Act of 1935. In its White Paper, the India Committee had stated that the 'safe hypothesis' on which they were proceeding was that the future Government of India would not be a new creation, but would rather imply a natural evolution of an existing government structure and the extension of its past tendencies.[12] In its essentials the 1935 Act merely envisaged extending a measure of self-government, in what amounted to an astute move by the British to divert Congress energies and distract the nationalists from the path of further direct action.

The British fondly, and perhaps realistically, hoped that the Congress would split on the issue and that conservative social and political forces would emerge from within the Congress willing to collaborate with the Act.[13] At the same time, in the British view, the possible dangers inherent in the area of administration under the control of elected representatives might be checked by substantial financial and administrative safeguards provided under the Act.[14] In particular, regarding the provincial section of the 1935 Act, the governors were given considerable powers to veto legislation. The federal part of the Act was to be introduced much more gradually than the provincial

[11] B.R. Tomlinson, *The Indian National Congress and the Raj*, pp. 17-20.

[12] Samuel Hoare in his speech on the India Bill in the House of Commons, 8 February 1935, quoted by Zetland to Linlithgow, 24.1.1939, Linlithgow Papers, Mss Eur F 125/7.

[13] Zetland to Linlithgow, 3.5.1937 in Linlithgow Papers, F 125/4.

[14] See R. Coupland, *Indian Politics 1936–42*, pp. 134-6, and B.R. Tomlinson, *The Indian National Congress and the Raj*, p. 30. The Act consisted of two main sections, the Federal section for the projected All India Federation of British India and the Indian States and the provincial section for the provinces in British India. The Federal Section envisaged giving special and discretionary powers to the Governor-General. No less than 94 sections of the Act conferred (i) discretionary, (ii) deferred and (iii) special powers on the Governor-General which permitted him to overrule the elected legislature in cases where the peace and tranquillity of India were threatened, British rule and the financial stability of India were endangered, and where the interests of minorities, public servants and states were thought to be jeopardized. Thus, sovereignty did not lie in the Federation but in the British ruling power, outside the Federation in the British crown. A most serious objection to the 1935 Act was that it envisaged the creation of a bloc of princely states in the Federation, which would have almost equal representation in the Council of State and the Federal Assembly and could effectively block any legislation.

section. These proposals fell far short of Congress demands and, at the Calcutta session of the Congress in April 1933, the party soundly condemned the White Paper on the Act and reiterated its faith in civil disobedience. In reality, however, the leadership was already placed in a difficult position. It was rapidly becoming clear that the civil disobedience programme was beginning to grow politically stale and inappropriate. On 8 May 1933 Gandhi was forced to make a statement advising the temporary suspension of civil disobedience.[15] He admitted that fear had seized the common masses and that the government ordinances had successfully intimidated them. He therefore advised the president of the Congress, Madhavarao Aney, to officially declare a suspension of the civil disobedience movement for one full month or six weeks.[16] The ending of the mass movement gave no clue as regards the future programme of the Congress. Gandhi, however, realizing the need to continue with mass activity, was interested in formulating a constructive programme. In his view, spinning, harijan work, prohibition and other social activities and interventions would keep Congress workers in touch with the masses.[17]

The bulk of the leadership, however, was not enamoured of this constructive programme. Some Congress workers were clearly thinking on entirely different lines. In October 1933, Satyamurthi, a Congressman from the south, founded a provincial Swaraj party to contest elections to local boards and the legislatures. His initiative was taken up in 1934 by D.A. Ansari and Dr B.C. Roy, among other national leaders, who also felt that a programme of entry into legislatures would help to chart a way out of a political stalemate. They felt that, given the historical circumstances, the Congress could not hope to transform the social and cultural conditions of India through the 'constructive' programme. Instead they saw the colonial state as too

[15] Gandhi's statement in *Indian National Congress Resolutions 1930-34*, Appendix No. 3, p. 201.

[16] On 9 May, M. Aney finally suspended the civil disobedience movement. Within a short time, however, the movement was restricted to an individual civil disobedience campaign and the mass agitation had to be called off completely. Pattabi Sitarammayya, *History of the Indian National Congress*, vol. I (1885-1935) p. 562.

[17] Gandhi had always laid the highest emphasis on his constructive programme. He believed that activities such as khadi spinning, and harijan work would transform India's villages. The quest for modernization and industry, he thought, would only increase the exploitation of the villagers and increase poverty. See Discussion with Frydman, CWMG, vol. lxiii, p. 241.

great an impediment to all such attempts at transformation. Hence, rather than wasting energy on allegedly futile projects such as constructive work in the villages, it was felt to be preferable to concentrate on the immediate task of winning self-government. The acceptance of a parliamentary programme in the context of the 1935 Act was thus seen to be a step in the right direction. The new group sounded Gandhi out as to his views on a putative Congress parliamentary party. On 18 March 1934, Ansari received a letter from Gandhi giving his blessing to the notion of forming a parliamentary party without suspending individual and civil resistance. The reason behind Gandhi's approbation of the scheme, as he was to admit later, related to a realization that the constructive programme alone would not satisfy the Congress majority. He felt rather that a parliamentary wing of the Congress should be allowed to emerge. Nevertheless, in this period Gandhi did not see the parliamentary programme as being the dominant part of the Congress agenda which it was later to become. To him it would instead constitute a marginal activity indulged in by some Congressmen. Indeed, the constructive programme remained the only worthwhile policy of the Congress alongside which compromises might be made in the field of practical politics. He believed that 'the parliamentary work must be left to those who are so inclined. I hope that the majority will always remain untouched by the glamour of council work. In its own place, it will be useful. But . . . Swaraj can only come through an all-round consciousness'.[18] On 5 April 1934 Gandhi further noted that it was the duty of every Congressman who did not wish to take part in civil disobedience to enter legislatures to pursue a programme which they believed to be in the interest of the country. This he followed with a public statement calling off even individual civil disobedience and instead confining the action to himself.

The reactions of Congressmen to these changes in policy were

[18] Speech at the AICC Meeting at Patna, 19 May 1934, CWMG, vol. lviii, pp. 9–11. In April 1934, in Delhi, leaders such as Ansari, B.C. Roy, Dixit, K.M. Munshi and Bhulabhai Desai met to discuss a suitable programme for such a party and concluded that only a policy of contesting the elections would serve. The official statement of the party announced that the Swaraj party had been revived to enable Congressmen not offering civil disobedience to undertake parliamentary activity on the issues of (a) implementation of the country's mandate and repeal of all repressive laws and (b) rejection of the proposals of the White Paper.

varied. Nehru viewed the withdrawal from civil disobedience as a blunder and felt that direct action was a better tactic of the Congress. Council entry to him was not compatible with Congress policy. He warned that no real power would be devolved and that Congress policy would degenerate into reformism and compromise with the colonial authorities. Nehru's reaction was thus one of growing unease at the developing parliamentary mentality of the Congress. He pointed out that, while he was prepared to reconcile himself to the withdrawal of civil disobedience, the suggestions for future work upset him. Nehru went on to criticize the factional activities of the Congress parliamentary board, stating that the real socio-economic objectives of the Congress were being neglected by the kind of politics followed by the Congressmen who made up the board. He was not alone in his criticism. By this stage a section of Congressmen with leftist views, dissatisfied with the policy of the Congress, had already created a separate party within it. This was the Congress Socialist party, formed in May 1934.[19] At a conference held to discuss the political situation in the country a majority of the delegates, who included Jayaprakash Narayan, Asok Mehta, Achyut Patwardhan, Yusuf Meherally and M.R. Masani, expressed misgivings about the newly-revived Swaraj party, whose policy, they felt, would degenerate into responsive co-operation.[20] As the parliamentary mentality in the Congress strengthened, this left-wing section intensified its critique of the new policy. They asserted that acquisition of partial control of the state structure within the constraints of the imperial constitution would tie the hands of the Congress, that the mass programme would be hindered and little would be achieved. The emergence of the Socialist

[19] The formation of this party was not a sudden step. In the late 1920s much of the dissent and dissatisfaction with the old Congress had expressed itself in socialist terms, especially in youth and student politics. In the 1930s the Youth League in many provinces had already demanded *purna swaraj*, had launched a series of demonstrations and had demanded reform. In 1931 the Bihar Socialist party was organized to propagate socialistic ideas with the specific programme of the abolition of zamindari and redistribution of land. Their programme went further than the Karachi resolution of the Congress which talked of rent reduction for the peasantry and minimum wages for workers in general terms only. See A.K. Chaudhuri, *Socialist Movement in India, the Congress Socialist Party (1934-47)*, Calcutta, 1980, p. 17 and *Indian National Congress Resolutions 1930-34*, pp. 66-8.

[20] Details of the All India Socialist Conference, Patna, 17 May 1934, *Indian Annual Register, Vol. I, 1934*, pp. 341-3.

party marked the beginning of a process of polarization within the Congress party along ideological lines, and along lines that challenged the 'bourgeois' vision of the Gandhian group.[21] In a broad programme, the CSP foresaw the development of India on essentially socialist lines. In this vision all power would be transferred to the producing masses and a progressive socialization of all instruments of production, distribution and exchange would take place.[22] The self-appointed task of the CSP in the national struggle was made clearer in 1936, when it was pointed out that the Congress was the principal existing mass organization of diverse elements seeking national liberation. It had already succeeded in uniting large numbers of the Indian people for the national struggle. For this reason the CSP felt that it had to work within the national Congress and not outside it. It would be the task of the CSP, therefore, to wean the anti-imperialist elements in the Congress away from the present bourgeois leadership and to gradually bring them under the influence of revolutionary socialism.

Other Congressmen opposing council entry included men such as Pattabi Sitarammayya (from the old 'no-changer' groups) from Andhra and the Punjab and Bengal Congressmen. These latter groups felt that, since the Congress had little hope of forming the ministries in these provinces, they personally would gain little from the new entryist policy. The majority of Congressmen, however, wanted the parliamentary programme to become a plank in the Congress platform. A widespread demand thus developed that the parliamentary programme be controlled by the All-India Congress Committee and the provincial legislatures, and not by the narrow circle dominated by Ansari and Roy. Within the Congress a group of people was now emerging who realized that the Act could be utilized to gain governmental power and partial control over the state apparatus, a strat-

21 The Gandhian group saw the development of India on broadly bourgeois lines. They were the dominant or the hegemonic group within the Congress in this period. They were in the Gramscian sense 'intellectuals already naturally condensed by the organic nature of their relation to the social groups whose expression they were . . . in other words they were the real organic vanguard of the upper classes, to which economically they belonged'. A. Gramsci, *Prison Notebooks*, p. 60.

22 In 1936 Jayaprakash Narayan the main socialist leader in Bihar published *Why Socialism?* in which he strongly criticized Gandhi's philosophy, particularly concerning trusteeship.

egy with enormous political significance. At the AICC meeting in mid-May 1934 a new parliamentary board had been formed to control election matters, with Ansari as president.[23] For his part Gandhi cautioned that, while the discipline of the AICC would prevail over the board, the AICC was not an exclusively parliamentary body. Instead, it should be required to represent and regulate all the national activities of the Congressmen, especially in the constructive programme. On this basis, Gandhi attempted during 1934 to re-infuse the Congress with his idealism by extending the constructive programme and later by proposing changes in the Congress constitution with a view to making it a more disciplined organization.[24] These activities, however, failed to attract the majority of Congressmen whose real interests lay in extending the parliamentary programme. A growing feeling was developing among them that constructive work alone would lead the Congress nowhere and a parliamentary programme was necessary to keep up Congress morale. The issue was to be recurrently and vehemently debated in the following years.

When the resolution on the formulation of the parliamentary board was put forward at the AICC session, the socialists moved an amendment stating that the decision of the Lahore Congress imposing a ban on council entry should not be rescinded, except by an open session of the Congress, and that elections should be run on the basis of a socialist programme. The amendment was lost by 35 votes to 86 at the AICC meeting. Nevertheless, the socialists had made their pres-

[23] While moving the resolution to form the board Gandhi noted that it was 'in the fitness of things' that he had been asked to move the resolution, as the idea of boycott of legislatures had originated with him and, though he retained his disbelief in the legislature as an instrument for obtaining swaraj, he felt that those Congressmen who had faith in council entry and were discontented for want of political work could enforce their desire to do so for 'Congressmen, as a whole, have never taken kindly to purely constructive work such as the spinning wheel'. CWMG, vol. lviii, p. 10.

[24] In June 1934, at Wardha, the working committee, in accordance with Gandhi's wishes, laid down for the newly organized Congress committee a constructive programme with the promotion of khaddar, removal of untouchability and enforcement of alchohol prohibition as the main points. Gandhi also proposed to discipline the Congress organization by introducing a spinning qualification for delegate and executive elections and also attempted to make the executive bodies of the Congress more compact and efficient. At the Bombay session of the Congress in October 1934 Gandhi's most radical proposals were watered down. CWMG, vol. lix, pp. 10–12.

ence felt. For their part, the right wing of the Congress did not take to this show of spirit by the CSP at all well. In Bombay, in June 1934, the working committee (which was predominantly right wing), directly criticized the socialist group for their 'objectionable' behaviour and seemed worried about the implications of the socialist programme.[25] Gandhi, too, seemed unhappy at the rise of the socialists in the Congress. In September 1934 he stated that he had 'welcomed the formation of the socialist group. Many of them are respected and self-sacrificing co-workers. With all this I have fundamental differences with them . . . If they gain ascendancy in the Congress, as they well may, I cannot remain in the Congress'. Gandhi's attitude to the socialists was determined by his distrust of socialist class-war notions which, he believed, were 'foreign to the essential genius of India, which is capable of evolving a form of communism broad-based on the fundamental rights of all and equal justice to all'.[26] Though he was extremely critical of the social and moral depredations of advancing capitalism, an exclusively socialist solution, Gandhi felt, would lead to unjustified violence. Furthermore, he was also wary, at a basic level, of the nature and ambitions of a potentially socialist state.

After the Bombay Congress of 1934 Gandhi publicly retired from the Congress, ostensibly because he felt that the party had no further use for his tactics. Indeed, he pointed out in a statement issued at Wardha that the Congress intelligentsia was tired of his statements and views.[27] In fact, this Wardha statement represented a political ploy and a reprimand to the Congress for failing to give importance to the constructive programme and to the ideals for which he was striving. It did not mark a genuine withdrawal. Instead, Gandhi remained a pragmatic politician, concerned with the practical organizational development of a political movement. He did not sever

[25] The working committee resolution noted 'whilst the working committee welcomes the formation of groups representing different schools of thought, it is necessary, in view of loose talk about confiscation of private property and the necessity of class war, to remind Congressmen that the Karachi resolution as finally settled by the A.I.C.C. at Bombay in 1931, which lays down certain principles, neither contemplates confiscation of private property without just cause of compensation, nor advocacy of class war. The working committee is further of the opinion that confiscation and class war are contrary to the Congress creed of non-violence'.

[26] CWMG, vol. lix, p. 6, and vol. lviii, p. 248.

[27] CWMG, vol. lix, p. 4.

connections with the Congress in any way, and general Congress policy continued to be of his shaping—as he himself admitted to Nehru in 1935.[28] Moreover, his retirement did not mean his dissociation from the parliamentary programme. Given the general response to such a programme, Gandhi had realized that it was inevitable. Furthermore, he was prepared to support it as the only alternative to a mass movement on civil disobedience lines and as a policy that attracted a majority of Congressmen.

Nevertheless, Gandhi remained in a personal quandary. This is clear from a letter to Nehru in October 1935 in which he noted:

> As to our attitude on the present world situation, I do not think there is want of appreciation of it. But it is our helplessness which imposes silence on us. There is no weakness either. It is a matter, if you will, of tactics. . . But I know that I cannot speak with effect at this juncture. . . But I have great faith in you in these matters. . . Therefore you may be able to evolve a dignified formula for national self expression in speech as well as in action assuming of course that at the moment direct action is ruled out.[29]

He himself was thus indifferent to and even wary of participation in government while aware of the strong current of opinion in favour of it. Gandhi remained much more suspicious, however, of the socialist group and their revolutionary programme and tactics. Indeed, this was perhaps the deciding factor behind his support for the parliamentary programme.

Gandhi's tacit support for the parliamentary programme did not prevent him from appealing to Nehru to accept the presidency of the Congress. He had realized that Nehru would probably help evolve a consensus on the twin issues of council entry and office acceptance. The other right-wing Congress leaders viewed Gandhi's choice of Nehru as president with trepidation. While in London Nehru had condemned the Government of India Act in the strongest terms and, though he had pointed out that he would accept council entry to avoid disruption within the Congress, he had also stressed he would have no truck with office acceptance.[30] Meanwhile, provincial opinion seemed strongly in favour of both council entry and office acceptance.

[28] Gandhi to Nehru, 22.9.1935, CWMG, vol. lxi, p. 439.

[29] See Gandhi to Nehru, 17.10.1935, CWMG, vol. lxii, pp. 39-40.

[30] Review of Nehru's correspondence, 4.12.1935, in Home Political, National Archives of India (hereafter NAI), File no. 4/13/35.

At a parliamentary board meeting in February 1936, the issue of council entry and office acceptance was discussed. Almost all members of the board representing the general opinion in their provinces announced that they were in favour of contesting elections. A majority also reported that their provinces were demanding office acceptance too.[31]

On the eve of the Lucknow Congress the varying views in the Congress had crystallized. The anti-office group included some no-changers such as Sitarammayya, the Bengal and Punjab Congressmen and, more importantly, Nehru and the left-wing Congressmen and the CSP. Other left-wing groups, such as the Communist Party of India, accepted the CSP position officially, while retaining significant differences with it. The anti-office group thus formed a group larger than the Congress left wing, but deriving the core of its strength from the left wing. The pro-office group included the Gandhian group led by Prasad and Patel, constitutionalists such as K.M. Munshi and Satyamurthi, and the bulk of provincial opinion. Gandhi had, in the circumstances, given his tacit support to the pro-office group, while reiterating his disbelief in the use of legislatures as instruments for obtaining Swaraj.

At the 'subjects committee' meeting of the Lucknow Congress Nehru suggested a compromise position. He argued that the Congress resolution on the Government of India Act could permit elections but that no decision on actual office acceptance should be taken. He emphasized the difference in outlook among sections in the Congress and the difference between what he called the reformist and the revolutionary mentality. The principal benefit of contesting elections would be to carry the Congress message to the millions of the Congress voters and to the millions of the disenfranchised. Having once entered the legislatures it would be foolish to ally with imperialist machinery, for this might mean that the programme on which the elections were fought would have to be faithfully carried out. This would be difficult given the constraints of the situation and the fact that the Congress would be operating within an imperial constitution. Were the decision to be for office acceptance it would imply responsibility without power. The Congress, Nehru prophetically suggested, would then find itself responsible for the administration, the apparatus

[31] Kripalani to Prasad, 15.2.1936, Rajendra Prasad Papers, File no. III/36, col. 1.

of imperialism, the deficit budget, and the suppression of labour and the peasantry. He therefore suggested that the Congress go to the people and make the constraints of the constitution clear to them by indicating that the legislative platform could be used to handicap the imperialist bodies by creating deadlocks within them.[32] The CSP adopted a similar stance. They accepted S. Caveesher's amendment of the resolution which clearly stated that Congressmen, even when in a majority in the provincial legislature, would not accept office and would endeavour to bring the constitution to a standstill. Sampurnanand, the CSP leader from UP, seconded the resolution, claiming that little would be achieved in the councils. M.R. Masani, another socialist, pointed out that, even were the Congress to accept ministership within the constitution, it would be unable to fulfil its promises to the people. All in all, it was strongly believed that acceptance of the principle of Congress ministries would lead to a compromise with the colonial government and thus to a dispersal of mass energies into apathy or reformism.[33]

The Gandhian group contested this apprehension by arguing that, were the Congress not to take office, other parties would do so, and bring about an erosion of Congress influence. In contrast, the acquisition of the control of the provincial state structure would tilt the political balance in favour of the Congress. Prasad argued that the danger of the Congress settling down to routine administration could be avoided by using the Act in the way in which Congress wanted. They could still be revolutionary by accepting office and then rigidly prosecuting the Congress programme. He felt that the sight of Congressmen in authority and controlling the state apparatus would restore the flagging morale of the people and help create the context for improving civil liberties. He argued that it might not be possible to avoid office acceptance (1) when other parties or individuals were willing to work the constitution in a way detrimental to the country, (2) when people expected them to do as much good as possible through the constitution, or (3) when it was possible to strengthen and support the movement of freedom by accepting office. At the general session he stated that, having considered the pros and cons of the case, he felt that if circumstances so necessitated, there was no harm in

[32] Nehru's presidential speech at Lucknow, April 1936 in *Selected Works of Jawaharlal Nehru*, (hereafter SWJN). vol. 7, pp. 182-8.

[33] *Report of the Indian National Congress 1936-48*, pp. 52-4.

accepting office.[34] The constitutional group, led by K.M. Munshi, expressed this viewpoint with more sophistication. Munshi argued that, while office might change Congress activity, the mentality of resistance would be maintained just as much through an uncompromising administration as through extra-parliamentary opposition. In effect this would mean transforming the institutions of the colonial state into institutions of the new nationalist bourgeoisie. By this process colonial authority could be eroded further and the civil services, in particular, be re-situated in a different power structure. Control over the state apparatus would change the political balance in favour of the Congress. In effect, the experiment in office would further legitimize the Congress, for it would be seen as the successor state.

The Lucknow Congress also saw the Congress Socialist party pressing for the adoption of a more concrete mass programme. The original socialist suggestion for direct representation of workers' and peasants' organizations in the Congress was side-stepped.[35] Instead, the resolution recommended the formation of a sub-committee, consisting of Rajendra Prasad, Jairamdas Daulatram and Jayaprakash Narayan, to suggest ways in which, by amending the constitution, the Congress might develop a closer association with the masses.[36] Regarding the agrarian programme resolution, the original socialist resolution (drafted by Nehru and P.D. Tandon), which included recommendations for the abolition of zamindari and a 50 per cent moratorium on agricultural debt, was seen to be premature. The final

[34] See Rajendra Prasad to Bhulabhai Desai, 19.12.1935, Rajendra Prasad Papers, File no. 111/35, col.5.

[35] For a recent article on the subject, see D.A. Low, 'Congress and "Mass Contacts" 1936-37: Ideology, Interests and Conflict over the Basis of Party Representation' in R. Sisson and S. Wolpert (eds.), *Congress and Indian Nationalism*, London, 1988. Low points out that the clash in the mass contacts committee in 1936 between Rajendra Prasad and Jayaprakash Narayan was the most crucial debate in the whole history of the Congress on the fundamental issue of class vs non-class in the structuring of the Congress organization. See pp. 149-50.

[36] New ideas regarding Congress mass contact were first mooted at the Karachi Congress in 1931 where a concrete socio-economic programme was adopted to define the Congress conception of independence in economic terms. These ideas were further systematized in 1936. See 'Congress and Mass Contact' in A.M. Zaidi (ed.), *The Encyclopaedia of the Indian National Congress*, vol. ii, 1936-8, p. 59. See also Congress and Mass Contact, Rajendra Prasad Papers, File no. ix/1936.

resolution merely described the problem in general terms and laid out a questionnaire to PCCs by which recommendations might be proposed on the subject. From hindsight, the failure to address the agrarian problem was highly significant.[37]

Nehru's unprecedented action in speaking against the resolution on the Government of India Act at the subjects committee meeting, as well as the renewed socialist show of spirit, did not go down well with the right-wing leadership. Nevertheless, Gandhi remained confident. He himself had taken no part in the debate but had gradually come around to accepting the parliamentary programme in view of the circumstances and of the fact that direct action was now practically ruled out.[38] Moreover, he regarded Nehru's stance on office acceptance as one that would not endanger unity in Congress ranks.[39]

Despite Gandhi's faith in him, the right wing continued to hold reservations about Nehru and were strongly suspicious of the growing influence of the socialist group. This difference was bound to create tension in the working committee. In a letter to Gandhi, Nehru complained that the committee was not his child and that he strongly resented certain inclusions. On one occasion in June 1936, Rajendra Prasad, S.R. Deo, J. Bajaj, C. Rajagopalachari, Acharya Kriplani and Daulatram threatened resignation. In a strongly worded letter to Nehru they complained that they had hoped it would be possible to evolve a common line of action. Their wish was not to be fulfilled.[40] The rift

[37] It portended the future betrayal of poor peasant interests once the Congress came to power.

[38] At Faizpur he was to point out, 'The parliamentary programme is in the air. It has come to stay and rightly. But it cannot bring us independence. Its function is strictly limited though quite necessary. Its success will prevent the government from claiming that ordinance rule or any measure restricting our progress to the goal was sanctioned by popular representatives'. Gandhi's speech at Faizpur, 27.12.1936, CWMG, vol. lxiv, p. 193.

[39] He wrote to Agatha Harrison in April 1936: 'His address is a confession of his faith. You see from the formation of his cabinet that he has chosen a majority of those who represent the traditional view i.e., from 1920. Of course, the majority represents my view. I would like to kill the new constitution today if I can. . . But Jawaharlal's way is not my way. I accept his ideal about land. But I do not accept practically any of his methods . . . So far as I know him, he will not precipitate a conflict nor will he shirk it if it is forced on him.' Ibid., Gandhi to Agatha Harrison, 30.4.1936, CWMG, vol. lxii, p. 353.

[40] They felt 'the preaching and emphasizing of socialism particularly at this stage, by the president and other socialist members of the working committee is prejudicial to the best interests of the country. . . The effect of your propaganda

between the members of the working committee was, however, temporarily patched up by Gandhi.

At the AICC meeting in Bombay in August 1936, the election manifesto was passed, but the question of acceptance of office was postponed until after the elections.[41] It was clear, however, that the Gandhian group, headed by Prasad and Patel, were all in favour of a parliamentary programme culminating in office acceptance. Gandhi was anxious that Nehru be allowed another term in office to help evolve a compromise; but Nehru's assertion that his re-election would amount to support for his programme worried Patel. He feared that non-acceptance of office would be a betrayal of the commitment of the Congress to its financiers. Patel therefore decided to contest the election himself. As it turned out, Nehru confirmed that the views he had put forward were personal, and would not bind the Congress. The potential crisis was therefore defused.[42] He further underlined the necessity of united joint action, whatever differences of opinion remained. It was this basic problem of the need for a united anti-im-

. . . has been very harmful and we feel that in the situation created we cannot shoulder the responsibility of organizing and fighting the coming elections'. Things were, however, patched up with the intervention of Gandhi and attention was focused on the functioning of the parliamentary committee.

41 The manifesto noted that the new legislatures, hedged and circumscribed by safeguard and special powers, would be incapable of solving India's problems. However, as there was danger of these legislatures being used by British imperialism for its own purposes, and to the disadvantage of the Indian people, the Congress would enter legislatures to take all steps to end various regulations, ordinances and action that oppressed Indians and would work for the establishment of civil liberty, release of political prisoners and other matters. It underlined that the main work of the Congress still lay outside the legislatures but it was to prevent the operation of the forces calculated to strengthen alien domination that the Congress had decided to contest seats. The manifesto also promised to stand by the Karachi resolution relating to fundamental rights and duties and stated that it would give its first attention to prohibition, land reform, reduction of the burden on land taxation and rent, renewal of intermediaries, scaling down of debts and provision of cheap credit facilities. In the domain of industrial labour a decent standard of living, regulated hours of work, the satisfactory settlement of disputes, the right to form unions, and relief against old age, unemployment and sickness were promised. The other reforms included upliftment of harijans, readjustment of communal claims, improvement of the treatment of political prisoners and the encouragement of khadi and village industries. See Election Manifesto, *Indian National Congress Resolutions, 1936-37*, pp. 2-8.

42 Nehru's statement from Hargaon in SWJN, vol. 7, p. 594.

perialist struggle that Nehru and the left wing in the Congress came up against, time and again. Finally it was to completely tie their hands, especially where the participation of the CSP in office was concerned.

Notwithstanding the debate on office acceptance, the election campaign in 1936 evoked genuinely joint action by the opposing groups in the Congress. Although the selection of candidates did cause some problems, the joint efforts of the left and the right wings in the party eventually paid off, securing the Congress a healthy majority in most provinces. There was no doubt that the election campaign succeeded in bringing the Congress message to the people and they apparently gave their wholehearted support to the Congress appeal. In fact, the elections brought sweeping success to the Congress in all the general seats. In the Muslim seats, however, the Congress did not fare very well. However, as these seats were contested by many splinter Muslim parties, the Congress remained the single largest party in a majority of provinces, namely Madras, C.P., UP, Orissa, Bihar and Bombay. Despite the limitation in the franchise the Congress success showed that the party had generalized mass support and, as Linlithgow noted, the elections had enabled the party to establish contact with the rural population on a wide scale under a pretext which made it difficult for the British to interfere with them.[43] The provincial leaders now pressed harder for office acceptance. Significantly, Linlithgow already believed at this time that the Congress would accept office and that there would be no widespread attempt to wreck the Congress from within.[44] This time his perception was accurate in the sense that, whatever political differences remained, the notion of a unified movement had rarely given way to the belief that ideological differences would lead to a split in the alliance. The reasons why this did not happen may have been due to the intellectual, moral and political hegemony of the Gandhian group within the Congress. In

[43] Linlithgow to Zetland, 5.3.1937, Linlithgow Papers, Mss Eur F 125/4. Nehru, on whom, as Congress president, fell the lion's share of the party election campaign travelled 50,000 miles and addressed 10 million people. For details of the election campaign and results in Bihar, see the next section.

[44] Zetland to Linlithgow, 21.2.1937, Linlithgow papers, no. F 125/4. The British had fondly hoped that, as the debate on office acceptance grew sharper, the Congress would split up into many provincial parties. The British aimed to disengage provincial pro-office groups from the leadership's influence and thus to realize the provincialising aims of the Act and so erode a national centre of opposition to their rule.

many ways the Gandhian group sought throughout the period to incorporate the socialist opposition.[45]

Within the working committee, Nehru, Narendra Dev and A.P. Patwardhan still opposed office acceptance. Gandhi had come around to accepting office, while stressing the limited role it could play. The Congress was going into the legislatures, he said, not to give but to take co-operation. Granting that the council programme was 'full of temptation' he confirmed his support for parliamentary democracy, while insisting on 'the pursuit of truth and non-violence'. If the constructive programme could be promoted through the legislatures, Gandhi asked, why should they not use these bodies while still working outside? At a decisive AICC meeting in March 1937 the office acceptance resolution was passed in the face of grave criticism from Nehru and the CSP. Ultimately Jayaprakash Narayan's resolution was lost by 135/70 votes, leaving the working committee resolution to be passed 127/70. The victorious resolution declared that the Congress acceptance of office would be based on the policy of combating the new constitution and seeking to end it. This would be achieved by carrying out the Congress programme in office and by co-ordinating mass activity outside the legislatures with activity inside the legislatures intended to create public opinion for Congress demands.[46]

Gandhi added a rider to the office acceptance resolution. This stated that the Congress would form ministries on condition that the governors would not employ the special powers vested in them by the

[45] In the case of the Socialist party in 1937, such an incorporation had yet to happen. It did happen by the late 1940s. Until then, however, the socialist opposition was to grow, though they never formally broke away from the Congress.

[46] Other details of the programme included (a) resisting the introduction and the working of the federal part of the Act and demanding a constitutional assembly, (b) refusing to co-operate with any function designed to enhance British prestige, (c) carrying out the Congress programme as enumerated in the election manifesto in particular [(1) a substantial reduction in rent and revenue (2) fixity of tenure (3) relief from the burden of rural debt and arrears of rent and revenue (4) repeal of all repressive laws (5) release of political prisoners, internees and detenues, (6) restoration of lands and property confiscated or sold by the government during the civil disobedience movement (7) institution of an 8-hour day for industrial workers without reduction of pay (8) introduction of unemployment relief] (d) not avoiding deadlocks with the British government in pursuance of Congress policy.

Act, which permitted interference in the constitutional activities of ministers. This rider can be seen as a palliative to Nehru and the CSP. More importantly, Gandhi's demand hypothesized a genuine attempt to paralyse the authoritarian features of the constitution and enlarge the space for popular control under the Act. It also constituted an ingenious attempt to test the sincerity of the British policy-makers, since it demanded a demonstration of British intentions. If they were indeed to ultimately transfer power, the British would have to demonstrate it by not interfering with the electoral mandate of the Congress. Gandhi insisted that once this demand was accepted, a peaceful transfer of power would be made possible. Office acceptance, he now believed, would be the mechanism through which independence and a constituent assembly could be attained.

This debate had also successfully reinforced a new Congress self-image of power that placed a potential Congress government on an equal footing with the colonial government. This was most evident in Gandhi's suggestion that a tribunal be set up to solve the disagreements between the British authorities and the Congress, with one representative from each side and a third independent one to decide the issue. The British realized that Gandhi's demand concealed a strategy to increase Congress legitimacy, and they feared that conciliating the Congress would further weaken their institutional supports. In response to this political impasse Zetland and Linlithgow finally offered a conciliatory statement which actually conceded very little. However, the Congress could not afford to dither longer.[47] At the provincial level mass aspirations, fuelled by the Congress election campaign, had resulted in rural and urban unrest in UP and Bihar. The provincial leadership in the Congress majority provinces was now urging the Congress to accept office, arguing that the electorate expected the Congress to implement its electoral promises. On 5 July 1937 the working committee permitted the Congress government to accept office, wherever they were able to form the government.

The pro-office group had thus had its way. They firmly believed that the control of the state structure would restore the flagging Congress organization and revive the morale of the people. The mentality of resistance would not be compromised and the institutions

[47] For the debates at the level of British policy, see Zetland to Linlithgow, 5.4.1937, and 12.4.1937 and Linlithgow to Zetland, 22.4.1937 in Linlithgow Papers, Mss Eur F 125/4.

of the colonial state would be transformed by the new nationalist bourgeoisie. In particular, the services would be re-situated in a different power structure and their loyalty to the new national state would be ensured. In this process the colonial state would be destabilized and the Congress would be further legitimized. In theory this can be seen to have constituted a valid strategy. Indeed, given the circumstances, the choices open to the Congress were very limited. On the other hand the possibilities of using office acceptance to enhance the credibility of the Congress were good. It can be argued, therefore, that the Gandhian group was realistic in its evaluation of the political circumstances. Office in itself was not a bad thing. What was more important was whether the mentality of resistance would be maintained through an uncompromising administration.

Nationalism now had to constitute itself as a state ideology. It had accepted the realities of power. It remained to be seen how the institutions of the new national state structure would operate. Nehru remained critical to the very end, arguing that by accepting office,

> we can take upon ourselves the odium and responsibility of keeping the imperialist structure functioning. We can indirectly become responsible for the repression of our comrades. We can take away the initiative from the masses and tone down their fine temper, which we ourselves have helped in building up.[48]

The experience of office showed that Nehru's bitter warnings of the consequences of office acceptance would be borne out. It was here that the ideological predilections of some members of what one may term the Congress 'high command' of Prasad and Patel would have their greatest impact. In office they compromised time and time again with the dominant groups in Indian society, considerably diluting the benefits of their rule and enabling a widespread movement of opposition to develop to the Congress in Bihar. This opposition directly attacked the legitimacy of the Congress and challenged it to meet its promises to the people. The Congress in power thus failed to fulfil many of the proclaimed aims of office acceptance set out in its election manifesto. However, its policy with respect to the colonial state was more successful. Control of the state institutions did gain

48 Nehru's presidential speech at the All-India Convention of Congress Legislators on 19.3.1937, SWJN, vol. 8, p. 70.

for it the loyalty of the subordinate bureaucracy, thus effectively destabilizing the colonial state.

Before one examines these issues more closely, however, it is worth tracing the development of the Congress organization in Bihar and identifying the dominant groups that had emerged within it by the 1930s.

BIHAR AND THE CONGRESS ORGANIZATION IN BIHAR IN THE 1930S

The main task of the following section consists in examining the nature of the political developments in Bihar in the period prior to office acceptance. To do this effectively demands a careful investigation of the complex character of the national movement in Bihar. In particular I have attempted to analyse the diversity of the social composition of the movement and its regional distribution. The period between 1920 and 1930 was a critical one for the political development of Bihar. In the wake of the non-cooperation movement a whole set of classes and castes entered the nationalist arena for the first time. They gave the Congress its mass base and linked it up firmly with people in rural areas. The Congress in Bihar thus became a mass party able to establish a highly credible position *vis-à-vis* the colonial state.

The significance of this important development needs to be assessed in the context of Bihar society as a whole. The following outline of some of the basic features of the economic life and social structure in the province is set out with the object of elucidating the political context and in order to relate it to the historical narrative of the rest of the work.

Bihar, a socio-economic survey

The province of Bihar in our period was divided into four administrative divisions. These were Tirhut, Bhagalpur, Patna and Chotanagpur.[49] These divisions were again divided into districts, as follows:

[49] Bihar, which stretches up to the eastern part of the Indian subcontinent, is a landlocked area which in our period extended to 69,745 square miles. It forms the north-eastern part of the table land of central India. Topographically the state had three natural divisions. The northern Gangetic plain, divided by the river Ganges, makes two natural regions, namely north Bihar and south Bihar. The Chotanagpur plateau is the third division and provides a great contrast to the

BIHAR, ADMINISTRATIVE DIVISIONS, 1937

	Districts	*Administrative divisions*
North Bihar	Saran	
	Champaran	Tirhut
	Muzaffarpur	
	Darbhanga	
	Bhagalpur	Bhagalpur
	Purnea	
South Bihar	Patna	
	Gaya	Patna
	Shahabad	
	Monghyr	
Chotanagpur Plateau	Hazaribagh	
	Ranchi	
	Chotanagpur	
	Palamau	
	Manbhum	
	Singhbhum	
	Santhal Parganas	Bhagalpur
	Feudal States	

The population of Bihar and Orissa, according to the 1931 census, was 42,329,583.[50] It had increased after the previous census of 1921 by 11.5 per cent. It was, however, overwhelmingly rural and only 3.7 per cent of the population was located in urban areas. The capital city

Gangetic plain. Gradually the fertile riverine area gives way to the rugged hills of Chotanagpur. It is sparsely populated and has great mineral wealth, which in our period was largely underdeveloped. Significantly, the nationalist movement had its greatest strength in the densely populated districts of north and south Bihar.

[50] This figure includes Orissa which was part of Bihar till 1936.

Map Showing the Districts of Bihar in 1938

of Patna was the only place with a population of over 100,000 while the population of the other major towns in the province was small, averaging about 20,000. Economically, the region was relatively underdeveloped even in the 1930s. The majority of the population being rural, 77.8 per cent of the working population was engaged in agricultural pursuits.[51] This dependence on land was not really relieved by the presence of large industries. As the population increased there was no matching urban or industrial development and only about 7.7 per cent of the working population was engaged in industrial activity.[52]

There were marked regional variations between north and south Bihar. Most of north Bihar required regular drainage and was prone to constant flooding while south Bihar had a greater incidence of drought. The south Bihar districts of Gaya, Patna, Shahabad and Monghyr were more fertile, better irrigated and more populous. The population pressed most heavily on the land in the districts lying on either side of the Ganges. Despite the fertility of the soil these areas were deficient in foodgrains, due to the density of the population and the extensive production of cash crops such as sugarcane, tobacco and chillies, which all yielded big returns in the rich alluvial soil.[53] This had enabled a section of the tenantry, exemplified by the *guzashtadars* of Shahabad, to prosper and thrive.[54] The north Bihar districts, on the other hand, (notably Bhagalpur and Purnea) had poorer soils, were little irrigated and were less densely populated. In Purnea the infertile soil gave few returns and the majority of the peasants were poor with

[51] Census of India, 1931, vii, part 1, p. 185.

[52] The nature and extent of its economic backwardness in comparison with Punjab or western UP may explain the course of its political development. However, such a conclusion requires a comparative survey, a task not tackled in this study. For the impact of economic variations and tenurial relations in the different regions of India, see K. N. Raj, *et al.* (eds.), *Essays on the Commercialisation of Indian Agriculture*. For example, Amit Bhaduri argues for western UP that its land revenue system had some ryotwari features which resulted in the development of a genuine class of peasant proprietors quite unlike in eastern UP. See A. Bhaduri, 'Class Relations and Commercialisation in Indian Agriculture: A Study in the Post Independence Agrarian Reforms in U.P.' See pp. 312-15.

[53] See K. Prasad, *The Economics of a Backward Region in a Backward Economy*, vol. 1, Calcutta, 1967, p. 247.

[54] *Final Report on the Survey and Settlement Operations in the District of Shahabad 1907-1916* by J.A. Hubback.

little chance of bettering their lot. Lack of nutrients had caused large-scale deficiency diseases among the people. An early gazetteer indicates that the majority of the people in certain tracts had large goitres and were often weak with disease.[55]

This regional variation did not, however, indicate a greater degree of industrialization in the south Bihar districts.[56] The overwhelming dependence on the land remained the same throughout Bihar and greatly affected the political economy of landownership. Within the constraints of an ill-conceived and oppressive land-tenure system, this degree of agrarian dependence manifested itself in a highly stratified rural society, in which the process of rent-paying lay at the heart of rural social relations. The Permanent Settlement had resulted in a complex system of tenurial rights. Apart from the landlords, who paid land-revenue to the colonial state, there were generally one or more subordinate layers of holders of rights in land, paying rents to their immediate superiors for the land they cultivated as tenants or sub-tenants. Beneath these groups subsisted the mass of landless or near-landless labourers. Since the beginning of the twentieth century the poorer sections of the peasantry had, in essence, been rack-rented to such an extent that they faced a growing and basic subsistence crisis by the 1920s. This poor majority had become enmeshed in a whole set of new monetized transactions. Such changes were rarely to their advantage as they were tied to the market by various debt and rent obligations.[57] In some south Bihar districts, however, a section of the tenantry had thrived as a result of the improved irrigation and had managed to corner some of the advantages of the increased commercialization of agriculture for themselves.

As elsewhere in India the population of Bihar was divided on the basis of caste and religion. Of the total population in the province 83.2 per cent were Hindus. In fact, Hindus formed the absolute majority in

[55] Purnea District Gazetteer, 1911 and Bhagalpur District Gazetteer, 1910.

[56] Most of the towns in north and south Bihar were overgrown villages and not real urban conurbations. The Bihar Unemployment Committee in 1954 estimated that only 20 per cent of the population was gainfully employed and the load of dependency was thus extremely heavy in the state. K. Prasad, *The Economics of a Backward Region in a Backward Economy*, vol. 1, p. 67.

[57] See Gaya Survey and Settlement Report, Patna Survey and Settlement Report. See also J. Pouchepadass, 'Land, Power and Market; The Rise of the Land Market in Gangetic India' in P. Robb (ed.), *Rural India, Land, Power and Society Under British Rule*, London, 1983, pp. 93-5.

the population of every district of the province, except in the Singhbum and Santhal Parganas. The number of Muslims in the province amounted to only 10.1 per cent of the total population. They constituted a substantial minority, however, in the Bihar towns. On the Chotanagpur plateau tribal religions dominated among the indigenous groups which formed 5.7 per cent of the population of Bihar. Christians were very few in number, there being only 416,726 in the whole province, of whom 95 per cent were resident in Chotanagpur. Among the Hindus, distinctions between one class and another were, of course, bolstered by caste. Bhumihars, Brahmins, Rajputs and Kayasthas were found at the top of the hierarchy and were mainly landlords or wealthier tenants. The Bhumihars were chiefly distributed in the districts of Monghyr and Muzaffarpur as well as parts of Darbhanga, Saran, Gaya and Patna. Economically, they were the strongest of all castes, Educationally, however, they were held to be more backward than the three other advanced castes. Kayasthas and Rajputs were normally placed next to the Bhumihars in status. Their strongholds were in Saran and Shahabad, where they were known to be harshly exploitative petty landlords.[58] Kayasthas, according to the 1931 census, were numerically weaker than many other castes at 1.15 per cent and their population was unevenly distributed, more in towns than in villages. Traditionally listed as 'writers', according to the 1931 census, they returned themselves both as cultivators and writers, there being about 372 literate Kayasthas for every 1,000 individuals. Educationally, they were far ahead of the other castes, as well as having been the first caste to receive an English education. These four upper castes were followed by the middle-ranking castes. Of these, Ahirs and Kurmis were the two most socially and politically prominent caste groups. Occupationally they can be termed as 'middle' and 'poor' peasants. Some had benefited from agrarian price inflation since the early part of the twentieth century and had improved their economic status. They had also managed to improve their ritual status, often returning themselves as higher castes in various census operations.[59]

[58] R.R. Diwakar, *Bihar through the Ages*, Calcutta, 1958, p. 693. See also Kisan Enquiry Committee Report 1936 (unpublished Congress report on peasant grievances in Bihar).

[59] Recent studies have argued for a strengthening of caste barriers in nineteenth-century India as a result of the engagement with colonial institutions. This process was evident in an intensified concern with the purity of lineage

Some were active in cow protection Sabhas, joining hands with upper castes in an attempt to acquire upward social mobility. Among the lower castes Chamars were numerically the strongest. Along with Dusadhs, Musahars, Doms, Dhobis and Pasis they formed the bulk of the population of agricultural labourers. Their position at the bottom of the agrarian hierarchy had been structured by the traditional caste norms, whereby untouchables were debarred from holding land. They lived in extreme poverty and were harshly exploited by the upper castes. Among the Muslim castes the only community for which the Census has recorded statistics were the Jolahas. Traditionally weavers, they formed a low caste among Muslims.

The religious differences between Hindus and Muslims in rural society did result in the formation of some kind of separate identity. Ahirs and Kurmis had been active in cow protection activities since the early part of the twentieth century and this served to reinforce the worsening relations between some sections of the Hindu and Muslim peasantry at the village level. At that period, however, there was nothing to compare with the communal divisions and concomitant communal violence that developed in the 1930s and 40s. The latter was based on the active communalist interventions of political parties like the Hindu Mahasabha and the Muslim League. On the contrary, at the level of rural local society, relations between the different religious groups were relatively relaxed and only rarely hostile. Nonetheless, fears and suspicions were gradually building up by the 1920s, fed by elite propaganda and landlord machinations and culminating in rural communal violence both in 1937-9 and in 1946.[60]

In the village, the local landlord was located at the apex of the power structure. In one recent study of a north Bihar village, the kayastha landlord and his sub-infeudaries were found to have maintained their hold on the land with relatively little disturbance for over

---

among the upper castes like the Rajputs, the closing-off of social boundaries among the middle-level castes like the Ahirs and Kurmis and the loss of status suffered by pastoralist and tribal groups and their assimilation into the expanding ranks of low-caste agricultural labourers. See Nicholas B. Dirks, 'The Invention of Caste: Civil Society in Colonial India' in *Social Analysis*, no. 25, September 1989; and R.O'Hanlon, 'Cultures of Rule, Communities of Resistance: Gender, Discourse and Tradition in South Asian Historiographies' in *Social Analysis*, no. 25, September 1989.

[60] See chapter 6.

three centuries.[61] Beneath them were the few settled occupancy tenants who had been able to take advantage of the price rises and improve their economic condition. The majority, however, were agricultural labourers or *khatbeys*. The latter were de-tribalized lower castes with houses on the fringes of the village. In the summer months, the khatbeys were allotted miniscule plots to laboriously cultivate ragi on a sharecropping basis, the share of the landowner not being used for consumption but for disbursal as wages. For the rest of the year they foraged for food, even eating mice and other small edible animals. The bulk of the social groups were thus living at an extreme margin of subsistence. Any small economic uncertainty, reduction in rainfall, changes in access to common rights or price fluctuation was sufficient to despatch a proportion of them over the boundary separating subsistence from famine. This overwhelmingly rural and divided society, faced with a steadily growing competition for resources, provided the socio-economic context for agitation and turbulent political change in the 1920s and 1930s.

There has been relatively little research carried out on the nature and extent of these changes in twentieth-century Bihar. A partial reading of the sources reveals a growing tendency towards the alienation of peasant lands (with peaks in incidences of alienation at particular times), the whittling-down of common property resources over a long period and a dramatic increase in prices during the two world wars. This progressive sharpening of economic deprivation stimulated the emergence of an embryonic political consciousness of the peasantry and provided the groundwork for wider political action.[62] It is important to note here that rural agitation in the 1920s and 1930s was not simply a product of divisions between the haves and the have-nots. Instead, a closer study reveals a more complex picture in which sections of the better-off peasantry joined hands with the more deprived sections of the poorer peasantry to challenge together the overweening power of landlords over resources, especially at those times when their traditional rights were under threat.

[61] See A.N. Das, 'Changel: Three Centuries of an Indian Village', *Journal of Peasant Studies*, vol. 15, no. 1, October 1987, p. 11.

[62] The nature of the changes at specific periods have been highlighted at relevant points in the study. By this I do not mean that these were successively more important issues, one relating to political change in the 1930s, one to change in the 1940s and so on. I only seek to delineate the conjuncture of forces that existed at specific moments of popular action.

POLITICAL DEVELOPMENTS AND THE NATURE OF THE BIHAR CONGRESS, 1917-37

In the years immediately following the First World War prices began to climb steeply. This hit most peasants hard for, as indicated above, even the poorer peasants had to enter into market relations, selling cheap and buying dear, in order to pay off rents, revenue and debt. In the cities the economic dislocation caused by the war and the following sharp price rise had affected all social groups. Thousands of soldiers had been demobilized and they, too, added to the discontentment. The price rise particularly affected the service classes in the towns, including the majority of the urban poor. Clerks, industrial workers, rickshaw-wallahs, and slum-dwellers and beggars were all drastically affected. This gradually worsening economic situation in both country and city provided a fertile ground for the sowing of a nationalist message under Gandhi.

Until that time Congress activity in Bihar had been confined to the ranks of the western-educated, a growing group since the end of the nineteenth century.[63] Lawyers were the most vocal of this new element and many of the Congress leaders in this period were, of course, lawyers by profession. They came mainly from upper-caste and landlord families whose politicization was not well developed. As late as 1917, however, Gandhi was to note that the Congress was unknown in many parts of Bihar.[64] With Gandhi's advent in Champaran this political picture started to change. The Congress base quickly widened, especially in rural areas. By taking up issues relevant to agrarian society, Gandhi was able to break out of the earlier parameters of Congress politics which had hitherto been largely insensi-

[63] This is not to say that political activity was confined to these sections. As Bayly has recently emphasized, the cultural and political activity of Brahmin and Rajput aristocracy, and Kurmi and Ahir middle peasants had been growing in rural Bihar and UP since 1800. Bayly, in a very cogent critique of nationalist historiography, argues against the view that the beginning of rural politics can be traced to Congress activity in rural areas after the First World War. Rather he argues that it is the persistence of these old patterns of rural politics much more than the relative radicalism of the Congress leadership that explains the evolution of national parties in the countryside after 1918. See C.A. Bayly, 'Rural Conflict and the Roots of Indian Nationalism, Allahabad District since 1800' in P. Brass and F. Robinson (eds.), *Indian National Congress and Indian Society 1885-1985*, Delhi, 1987, p. 229.

[64] Mahatma Gandhi, *My Experiments with Truth*, p. 411.

tive to basic rural problems. He was first able to do this effectively in Champaran in 1917 where agrarian unrest on indigo plantations, which were mainly European-owned, had been endemic since as early as 1820. The forcible cultivation of indigo had made radical interventions in the peasant production process, commoditizing almost every aspect of it. The new mode of production did not, however, bring the peasant any particular benefits. For poorer peasants their ecological parameters of survival were easily threatened in so far as the demands imposed by indigo cropping prevented the growth of a variety of food crops.[65] The better-off peasants, on the other hand, resented the fact that they could not participate in the indigo trade.[66] Recurrent peasant agitation in Bettiah and Motihari between 1908 and 1916 originated in these dual grievances of ecological and commercial exclusion. In 1916 some representatives of the middle and rich peasantry, Raj Kumar Shukla, Kedar Prasad Rai and Sant Lomraj Singh, appealed to Gandhi to visit Bihar.[67]

The nature of Gandhi's activities in Bihar emerged as being very different from that of earlier political leaders. He proceeded to meet the planters and investigated in person the conditions of the raiyats. A small body of nationalist intelligentsia gathered around him at once. Men like Rajendra Prasad, (a high court *vakil* in Patna) Braj Kishore Prasad, and other newcomers became his active workers. Gandhi preferred to work more with these men than with the local Congress organization. As Gandhi and his helpers toured the countryside,

[65] Indigo occupied the land for the entire year, preventing the peasant from cultivating his traditional *bhadoi* or *aghani* crops. The hated Tinkathia system demanded that the peasant compulsorily cultivate three-fourths of his land with indigo. The indifference or outright hostility of peasants and tribal people towards participation in the market economy has been a common phenomenon the world over. A compelling anthropological study of peasant attitudes and hostility to wage labour and capitalist development in the sugar-cane plantations of the Cauca valley is provided by M. Taussig, *Devil and Commodity Fetishism in South America*, Chapel Hill, 1980.

[66] For details, see B.B. Mishra (ed.), *Select Documents on Mahatma Gandhi's Movement in Champaran, 1917-18* and Colin Fisher, 'Planters and Peasants: The Ecological Context of Agrarian Unrest on the Indigo Plantations of North Bihar' in Dewy and Hopkins (eds.), *The Imperial Impact*, London, 1978, pp. 117-18. Also see J. Pouchepadass, 'Local Leaders and Intelligentsia in the Champaran Satyagraha (1917)', *Contributions to Indian Sociology (new series)*, 8, 1974, pp. 67-108.

[67] B.B. Mishra, *Select Documents*, p. 21.

peasants thronged to see him in large numbers, hailing him as a 'saviour' and 'liberator'.[68] Reinforced by this emotional political backing, Gandhi was able to demand an official enquiry into the indigo problem. Before long, then, the Tinkathia system had been abolished and the enhanced rents reduced.[69]

The Champaran episode was important both for Gandhi's style of politics and for the development of the Congress organization in Bihar. Instead of reaching down into the community with political ideas, Gandhi effectively worked from below with relevant local issues. Indeed, the first obvious feature of his influence was the crowd of peasants that showed up wherever he went. He was able to attract a genuinely wide provincial following and, as he grew to be associated more directly with the Congress by the 1920s, this popularity secured for the organization a firmer mass base in Bihar than in many other regions. This increasing influence in the province became more apparent when Gandhi launched the 'Rowlatt' satyagraha in 1919.[70] He then made use of the organization of the Home Rule League to communicate his plans. The nation-wide *hartal* organized by Gandhi as part of the agitation was as successful in Bihar as it had been elsewhere, although largely confined to such large towns as Patna, Muzaffarpur, Chapra, Monghyr and Gaya. The Rowlatt satyagraha reinforced Gandhi's claim as a leader with a more effective style of politics and helped him to gain control of the Congress organization. The year 1920 saw the development of a broad-based and more or less permanent organization. A Congress working committee was created which made the national Congress an active and functioning body throughout the year. Linguistic provinces were introduced and the organization now fanned out from the village level upwards with an open membership and a fee of four annas per annum. Much of the impetus for these changes came from Gandhi. He was thus able to reshape the Congress to suit his new agitational programme, in the form of the non-cooperation movement. It is worth noting that this programme was, from the outset, wholly supported by the Bihar and the Gujarat PCCs, where Gandhi had built a local reputation. In

[68] It is interesting that those first influenced by the Gandhian ideology were mostly vakils (lawyers). Rajendra Prasad, *Mahatma Gandhi and Bihar*, p. 170; and B.B. Mishra, *Select Documents*, pp. 102, 342.

[69] Ibid. Text of the Champaran Agrarian Bill 1917, pp. 518-20.

[70] H.F. Owen, 'Organizing for Rowlatt Satyagraha in 1919', in Ravinder Kumar (ed.), *Essays in Gandhian Politics*, pp. 65-6.

contrast, the other PCCs expressed more reservations with respect to the movement.[71]

Despite this advantage, in practical terms the non-cooperation movement in Bihar began very slowly. Some important senior leaders of the Bihar Congress, such as Sachidananda Sinha, resigned from it at an early stage in disagreement with the changed policy of the party. Once the movement was underway, however, it proceeded with vigour. The party programme, which included boycotts of schools, colleges and courts, was initially much more effective in urban than in rural areas. Muslims were also readily mobilized on the Khilafat issue.[72] The support of the general public can be gauged from their response to the question of council boycott in November 1920. Keeping in mind the limitations of the franchise one finds that 27.3 per cent of the Hindu voters and 28 per cent of the Muslim voters went to the polls in the towns. In rural areas, on the other hand, the equivalent figures were 41.8 per cent for the Hindus and 28.2 per cent for the Muslims.[73] By January 1921, an anti-liquor temperance movement was in full swing in the province, causing government excise revenue to fall by ten million Rupees during 1921-2. The tightening-up of the organization under the Nagpur Congress further aided the non-cooperation movement. The extension of the boycott to schools, colleges and courts was also successful.[74] In rural areas the Congress concentrated on the withholding of chowkidari tax. Henningham notes that attacks on police and chowkidars were very common in the 1920s. However, it should be pointed out that Congress attempts at mobilizing the peasantry in this period were still only very partial. The party made no attempt to take up the issues that were of real importance to the majority of the peasantry or to tackle rural problems

[71] K.K. Dutta, *Freedom Movement in Bihar*, vol. 1, Patna, 1957, p. 301.

[72] The khilafat movement attempted to mobilize sections of the Muslim community against the British decision to abolish the caliphate at Turkey following a state of belligerency between the two countries. It thus fanned pan-Islamic sentiments which sought to protect the caliphate and maintain the integrity of Islam. See Mushir ul Hasan, *Nationalism and Communal Politics in India, 1916-1928*, Delhi, 1979.

[73] Judith Brown, *Gandhi's Rise to Power*, p. 288.

[74] By March 1921 there were 6,000 volunteers in Bihar. Bihar's contribution to the Tilak Swaraj fund amounted to Rs 7,10,628 while Congress membership figures were 350,000. See A.M. Zaidi, *The Encyclopaedia of the Indian National Congress*, vol. 8, p. 390.

in the way Gandhi had seemed to anticipate in 1917. These real and unaddressed grievances included landlord attacks on occupancy rights, erosion of common property resources and the gradual dispossession of the peasantry. This was well illustrated by the way in which, in Darbhanga raj, a kisan leader called Swami Bidyanand, who mobilized on these lines, was able to secure a substantial following. On the strength of this support he was able to oppose the council boycott resolution of the Congress in 1920 and secure four seats for his peasant candidates in north Darbhanga and north Bhagalpur.[75]

It is therefore clear that, despite the broadening of the Congress base in Bihar during the non-cooperation movement, a more effective mobilization of the peasantry had still to be achieved, while real peasant grievances had still not been taken up. Here, perhaps, one can observe the seeds of later Congress neglect of low-level peasant grievances. In contrast, in neighbouring UP, widespread landlord-tenant conflicts broke out relatively spontaneously under Baba Ramchandra, a peasant leader. In Bihar the official Congress leadership seemed to have retained control to such an extent that no large scale peasant movement developed at this stage. The British were well aware of this distinction and even took a decision not to establish 'anti-red' leagues or Aman Sabhas in Bihar, (as they had in UP), to curb non-cooperation propaganda. They believed that while there was a 'bolshevik' tendency within the nationalist movement in UP no such element existed in the nationalist camp in Bihar.[76] Part of the reason for this distinction lay in the very particular influence of Gandhi in Bihar. Gandhi's strategy was to mobilize the population on specific anti-British issues, thus securing a united platform of all classes for an anti-imperialist struggle. This particularist strategy of Gandhi and the Gandhian group was to come increasingly under scrutiny in the 1930s as the socialists and the communists in the Congress sought to integrate a new class struggle with the established national struggle.

The ending of the non-cooperation movement had marked the beginning of a new phase in Congress history.[77] The period of en-

[75] S. Henningham, *Peasant Movements in Colonial India, North Bihar, 1917-42*, Canberra, 1984, p. 77.

[76] K.K. Dutta, *Freedom Movement in Bihar*, vol. 1, pp. 387-8.

[77] Gyan Pandey has noted that the Congress organization in this period had two major related weaknesses. First, a completely open organization like the Congress relying on a general nationalist appeal and discouraged by the government of the time could maintain its strength only when nationalist feeling was

forced inactivity caused a section of Bihar Congressmen to suggest a programme of council entry which stood in contrast to the earlier Congress preoccupation with 'constructive work' activity.[78] While the Congress vetoed the idea, resulting in the 'Swarajist split', it was forced, particularly in Bihar, to contest municipality and district board elections in May 1923 in order to satisfy majority opinion.[79] It is important to note that while the franchise in these areas was limited to small landlords and rich tenants in rural areas, the Bihar Congress was still able to win a good majority. A Congress strategy which had been confined in Bihar to emotive appeals to nationalism, while avoiding issues of class conflict during non-cooperation, had clearly paid off. The party had correspondingly succeeded in gaining the support of more conservative social groups in rural areas, especially that of small landlords. These individuals were to form the core of Congress support in Bihar well into the 1940s.

Electoral politics was not without problems. Factions based on caste or status tended to grow and struggle for seats in local bodies. Indeed, A.N. Sinha noted in 1923 that many of the election speeches laid stress on the candidate's caste and other related matters rather than emphasizing the party programme. Council elections fought by the Swarajist party in Bihar were also plagued by factional and caste politics, especially between the Bhumihar and the Rajput-dominated factions. Rajendra Prasad openly condemned this development as 'dirty' and 'nauseating'.[80] The Swarajist election experience did not, therefore, prove especially rewarding. To add to these problems, the report of the General Secretary of the Congress noted that many PCCs were not functioning properly. In Bihar the Tilak Swaraj Fund collections amounted in 1924 to only Rs 9,326 while membership figures

at a high pitch. Rapid disintegration soon followed a mass movement. The other weakness with the organization was that it made no specific provisions for the different sections of the Indian masses whom the Congress sought to represent. Even Gandhi's constructive programme failed to break through to classes below the rich peasants and petty bourgeoisie except in cases of the khadi promotion campaign and the programme of harijan upliftment. See Gyan Pandey, *The Ascendancy of the Congress in U.P.*, Delhi, 1978, p. 30.

[78] A.N. Sinha, *Mere Sansmaran*, Patna, 1942, p. 73

[79] A.N. Sinha, *Mere Sansmaran*, p. 73; Rajendra Prasad, *Autobiography*, Bombay, 1957, p. 205. Rajendra Prasad himself became the chairman of the Patna municipality.

[80] Rajendra Prasad to Motilal Nehru, 14.10.1926, AICC papers (Nehru Memorial Museum and Library, hereafter NMML), File no. 21, Part II/1926.

stayed as low as 11,795.[81] Finally the 1920s also saw the disintegration of the Muslim League/Congress alliance and an increasing growth of provocative communal propaganda which soon led to riots in several places.[82] This, then, was the picture that the Congress presented in the post non-cooperation period, at a time when it clearly required a new agitational programme to properly galvanize the movement.

The shortcomings of the Simon Commission, which had come to India in 1928, provided a much-needed background for the new struggle. The reorganization of the Congress was now undertaken with vigour. At the AICC in Bombay, Gandhi had moved a resolution calling on the provinces to reorganize themselves and to enrol at least one quarter of the total population as original members. Efforts in this direction in Bihar were quite rewarding and the Congress party exceeded its quota of membership of 72,558 members for the year by an actual enrolment of 90,525. The membership was by and large drawn from small landlords, the upper echelons of the peasantry, and the trading sections in urban areas.[83] This renewed burst of life in the Congress was soon followed, in December 1929, by Gandhi's resolution declaring independence as the chief aim of the Congress, to be achieved by a programme of civil disobedience.

There being no all-India blueprint for the initiation of civil disobedience, the working committee permitted extensive provincial flexibility.[84] In Bihar, once again, Rajendra Prasad decided on an attack on the chowkidari tax. This was the main and admittedly very limited strand of the Congress civil disobedience campaign in the provinces. It urged people to stop paying the taxes levied for the support of their local chowkidars and urged chowkidars themselves to relinquish their posts. The campaign also provided a totemic issue around which all sections of the population could agitate jointly. Significantly, the leadership felt that, unlike rent or revenue matters, the chowkidari issue was attended by few risks of internal division. Once again the peasants had to wait for a more committed political

[81] A.M. Zaidi, *The Encyclopaedia of the Indian National Congress*, vol. 8, p. 693.

[82] A.N. Sinha, *Mere Sansmaran*, p. 71.

[83] A.M. Zaidi, *The Encyclopaedia of the Indian National Congress*, vol. 9, 1926-9, p. 678.

[84] Obviously the agitation against the salt tax by collecting salt from the sea could not constitute an issue for mobilization in landlocked Bihar.

leadership to articulate their most pressing grievances. For the time being, however, the movement seemed to be reasonably popular, while in rural areas in Monghyr and Tirhut divisions, large sections of the population assumed a militant stand against the administration.[85] Police retaliation worsened the relations between the people and the provincial administration, while the brutality of the state response gave valuable publicity to the Congress movement.

By May 1931 all the potential of these activities had been exhausted and there were no signs of the movement in rural areas developing into a substantial no-rent campaign and thus securing a new lease of life.[86] Clearly the real and pressing issues were not being addressed by the Congress. Instead, a growing and uncoordinated peasant resentment was developing in the countryside against the landlords. In rural areas this resentment threatened to reach a crisis following the impact of the depression, due to the inability of large sections of the tenantry to pay the cash rents which had been commuted when prices were higher. Despite their urgency these were issues that the Gandhian group did not and would not incorporate into the national struggle. On the contrary, against this background, the civil disobedience movement was formally called off in 1934. In reality it had ceased to effectively articulate real grievances or constitute a mass movement long before that time.[87] Despite all the limitations of the Congress programme, however, one finds that by 1934 the Congress in Bihar had in fact retained its stable base. The two mass movements of 1920 and 1930 had shown that large sections of the people hitherto outside the nationalist struggle were willing to rally to the Congress call. Though the mobilization was still partial, the Congress in Bihar had succeeded in creating a semi-permanent and a broad anti-imperialist front.

The ending of the mass movement gave no clues as to the future programme of the Congress as, after the non-cooperation had ended, the leadership was quite simply unsure how to resuscitate the party

[85] See S. Henningham, *Peasant Movements in Bihar*, pp. 117-19.

[86] In neighbouring UP, on the other hand, the civil disobedience movement became for a time a non-rent campaign. See P.D. Reeves, 'Landlords and Party Politics in the United Provinces, 1934-7' in D.A. Low (ed.), *Soundings in Modern South Asian History*, London, 1968, p. 264.

[87] This inability of the Congress to develop the movement further by taking up the real issues of the peasantry, coupled with draconian government action, diluted the effectiveness of the agitation as early as 1931.

organization. Relying as it did on a general nationalist appeal, and being a completely open organization, the Congress could maintain its strength only in periods of intense mass struggle. As noted earlier, the 'constructive work' programme had not really succeeded in dominating Congress activities. The British, who had at first viewed the programme with suspicion as an attempt to renew the civil disobedience movement, found it amounted only to a damp squib in Bihar.[88] They noted that the programme existed, effectively, only on paper and that little work was being done except in stray, isolated instances. A new programme to revitalize the party organization had to be found. It is significant that this rebuilding of the Congress appeared, in reality, to relate almost entirely to the task of contesting elections as soon as a parliamentary programme showed signs of taking shape.[89] A growing section in the party was attracted by the idea of such a programme as a counterweight to a frustrating policy of 'wait and see' in a period of low mass morale.

The distinctive character of the Bihar Congress had by this period become fairly clear. The repeated resurgence of the Congress organizations had been made possible by the establishment of a firm infrastructure in 1920, a network of Congress branches and offices run by a nucleus of active workers, and the presence of a general mass sympathy for the Congress. The continuity of leadership also ensured a certain stability. The bulk of the non-cooperation leadership in Bihar stayed on to run the Congress organization in the years that followed while men like Rajendra Prasad and S.K. Sinha were rapidly rising in the hierarchy of the Congress structure. These were upper-caste men, lawyers by profession who came from petty zamindar families. The leadership was thus provided by men who had some degree of economic independence and political sophistication. The middle-level leadership that had emerged in the context of the mass struggles, however, tended to come from a more economically deprived background. They came from middle or lower castes and were sons of less well-to-do peasants or petty businessmen. These included men such as Mahamaya Prasad, Suryanandan Thakur and Satyanarain, who rose to control the District Congress Committees in the 1930s.

Thus, despite the extension of membership and the widening of the

[88] Home Political, NAI, File no. 11/3/35.

[89] AICC resolution, May 1934 in *Indian National Congress Resolutions 1930-4*, p. 176.

Congress base, as seen in the mass movements, the leadership of the Bihar Congress remained staunchly conservative. A study of the socio-economic background of the Bihar Provincial Congress Committee in 1934 has shown that about 70 per cent of the committee were upper-caste men, mainly lawyers, most of them belonging to small landlord families.[90] This should not, however, be held as evidence to indicate the attitude or character of the Bihar Congress.[91] The influence of ideological factors was far more important in determining historical action. Indeed, it is possible to determine the dominant ideology in the Bihar Congress in this period as being closely identified with that espoused by such relatively conservative individuals as Rajendra Prasad, A.N. Sinha, S.K. Sinha and others. These men believed themselves to be staunch Gandhians and had been drawn within the Congress during the Champaran satyagraha days. The views of this group of 'Gandhian rightists' differed markedly from those of Gandhi himself on most matters pertaining to the nature of the state, and modern civilization.[92] However, they were well aware that Gandhian ideology could be used to mobilize large sections of the people into the national movement. They were thus actively involved in the mass movements of 1920 and 1930 and believed that the mass-mobilization activity of the Congress should be specifically geared to the creation of a broad anti-imperialist united front of all classes. Their populist rhetoric was thus directed towards the construction of a united national identity. For these reasons socialist ideology, with its emphasis on developing class struggle, was seen as a direct threat to the building of a united front by the Congress.[93] Instead, the Congress vision of the future development of India was conceived of in broadly capitalist terms, so that any changes in land structure or industrial policy were to be achieved through compromise

[90] Ramashray Ray, 'A Study of the Bihar Pradesh Congress Committee, Bihar India', (unpublished Ph.D. thesis, University of California) pp. 272, 281.

[91] Indeed many of the Congress socialists, and even the communist leadership in Bihar, came from similar backgrounds.

[92] They viewed the future development of India along bourgeois lines and, though they paid lip-service to the abolition of zamindari and adopted a concrete socio-economic programme to improve the conditions of the poor in 1931, they were only too keenly aware of their own power base among the petty landlords and urban capitalists.

[93] A.N. Sinha noted that he saw the rise of the socialists as harmful to the interests of the country and he felt that socialist ideas should not be allowed to influence 'day to day politics'. See A.N. Sinha, *Mere Sansmaran*, p. 190.

and bourgeois reformism. Whatever its drawbacks, this approach certainly had great pragmatic validity in the context of the anti-imperialist united front. Indeed, by the 1930s, the links between the Congress organization and the capitalist classes had become transparently clear. Large industrialists such as Dalmia had begun to contribute to the party coffers on a large scale. Many petty zamindars also funded the organization generously. All these developments ensured that the Bihar Congress leadership developed a growing stake in preserving the prevailing social order. Even so, the views of the right wing were to be further challenged as the socialists began, on the basis of the agreed programme, to play as major a role in Congress politics in Bihar as elsewhere.

The rise to influence of the socialists in Bihar can be traced back to 1931, when a section of Congressmen started to express their dissatisfaction with the mass-mobilization approach of the Congress and decided, unilaterally, to form the Bihar Socialist party. It was first formed with Abdul Bari as president, Phulan Prasad Verma and Ganga Sharan Das as secretaries and Ambica Kant Sinha as treasurer. Once established, it was able to mark out the ground for the creation of a national-level socialist party for the furtherance of socialist aims. The first All-India Congress Socialist Conference was, significantly, held in Patna on 17 May 1934. On this occasion it was stated that, while the Congress constituted the foremost anti-imperialist organization in the country, its mass programme was still very limited in practical scope. It should be the task of the socialists, it was said, to transform the organization through patient propaganda and contact and to steer it towards adopting a more radical mass programme.[94] While the long-term objective of the party was the creation of a socialist state 'devolving all power to the producing masses', the short-term strategy was to work towards such a state by broadening the anti-imperialist front through the development of separate new organizations of workers and peasants. The focus of activity was to be inside and within the Congress. Those prominent in the socialist movement at that time included Achyut Patwardhan, Jayaprakash Narayan, M.R. Masani, and S.M. Joshi. Their early strategy was developed much further in the Meerut and Faizpur theses of the CSP in 1936.[95]

[94] Patna Socialist Conference, 17.5.1934, *Indian Annual Register*, January to June 1934, p. 343.

[95] A.K. Chaudhuri, *Socialist Movement in India*. For the Meerut and the Faizpur thesis, see appendix I and II. The Faizpur thesis pointed out that, while

The rise of the socialists in the Bihar Congress during the 1930s gradually gave rise to a growing conflict between the left and the right wings in the party. Their programme was as yet very moderate and was aimed only at broadening the anti-imperialist front by giving adequate representation to peasants' and workers' organizations. Yet the right wing greeted their growing influence with hostility and all through the succeeding period sought to dilute socialist opposition. As at the national level, the main issue on which differences emerged was the office acceptance programme. In Bihar the socialists rejected the programme outright, preferring instead to work towards developing class organizations among the working classes and the peasantry. In particular, they took an active interest in the Kisan Sabha. It was the resulting association of the Sabha with the Congress that built up the reputation of the latter as a mass organization.[96]

Once the decision to contest elections was taken the socialists were left with little alternative but to accede to it. Thus, despite strong reservations, the decision to contest elections saw joint action by both

---

the Congress was the principal existing mass organization of diverse elements seeking national liberation, it did not yet embrace the broadest sectors of the masses and still stood aloof from their day-to-day struggle for the satisfaction of their pressing immediate needs. This was because of the fact that, while the Congress was a mass organization, its leadership was predominantly bourgeois. The task of the CSP, therefore, was to wean the anti-imperialsit elements in the Congress away from the bourgeois leadership and bring it under the influence of revolutionary socialism. This could be done by (a) working within the Congress organization by enlarging membership and extending Congress organization further, (b) placing an alternative programme before the Congress so that it could begin actively to develop the struggle of the masses taking their immediate demands as the basis for policy. This alternative programme would include the development of the class organization of the workers and peasants. This would be linked up with Congress committees, thus widening the anti-imperialist front of the Congress, while changing the entire structure and leadership of the organization.

[96] It can be pointed out here that the socialist elements were to dominate the left wing in the Bihar Congress *vis-à-vis* other left groups. They were a predominant element in the Kisan Sabha and the communists despaired of ever getting a hold in Bihar in this period. As such the communists were able to form the Bihar state unit of the CPI only in October 1939, when their United Front agreement with the CSP had collapsed. See *Indian Communist Party Documents 1930-56*, p. 44, and B. Jha, 'Birth and Growth of the Communists in Bihar', *Indian Left Review*, February 1976, pp. 8-16. For a study that examines the history of radical left politics in Bihar, see P. Brass (ed.), *Radical Parties of the Left in Bihar, a Comparison of the SSP and the CPI.*

right and left wings in the party in Bihar, as elsewhere.[97] The franchise was limited to only 10 per cent of the population. Much further research would be needed to determine the exact composition of the new electorate.[98] Nevertheless, it clearly included a large number of individuals who had hitherto had no contact with electoral institutions and, as British officials noted, the extension clearly implied an increase in the representation of tenant interests. In Bihar, as in Madras, Bombay, Bengal and the United Provinces, a second chamber with special representation was also created to provide a check on the first chamber.[99] The upper house had a stricter franchise qualification and also had a high percentage of nominated members. This ensured a conservative bloc. [The allocation of seats in the Bihar Provincial Legislative Assembly is given in Tables 1 and 2.]

Selection of candidates was not an easy task for the Congress. In Bihar the working committee in the province was converted into a 'parliamentary' committee to organize the elections and it was decided that the parliamentary committee alone would select candidates. In practice, however, many of the decisions were taken at the DCC level. The consequences of electoral participation became immediately noticeable in terms of the multiplying of new political groupings. Furthermore, the communal categories sanctioned by the constitution soon encouraged major alignments on precisely communal lines. The Muslim Independent party was created by a section of Congress Muslims as a result of which the Congress contested only a few

[97] Rajendra Prasad had been put in charge of organizing elections in Bihar. AICC papers, File no. E 1/1936.

[98] The franchise qualification was set on the basis of rent or cess in rural areas and the payment of a municipal tax in urban areas. The person was qualified to be included in the electoral roll for a territorial constituency if he was assessed during the previous financial year for income tax or an aggregate amount of not less than Rs 1.50 in respect of municipal tax or chowkidari tax of 9 annas. He was also qualified to be included in the electoral roll if he occupied a land or building for which he was liable to pay a rent of not less than Rs 6 per annum or local cess of 3 annas. See *Government of India Act 1935*, pp. 274-5. It is important to point out that only 0.06 per cent of women were enfranchised and the new political institutions gave them virtually no representation at all.

[99] It was pointed out by the Government of India that, though only 10 per cent of the population was enfranchised, this would cause a vast increase in the representation of tenant interests which would swing other interests and tend towards instability. Hence the second chamber to introduce the stable element. See Indian Franchise Committee, vol. iii, p. 109.

TABLE 1

ALLOCATION OF SEATS IN THE BIHAR LEGISLATIVE ASSEMBLY

Category	*Number of seats*
Total	**152**
General Urban	5
Rural	73
Scheduled Caste	15
Muslim Urban and Rural	39
Anglo-Indian	1
European	2
Indian Christian	1
Commerce	4
Land holder	4
University	1
Labour	3
Women	4

SOURCE : Indian Annual Register, (1937), vol. 1, p. 168.

TABLE 2

ALLOCATION OF SEATS IN THE BIHAR LEGISLATIVE COUNCIL

Category	*Number of seats*
Total	**30**
General	9
Scheduled Caste	–
Scheduled Tribe	–
Muslim	4
European	1
Indian Christian	–
Anglo-Indian	–
Commerce	–
Land Holders	–
University	–
Legislative Assembly	12
Nominated by Governor	4

SOURCE : Indian Annual Register, (1937), vol. 1, p. 168.

Muslim seats through an informal agreement with this new party. The tenuous links with the Congress were soon broken once the Muslim League sought to broaden its base among regional communal parties.[100] Among the 'depressed' classes the 15 seats awarded to them by the constitution resulted in the creation of the Depressed Classes League. In this case the League spokesman, Jagjivan Ram, was willing to negotiate with the Congress over specific depressed-class seats. The Congress was thus able to contest the seats in its own name and retain a hold over the organization.[101] Another party, the Tribeni Sangh, was also created to defend the interests of the depressed classes. It had a predominantly Kurmi and Koeri membership, but did not, however, contest many seats. At one stage the Tribeni Sangh leaders alleged that they had wanted an agreement with the Congress party, but that the latter had accepted none of their nominees. As a result the Tribeni Sangh was forced to put up its own candidates, who then lost heavily to the Congress.[102]

Candidate selection also encouraged factional groupings and the formation of corrupt cliques. In Shahabad district it was discovered that nomination papers had been given to the friends of existing DCC members. Other irregularities were also imputed, such as the failure to appoint an impartial returning officer, and the granting of permission to file nominations beyond the appointed date, all of which permitted further fraud. In Saran there were two factions among the Congressmen in the district. One was led by the president of the DCC, Mahamaya Prasad, and the other by Girish Tiwari. A serious tussle between the factions had to be defused from above by Prasad's intervention and the distribution of seats shared equally between the factions. Prasad himself recorded several cases in which members had actually been nominated because they belonged to a particular caste.[103]

Despite this evidence of factionalism based on caste and personal

[100] Rajendra Prasad, *Autobiography*, p. 435. See also Syed Mahmud to Rajendra Prasad, undated, Rajendra Prasad Papers, File no. 1/36, col. 1.

[101] Rajendra Prasad to Patel, 4.8.1936 and Rajendra Prasad to A.N. Sinha, 17.9.1936, Rajendra Prasad Papers, File no. 1/36, col. 1.

[102] See Kalyan Mukherjee and Rajendra Singh Yadav, *Bhojpur Naxalism on the Plains of Bihar*, New Delhi, 1980, pp. 29-30.

[103] Rajendra Prasad recorded a few cases of cliquism and nepotism in a letter to Patel. See Prasad to Patel, 17.11.1936, Rajendra Prasad Papers, File no. 1/36, col. 3.

politics, the most important element of Bihar politics concerned the growing ideological cleavage within the party between the left and the right wings. Socialist support for the Congress decision to contest elections had rested on the premise that the party could be transformed from within. The CSP therefore staked a claim to seats in areas where specific contacts had been developed with the kisans. However, CSP officials reported that complaints of discrimination against socialist candidates had been received from a number of constituencies where established groups prevented the nomination of socialists.[104] In Darbhanga, for example, Sahajanand was insistent on installing Jamuna Karjee and Dhanraj Sharma, both Kisan Sabha workers, in two of the six seats. However Satyanarain Sinha, the president of the DCC, was insistent on having neither of them. After much bitter argument it was finally decided to drop Dhanraj Sharma and take Jamuna Karjee, but for a different seat. Karjee's original seat was given to Suryanandan Thakur in accordance with Satyanarain's interests. Prasad noted on this occasion that Suryanandan was not a regular Congress worker at all but was a manager of a zamindari estate while Karjee was a devoted Kisan Sabha activist. For the sake of political peace, however, Satyanarain's wishes could not be overruled.[105] Sahajanand later stated in his autobiography that he had been disgusted by the way that nominations had been effected and that, because of his links with the Congress, he was forced to support the most blatantly sectarian zamindari candidates put up by the Congress. Overall, only twelve kisan candidates were selected in the province, while Sahajanand was unsuccessful in securing seats for a dozen more.[106]

Once the selection of candidates was completed, problems arose over the funding of the campaign. Candidates were asked to use their own resources, at up to Rs 2,000 per seat. This did not preclude frequent requests for money, especially for the Muslim and Depressed Class seats.[107] The Congress was thus forced to approach local in-

[104] *Searchlight*, 8 November 1936.

[105] Rajendra Prasad to Vallabhai Patel, Rajendra Prasad Papers, File no. 1/36, col. 3.

[106] In Muzaffarpur, over one such dispute where a Kisan Sabha worker, Kishori Prasanna Sinha, was dropped, Sahajanand threatened to resign from the working committee. See Rajendra Prasad Papers, File no. 1/36, col. 3, and Sahajanand Saraswati, *Mera Jeevan Sangharsh*, p. 481.

[107] Syed Mahmud to Rajendra Prasad, 30.9.1936 and 6.8.1936 and Krishna

dustrialists for funds.[108] One of the industrialists, G. Dalmia, indicated a willingness to contribute to the Congress coffers on condition that a socialist candidate, Budhan Rai Verma, was dropped.[109] The latter, it was claimed, had fomented an agitation in one of Dalmia's mills. To its credit, the Congress decided not to accept Dalmia's opportunist intervention. Prasad was particularly disturbed by this incident and it was Vallabhai Patel's placatory intervention that settled matters.[110] Budhan Rai was retained and nominated from Shahabad where the Kisan Sabha/ CSP lobby was strong and Dalmia continued to contribute to the Congress coffers. This incident highlights the ambivalent nature of the relationship between the Indian capitalist class and the Congress. By the mid-1930s the capitalists had drawn closer to the Congress, recognizing the dominant group in the organization to be right-wing in mentality and visualizing the future development of India on broadly capitalist lines. It would, however, be wrong to conclude that capitalists were, therefore, able to dictate terms to the right-wing leadership. In fact, during this period the leadership maintained their autonomy of action in the nationalist arena and responded relatively sensitively to pressures from a wider social context.

Dalmia's money, however, represented only a drop in the ocean and was used mainly to help campaigns for Muslim and Depressed Caste seats. The bulk of the funds were provided by the candidates themselves and by their local supporters. A.N. Sinha has observed that, in areas where the Congress campaigned for rich landlords to gain council seats, contributions had been made generously to the Congress coffers. At an obvious risk of compromise, therefore, the Congress was able to tide over a real financial problem.

The Congress election campaign enabled the organization to establish itself among the mass of the population on a very wide scale. The

---

Vallabh to Rajendra Prasad, 18.9.1936, Rajendra Prasad Papers, File no. 1/36, col. 1.

[108] It is important to point out here that the Congress was never totally dependent on big business funding even in the 1940s. While business groups gave material assistance to the Congress, their funds were by no means inexhaustible and were supplemented by frequent *ad hoc* collections from within as well as outside the mercantile community. See TOP, 111, enclosure to no. 595.

[109] Dalmia to Prasad, 15.11.1936, Rajendra Prasad Papers, File no. 1/36, col. 4; see also A.N. Sinha, *Mere Sansmaran*, p. 208.

[110] Vallabhai Patel to Durga Prasad, ibid., 20.11.1936, Rajendra Prasad Papers, File no. 1/36, col. 4.

extension of the franchise had brought about an increase in the representation of tenant interests in rural areas and of the lower middle classes and a small section of the industrial working population in urban areas. It is probable, too, that the electoral propaganda reached far beyond these enfranchised sections and relayed some kind of Congress or nationalist message to almost all sections of society. Furthermore, the association of the Kisan Sabha with the Congress led to an effective penetration of the campaign into the most remote rural areas. Vigorous activities in the countryside in 1935-6 meant, too, that the Kisan Sabha had created a secure base for itself among the tenantry. In spite of this, Kisan Sabha leaders consistently couched their appeal during the election campaign in moderate terms to avoid alienating the critical constituency of small landlords. Numerous kisan conferences were held in which kisan workers spoke in very general terms about the benefits of swaraj. Inaugurating the election campaign in Ranchi, for example, Sahajanand emphasized that, under swaraj, supplies of food, clothing and housing would be guaranteed to the masses. In Shahabad he toured villages with groups of workers, singing songs and holding tri-colour flags. In Nawadah he noted that, in spite of the winter season and the harvesting of the paddy crop, the kisans turned up in thousands to most meetings.[111] Fortnightly Reports, for example, noted that in rural areas the election campaigning had not only boosted the kisan agitation but also led to the diminution of *begari*.[112] The undoubted popularity of the Kisan Sabha resulted in an effective but moderate style of campaigning in areas such as Gaya and Shahabad; in other areas right-wing leaders were even more low-key. Men such as S.K. Sinha and A.N. Sinha talked in general terms of the benefits of Congress rule, while emphasizing the need for a harmonious relationship between zamindars and tenants. The British were aware of the implications of this intentionally moderate appeal for Congress legitimacy. Indeed, Linlithgow observed perceptively that the elections had enabled the Congress to establish control among the rural population in a manner which made it difficult for the British authorities to interfere.[113]

In popular perceptions, as a result, swaraj seemed imminent. The

[111] *Searchlight*, 18 December 1936.

[112] Fortnightly Reports of the Government of India (hereafter FR) (1), February 1937.

[113] Linlithgow to Zetland, 5.3.1937, Linlithgow Papers, F 125/4.

nature of the Congress/Kisan Sabha propaganda had allowed much of the population to build up a picture of the benefits of Congress rule. This further fuelled mass aspirations and increased popular demands. That Congress swept to power in the elections thus surprised no one. The British officials noted that the attendance at polling booths was unexpectedly good. Significantly, part of the reason for this enthusiasm can be related to the fact that, in many areas, the ballot box was directly associated in the popular mind with Gandhi.[114] The results gave testimony to the astonishing success of the Congress campaign. It had secured 75 per cent of the total votes cast and 65 per cent of the seats in the Bihar Assembly. In general urban constituencies, the Congress captured all five seats, while in general rural constituencies it was able to secure 68 out of 73 seats. In scheduled caste constituencies the Congress achieved success by capturing 14 out of 15 seats, 9 being opposed. The Congress also secured 2 out of three of the labour seats and 1 landholders seat. The position of the parties in the Bihar Legislative Assembly can be seen in Table 3.

TABLE 3

POSITION OF PARTIES IN THE BIHAR LEGISLATIVE ASSEMBLY

Party	*Seats*
Congress	98
Muslim Independent	15
Muslim United	6
Europeans	2
Constitutional	1
Anglo-Indians	1
Indian Christian	1
Loyalist	1
Ahrar	3
Independents	24

SOURCE : Indian Annual Register 1937

In the Legislative Council the Congress contested 6 out of 9 seats but did not win any. Having a majority in the Legislative Assembly, it was able to secure 8 out of 12 seats to be filled by the Assembly.

[114] FR (1), February 1937.

Thus in a joint session of both houses the party possessed a majority.[115] A survey of the election results, bearing in mind the limitations in franchise, shows that the Congress organization had successfully completed the transition from mass movement to parliamentary party.

The stage was now set for a new phase in the Congress mass-mobilization programme. In April 1937 the Congress ministry was formed in Bihar. The secretary of the BPCC noted that the ministers had been selected with great care, almost exclusively from the leaders of the right wing. Caste claims also affected the final distribution of posts. In particular, some dissatisfaction was expressed at the emerging strength of the Bhumihar lobby when S.K. Sinha, a Bhumihar, became the Chief Minister. This lobby had ensured that A.N. Sinha, a Rajput, was effectively denied the post.[116] Similarly, some Congress Muslims were disturbed that Syed Mahmud had not been asked to lead the party in Bihar. The final list, however, gave considerable representation to most of the castes as well as to the Muslims.[117]

It has been said that for a popular party in power its worst enemy is its own past propaganda.[118] Once in power, the Congress ministry had now to decide whether it would actually meet its promises to the people, and it was at this stage that the implications of office acceptance really started to make themselves felt. It remained to be seen how far the temptations of structure and office might promote the reinforcement of the contradictions with which Congress was already riven.

[115] Brief analysis of election results, Rajendra Prasad Papers, File no. 1/37, col. 2.

[116] FR (1), March 1937. See also S.Jha, *Political Elites in Bihar*, Delhi, 1972, p. 81.

[117] The four ministers were S.K. Sinha, A.N. Sinha, Syed Mahmud, and Jaglal Choudhury. The eight parliamentary secretaries were Shiv Nandan Prasad Mandal, K.B. Sahay, Jagat Narain Lal, Jimut Bhaban Sen, B.N. Jha, Jagjivan Ram, Syedul Huq and Sarangdhar Sinha.

[118] S. Orwell and I. Angus (eds.), *The Collected Essays, Journalism and Letters of George Orwell*, vol. 1, New York, 1968, p. 448.

CHAPTER TWO

Peasant Agitation, the Congress and the Kisan Sabha in Bihar, 1937-9

The 1930s in rural Bihar witnessed a large-scale agrarian movement and the development of a radical peasant organization to resist landlord encroachment on the traditional rights of the tenantry. Competition over agrarian resources worsened by the great Depression accounted for this rural unrest. The bakasht agitation, as it came to be known, affected many villages in Bihar and put severe pressure on the nascent Congress ministry to implement wide-ranging agrarian reform.

The historiography of peasant movements in India, which has grown extensively since the 1970s, includes a few studies of the Kisan Sabha movement in Bihar.[1] Of these, the works of Walter Hauser and Rakesh Gupta are most important. However their perspective yields only a partial understanding of the historical setting in which these agitations occurred. Several limitations still remain, notably with regard to their characterization of the peasantry in rural Bihar, the way in which tenant rights were being eroded during the nineteenth century and the political implications of the movement. With a view to overcoming these limitations, this chapter addresses the central question of the nature of the agrarian problem and peasant protest in rural

[1] These works include A.N. Das on the peasant movement in Bihar in general, Henningham's study of six kisan agitations in Bihar in the period 1917-42, Hauser's pioneering work on the Bihar Kisan Sabha and Rakesh Gupta's broad survey of the same movement. See W. Hauser, 'The Bihar Provincial Kisan Sabha 1929-42', (unpublished Ph.D. thesis, Chicago, 1961); Rakesh Gupta, *The Bihar Peasantry and the Kisan Sabha*, Delhi, 1982; A.N. Das, *Agrarian Unrest and Socio-economic Change in Bihar 1900-1980*, London, 1982 and S. Henningham, *Peasant Movements in Colonial India: North Bihar 1917-1942*, Canberra, 1984.

Bihar in the 1930s. It then moves on to examine the political repercussions of peasant discontent and the development of peasant protest organizations in a period when the Congress was changing from being a movement to a party in power and found itself hard pressed to deal with an intractable agrarian problem. The chapter attempts an integrated study of the bakasht movement in Bihar and examines several related issues.[2] I begin by looking at the nature of peasant agriculture in Bihar in the 1930s and the increasing pressures on the subsistence economy of the region. I then link this with the bakasht agitation. Finally, I investigate the participation of elite leaders and the effect of Congress interventions both at the local level and in the arena of legislative politics.

THE DIFFERENTIATION OF THE PEASANTRY AND THE EROSION OF TENANT RIGHTS IN BIHAR IN THE 19TH AND EARLY 20TH CENTURIES

Drawing on the popular model developed by Eric Wolf and Hamza Alavi to stress the potential radicalism of a land-owning 'subsistence middle peasantry', the main body of literature on the Indian peasantry has concentrated on categorizing the agrarian population of colonial India and then on arguing that certain sections of the peasantry had a greater propensity to resistance. Hamza Alavi, Theodore Shanin, and Daniel Thorner all attempted to define the term peasant rigorously and to classify them into different categories: rich, middle, and poor.[3]

The formulation of these categories was clearly essentialist in character with the middle peasant being defined strictly as the independent subsistence-oriented small holder while the rich peasant was quite clearly a tenant who had done well for himself. Further, distinct political roles were ascribed to different strata of the peasantry. This had led Alavi to conclude that the 'middle peasants' who occupied the second sector, and who were independent small-holders, tended

[2] The term bakasht refers to lands where the landlord had more direct control over cultivation and where tenant rights did not normally accrue. Throughout 1937-46 the landlords were attempting to convert tenant lands into their bakasht.

[3] See Theodore Shanin, 'Peasantry as a Political Factor' and Daniel Thorner, 'Peasant Economy as a Category in Economic History' in Theodore Shanin (ed.), *Peasants and Peasant Societies*, London, 1971, pp. 238-63. See also H. Alavi, 'Peasants and Revolution' in A.R. Desai (ed.), *Peasant Movements in India*, Bombay, 1979, pp. 671-715.

to be the most radical in any revolutionary situation. There are two problems with this argument. The first relates to the problem of definition. How strictly subsistence-oriented is the middle peasantry? In Bihar, for example, many middle ranking peasants grew commercial crops and were tied to the market by rent and debt obligations. Does a rich peasant have to be strictly a tenant? Many of the richer peasants in Bihar combined rentier interests with their primarily tenant status. Therefore, at what point does a rich peasant cease being one and become a landlord? The variations within these categories have not been examined by historians who have tended to rigidly define the categories.

The second problem arises when distinct political roles are ascribed to different strata of the peasantry with the middle peasants being identified as the most radical. This school of historians has implicitly ascribed passivity to other sections of the peasantry, a conclusion in many respects untenable. As Charlesworth has pointed out in his qualifying piece on the middle peasant thesis, 'peasants are a heterogenous entity comprising disparate individuals playing a wide variety of economic roles with frequently conflicting interests'. In his view 'the capacity of a distinct stratum of the peasantry to fight against specific sources of oppression tends to be limited. For peasant militancy to be effective fierce grievances with a wide appeal are necessary'.[4] Thus any broad and general conclusions about the response and involvement of specific groups in peasant agitations emerge only fitfully. It is difficult to argue, as D.A. Low has done, that the Indian national movement represents, predominantly, the assimilation of 'dominant peasants' into politics.[5] In fact, the studies in the volume he has edited in reality present 'a bewildering range of peasant groups as the protagonists of specific actions'.[6] It is quite clear therefore that, while the peasantry in Bihar as elsewhere was differentiated on the basis of income, any rigid definitions of what constitutes 'middle', 'rich' or 'poor' are not borne out by the existing historical evidence.

[4] N. Charlesworth, 'The Middle Peasant Thesis and the Roots of Rural Agitation in India, 1914-47' in *Journal of Peasant Studies*, 7(3), 1980, pp. 259-76. In Bihar it can be argued the bakasht agitation had this wide appeal. Although the movement attempted to resolve specific grievances i.e. those of the landed peasantry, it also subsumed grievances of landless labourers who used the agitation to demand higher wages as in some parts of Shahabad.

[5] D.A. Low (ed.), *The Congress and the Raj*.

[6] Charlesworth, 'The Middle Peasant Thesis', p. 260.

Further, the distinct political roles ascribed to different sections of the peasantry are also not borne out by historical evidence. In truth, such existing categorizations may actually prevent a realistic analysis of political mobilization and protest, particularly at the critical, and neglected, lower level of action. Instead it would seem preferable to take an analytical view based on more rigorous local studies of peasant action made at some depth.

In the period under consideration Bihar was designated a 'permanently settled' area under the Cornwallis settlement of 1793. The Permanent Settlement had aimed to put an end to the uncertainty involved in periodic settlements by making an agreement with the zamindars which provided for a permanent limitation of the government's revenue demand. This, it was felt, would ensure an increase in zamindari income so that landlords might be encouraged to become staunch allies of the government.[7] By concentrating proprietary rights in the hands of the former state revenue collectors, namely the zamindars, the Settlement wiped out at a stroke the peasant hold on the land, forcing them to pay rents which could be set at any level pleasing to the landlord.[8] Taking advantage of their poverty and illiteracy the zamindars had heaped impositions upon their tenants since 1793.

The main consequence of the Permanent Settlement was, however, to convert *raiyats* who had hitherto been enjoying customary rights of occupancy into tenants whose title to occupation was a lease granted by the zamindar.[9] Buchanan regarded the right of occupancy as originating from the lease and ending with its term. He observed in 1818 that the leases were seldom renewed so that tenants were movable at will, and though in the north of Bihar, in Gayaspur and Maner, tenants were fighting for their claim to hold at fixed rates, the district courts had already decided against them. In sharp contrast, in Shahabad, among the traditionally prosperous *guzastha* tenancies,

[7] R. Guha, *A Rule of Property for Bengal: An Essay on the Idea of the Permanent Settlement*, Paris, 1963.

[8] In the Mughal revenue system primary zamindars or local lords with superior proprietary rights over a part of a village or a whole village or several villages were responsible for revenue collection. Often above them was a class of *jagirdars* who were assigned the revenue of a whole territory to enable them to maintain a military force. For details, see Irfan Habib, *The Agrarian System of Mughal India*, London, 1963.

[9] W.W. Hunter, Bengal MS, vol. 1, p. 125.

Buchanan noted that he had heard of 'few or no illegal exactions, a great proportion of the tenantry being too high-spirited to submit to trifling abuse'.[10]

In the 1880s occupancy rights had been given to tenants who had held lands for twelve years, in an attempt to give long-standing tenants legal protection. This legal provision did not, however, prevent the landlords from continuing to erode the position of tenants.[11] They felt particularly threatened by occupancy rights and sought to attack them throughout the period. Guzastha tenants in Shahabad, however, continued to hold out against zamindar attempts to increase their rents.[12]

The early twentieth century saw a further increase in landlord attacks on tenant rights. Pressure on land had substantially increased and the exploitation of the tenantry was now exacerbated by the proliferation of petty zamindaris.[13] In examining the nature of the land structure and the extent of differentiation among the peasantry it can be noted that the hundred years or so preceding our period had seen a remarkable division of zamindari estates. Though there were still many large estates like Mahsudpur, Darbhanga, Hathwa and others, the increase in the numbers of petty zamindaris was marked. The problem was complicated by the multiplicity of tenure rights which had resulted in a proliferation of sub-infeudation and intermediate tenures. The zamindars had followed a policy of farming their estates to intermediaries called *thikadars* in order to ease their tasks. As rents

[10] F. Buchanan, *Eastern India*, London, 1838, p. 325. These tenancies were close to the Ganges and were in a *rabi*-growing tract whose plentiful harvests had enabled the settlers to prosper for many hundreds of years. Further south conditions were different, the lands poorer and the landlords were able to dominate the weak tenantry.

[11] *The Bihar Tenancy Act 1885 (as modified up to 1st August 1940)*, Govt. of Bihar, 1941.

[12] During the field season of 1908-9, when the cadastral survey and *khanapuri* of Arrah *thana* were in progress, the survey officers came across guzasthadars who asserted that their rents could in no circumstances be enhanced. Some settlement officers, being under pressure from the landlords, were loathe to admit the existence of guzastha tenancies. However, even they could not dispute it in certain cases. A memorial presented by the Landlord's Association in April 1911 argued that there was insufficient evidence to support a custom of guzastha in the strict legal sense of the term. *Final Report on the Survey and Settlement Operations in the District of Shahabad, 1907-1916*, Patna, 1918, pp. 74-5.

[13] Bengal Presidency, *Report on the Administration of Bengal, 1901-1902*, p. 76; *Final Report on the Survey and Settlement Operations in the District of Darbhanga, 1896-1903*, Patna, 1904, p. 28.

in a large part of south Bihar were based on the division of produce, the *thikadari* system facilitated rent collection over large farflung areas. Holding land on a fixed rent from a landlord, many of these tenure-holders were themselves landlords of some kind on *istimrari* or *mukarari* tenure.[14]

It was also observed that some of these petty landlords were increasingly taking to cultivating land on their own to take advantage of the rise in prices. But this was only a weak trend. In our period, rent still constituted the heart of rural social relations while the imposition of capitalism from above did not constitute a growing phenomenon in Bihar. Even during bakasht disputes, the tendency was for landlords to settle land with short-term tenants and sharecroppers in order to make better profits, rather than to cultivate it themselves or employ wage labour. In Shahabad district these lands were tilled by Dusadhs, Chamars, or Ahirs, who were usually 2-5 *bigha* sharecroppers. In some cases the petty landlords also leased out lands from other larger landlords and were themselves categorized as tenants on those lands. What is significant was their considerable local standing in the village and their relationship with the rest of the village community.

Below the strata of the landlord classes lay the vast mass of the peasantry. The 1931 census indicated that about 80 per cent of the cultivating population were tenants or agricultural labourers. In the case of the tenants they were not an undifferentiated social group. The official definition of the 1880s listed different legal categories of tenants on the basis of the nature of the tenure by which they held lands. Apart from the tenure holders, who were effectively zamindars, as we have seen, there were three main categories of tenants. These were (a) tenants on fixed rates, (b) occupancy tenants, and (c) non-occupancy tenants, who also included the under-raiyats of the raiyats holding the right to occupancy.[15]

Fixed rate tenants held lands from zamindars at such rates 'for all time' but made up a very small category and occupied only 2 per cent of the holdings. Only a few of the guzasthadars were lucky enough to be recorded as such. The second category of tenants (that is, occupancy tenants) were the most numerous class of tenants.[16] The tenancy

[14] These tenancies were more stable and the tenants were subject to a fixed rent.

[15] S.P. Srivastava, *A Commentary on the Bihar Tenancy Act*, pp. 11-16.

[16] Kisan Enquiry Committee Report.

measures passed in the 1880s conferred some rights upon them, but these rarely prevented the landlords from circumventing the law. Tenants who had occupied a piece of land continuously for twelve years were deemed to have occupancy rights in land.[17] Legally this meant that they could not be ejected by the landlord except on the execution of decree for ejectment passed on the grounds that the tenant had broken a condition consistent with a provision of the Bihar Tenancy Act or had used the land in a manner which rendered it unfit for cultivation. The occupancy raiyat was also given the right to a fair and equitable rent, though what this would be was not specified in the case of money rent. Rents could be enhanced legally only if the rent paid by the raiyat was below the prevailing rate paid by the other occupancy tenants or if there was a rise in the prices of food crops or rise in the productivity of the land. In each district about 60-80 per cent of the holdings fell into this last category.

The third category of tenants constituted the non-occupancy tenants who did not enjoy any occupancy rights on land. These were tenants-at-will, being mainly sharecroppers. They held a varying percentage of the total raiyati holdings. They also included a large section of under-raiyats, i.e. sub-tenants of occupancy holders, for in many cases wealthier peasants leased out land on rent to tenants. In Saran, for example, they occupied 7.39 per cent of the total number of holdings, although this may have been a gross underestimation.[18] In Purnea, where the intermediate tenure-holders of the Darbhanga raj were recorded as occupancy raiyats in the 1920s, most of the lands in the Kosi Diara region were sub-let to Santhal sharecroppers. In this area more than 20 per cent of the total holdings were categorized as belonging to under-raiyats.[19] The law left this tenant class entirely at the mercy of their landlords. Below them a vast mass of agricultural labourers struggled to survive, forming over 25 per cent of the populations. Some of these held marginal plots of land as their *bari* lands. Most of them were extremely vulnerable to the actions of landlords and to ecological change.[20]

It is important to point out that these official definitions of different categories of tenants do not give us a complete picture of any emerg-

[17] *Bihar Tenancy Act 1885 (as modified up to 1st August 1940).*

[18] J.H. Kerr, *Final Report of the Settlement Operations in Saran, 1893-1901.*

[19] *Final Report of the Survey and Settlement Operations of the Kosi Diara Operations in the District of Purnea 1923-1926*, p. 16.

[20] Census of India, 1931, p. 195.

ing differentiation among the peasantry, even within these categories. There were many intermediate types with features deriving from each of the categories. In the case of the category of occupancy tenants the legal provisions did not provide an adequate barrier against exploitation. Instead, contemporary reports of the early 1930s reported that landlords rarely gave rent receipts and often demanded illegally high rates of rent. If the tenant refused to pay, the landlord would often sue him for four years rental.[21] In this way ample scope existed for landlords to raise rents indiscriminately. It was only through determined local struggle that a small section of occupancy tenants were actually able to resist landlord demands and take advantage of the rise in prices that occurred during the late nineteenth and early twentieth centuries.[22] There is little doubt that there was a section of better-off cultivators emerging from among the occupancy tenants in our period. While the majority were at the mercy of the landlords, some sections were able to defend their traditional rights, as had the guzasthadars of Gaya, and to take advantage of changes in the market, price equalization and inflation.[23] It would be wrong, however, to see them as employing farm labourers on a large scale, or representing a strong trend of capitalist relations developing from below. In some cases these peasants may have cultivated the land through hired labour. In most cases, however, they leased out the land to under-raiyats or sharecroppers and were themselves landlords. The problem, then,

[21] In Gaya the Kisan Enquiry Committee recorded that 64.49 per cent of the peasantry was recorded as having occupancy status. See also Report of Lyall, Collector of Bhagalpur, in Kisan Enquiry Committee Report.

[22] J.A. Hubback, *Final Report on the Survey and Settlement Operations in the District of Shahabad 1907-1916* (Patna, 1928), pp. 55-7, and E.H. Coupland, *Final Report on the Survey and Settlement Operations in the District of Monghyr (North) 1905-07*, Patna, 1908, pp. 26-39.

[23] Between 1914 and 1925 the prices of food grains had doubled. In the late nineteenth and early twentieth centuries, a struggle by some sections of the peasantry took place to take advantage of this and to retain their surplus. These peasants cultivated relatively larger pieces of land, adopted improved methods of cultivation and transported their own grain to the market. In Champaran this section was clearly identifiable. See Colin Fisher, 'Planters and Peasants' in Dewy and Hopkins (eds.), *Imperial Impact*, London, 1978, p. 124 and *Report of the Committee on Agrarian Conditions in Champaran, vol. 2*, pp. 61, 67-70. Settlement reports noted a growing section of the well-to-do cultivators, the rich peasant, being able to retain his surplus and defend his rights. These sections were mainly either Bhumihars or Rajputs and enjoyed a high ritual position in local society. *Shahabad Settlement Report*, pp. 54, 56, 57; Monghyr Settlement Report, pp. 26, 31, 39, 56.

with the categorization into 'rich' peasants or 'dominant' peasants that Alavi and Pouchepadass have tried to make is that these terms are not structurally descriptive and tend to ignore the extensive historical variations developing within these categories.[24]

Below the level of the richer peasants was a stratum of middle-ranking peasants who were mainly occupancy tenants cultivating smaller areas of land, and tied to the market by various rent and debt obligations. To define them as Wolfian 'middle peasants' presents a problem. They were not strictly subsistence-oriented and the commercialization of agriculture had left its impact upon them. They can, however, be defined as intermediate in economic and social status. These were the Koeri or Kurmi small-holders, increasingly aware of their rights and, since the late nineteenth century, actively seeking to improve their low ritual status through participation in various social movements, particularly that of Cow Protection. They had benefited from the growth associated with railways, urban demand, and poppy and sugar-cane production; but they were hit hard by landlord attempts to whittle down occupancy rights. It is not often realized by economic historians that from the 1920s onwards many Bihar landlords were steadily increasing their *Khas* landholdings and that the effects of this consolidation percolated right down to the bottom of the agrarian hierarchy.[25] In some parts of Gaya district where a strong section of Koeri and Kurmi middle-ranking peasants had emerged they harshly exploited their *kamia* labour force.[26]

Below the sector of middle peasants lay the vast majority of the poorer peasants, including the weaker occupancy raiyats, the non-occupancy raiyats and under-raiyats. Those without occupancy status possessed no formal legal provisions to their tenure and were often reduced to mere tenants-at-will. Their numbers were increasing throughout our period as the landlords successfully attacked occupancy rights. The Santhal *bataidars* of Purnea belonged to this category. With small holdings such people found themselves unable to with-

[24] Jacques Pouchepadass, 'Peasant Classes in Twentieth Century Agrarian Movements in India' in E.J. Hobsbawm (ed.), *Peasants in History: Essays in Honour of Daniel Thorner*, Calcutta, 1980, p. 147.

[25] *Khas* holdings denote the homestead lands of the landlord. Some work has been done in this regard by B.B. Chaudhuri; see 'The Process of Depeasantisation in Bengal and Bihar, 1885-1947', *Indian Historical Review*, July 1975, vol. ii, no. 1, p. 142.

[26] Gyan Prakash, *Bonded Histories: Genealogies of Labour Servitude*, Cambridge, 1990, p. 181.

stand the onslaught of the landlords, and were rack-rented and left with little surplus.[27] As a consequence of this vulnerability the 1930s and 40s saw a gradual change take place at the bottom of the agrarian hierarchy. As occupancy rights were dissolved the percentage of landless increased.[28] Where earlier they might have owned small plots of land, they were now landless and forced to eke out a living on the lands of their better-off neighbours, the occupancy tenants, or on the lands of the zamindars. These essentially expropriated groups belonged to the low caste of Musahars or Dusadhs whose situation, though never enviable, was quite clearly getting worse. Traditionally these castes had a low status. They lived in a separate neighbourhood in the village and were known as *chhote log* (small people) or *Razeel*, while the upper castes were called *Asraf.* A local proverb recorded the essence of this low status at the village level. In these terms 'a kayastha does what you want on payment, a Brahmin on being fed, paddy and betel on being watered, but a low caste man on being kicked'.[29] They maintained service ties with the village as artisans, washermen and barbers and had combined their service occupation with small-scale cultivation. As the traditional *Jajmani* ties[30] in the village loosened, the plight of these untouchable service castes steadily worsened and they gradually became pauperized. Thus, with the changes taking place in rural society, the larger farmers and landlords were gradually stripping away many of the economic and social ties that bound them to the poorer castes. They were raising rents, evicting labourers from their minimal bari lands, paying them less grain for

[27] The Shahabad Settlement Report recorded that the most troublesome disputes in Arrah were between co-sharer proprietors and low-caste tenants. The disputes in 1910-11 were over local landlords claiming lands as their bakasht and tenants claiming occupancy rights; see p. 50.

[28] The landless also included those who were traditionally bonded. These attached workers were known as harwahas or charwahas in Saran and Muzaffarpur districts and as kamiyas and bhagats in Manbhum and Singhbhum districts. See *Report of the Commission of Agricultural Wages in India, 1952*, p. 68.

[29] G. Grierson, *Bihar Peasant Life*, Calcutta, 1885, p. 328.

[30] These ties denoted a system of patron-client relationships between the landlord and the service castes in the village and were culturally and ritually structured. For an interesting study of ritual practices reflecting social relations at the village level, see Gyan Prakash, 'Reproducing Inequality: Spirit Cults and Labour Relations in Colonial Eastern India', *Modern Asian Studies*, 2 (1986), pp. 209-30.

traditional services, and cutting back on their ceremonial and charitable obligations within the community. As the percentage of landless peasants increased, differences with regard to wages became more important.[31] Furthermore, landlords even began reducing the quantity of grain that the labourers were getting. Coercion was also employed to exact more work from the labourers by, for instance, compelling them to guard the fields instead of going home at night, and by forcing women to carry the harvested grain from the fields to the *Khalihans* for no extra remuneration.[32] Women labourers were particularly exploited as they were paid less and were often sexually harassed by the zamindars and his *amlas*. The rape of lower-caste women by Rajput and Bhumihar landowners was common in Shahabad where, by the 1930s, anger and resentment against the frequent violation of women were openly expressed on the platform of the Tribeni Sangh.[33] Control of landless labour through coercive means added a new dimension to an older relationship.

THE DYNAMICS OF THE SUBSISTENCE ECONOMY AND THE ECOLOGICAL BASIS OF PEASANT PROTEST

Peasant agriculture in Bihar was always precarious and susceptible to periods of famine and scarcity. Historians have, however, neglected to examine the detailed working of this part of the economy.[34] There

[31] In Saran district by the end of the nineteenth century, 15 per cent of the inhabitants in every village were landless and were drawn from the lowest rung of the social and ritual order. See A. Yang, *Limited Raj: Agrarian Relations in Colonial India Sarsu District, 1793-1920*, Berkeley, 1989, pp. 48-9.

[32] See *Survey and Settlement Report, South Monghyr*, Patna, 1908, p. 179. These labourers included the bonded serfs of Kamias and casual labourers. Kamias were bonded labourers who contracted their whole labour to meet debt obligations. They were most common in Gaya and where great estates rather than peasant cultivation predominated. See Gyan Prakash, 'Production and Reproduction of Bondage, Kamias and *maliks* in South Bihar, *circa* 1300 to 1930s', Ph.D. thesis, University of Pennsylvania, 1984, p. 49.

[33] The legend of Mahtin Dai, a woman who was violated by the local landlord at Bahea and in whose memory a temple was erected, symbolized this exploitation for the lower castes. See K. Mukherjee and R. Yadav, *Bhojpur Naxalism on the Plains of Bihar*, pp. 26, 35.

[34] There are exceptions to this general statement. Historians such as Elizabeth Whitcombe and Shahid Amin have succeeded in painting a rich picture of the UP countryside in the nineteenth century and have dealt extensively with peasant commodity production. Whitcombe's work in particular contains a

is a thus a great paucity of literature on daily peasant life, the harvest calendar and on local irrigation systems and the nature of the subsistence economy. The non-arable dimension of the agrarian economy is the least well-documented. In turning to these matters a basic premise accepted here relates to the critical part played by the marginalization of peasants in terms of their access to the non-arable parts of land. The importance of this dynamic of ecological transition in eliciting peasant actions cannot be underestimated.[35] An attempt has been made in this section to highlight the nature of the ecological transition in Bihar in the nineteenth and early twentieth centuries. It is quite clear that when combined with other causes, the erosion of common property resources and irrigation rights was very important in fuelling peasant discontent.

In every district in Bihar the rice crop was the main crop in the year. Other crops included barley, pulses, peas, gram and cotton. Poppy, which had been widely cultivated in the nineteenth and early twentieth centuries as a commercial crop and occupied up to 10 per cent of the land gradually came to be less important to the peasant economy.[36] There were three main harvests, the *bhadoi* (summer), *aghani* (winter) and the *rabi* (spring) harvests.[37] While the canal areas yielded 3 crops in a year the less fertile lands yielded only one rabi crop and some bhadoi, and the fields remained barren the rest of the year. Cultivation was organized according to the lunar asterisms. In

---

wealth of information on almost every aspect of country life, from the implements used by peasants to the survey plans of engineers working on the western and central UP canal system. Shahid Amin, drawing inspiration from Whitcombe's work, has concentrated in his work on the small peasant sugarcane economy in UP. See Elizabeth Whitcombe, *Agrarian Conditions in Northern India, vol. 1, the United Provinces under British rule, 1860-1900*, Berkeley, 1972; and Shahid Amin, *Sugarcane and Sugar in Gorakhpur: An Enquiry into Peasant Production for Capitalist Enterprise in Colonial India*, Delhi, 1984.

[35] The term 'ecological' has been used here very broadly to denote the constraints over water, forests, and other natural resources to which the Bihar peasants traditionally had access.

[36] In Gaya district the cultivation of poppy under government monopoly depended on small Koeri peasants who worked with family labour. These peasants grew poppy because of the attractiveness of cash advances given by government agents. With the rise in the prices of foodgrains and other cash crops, opium cultivation lost out and was replaced by sugar-cane. See *Gaya District Gazetteer*, p. 112.

[37] Grierson, *Bihar Peasant Life*, p. 183.

this system the Indian year was divided into twelve lunar months of 30 days, each of which was further divided into lunar asterisms.[38] Every agricultural operation commenced in a certain asterism. The occurrence of rainfall in particular lunar asterisms was considered important, while the timing of many of the agricultural operations was based on traditional systems evolved regionally among the Bihar peasantry. Cultivation commenced in *Jeth*, in the asterism of *Rahu*, the time when ploughing and sowing began. The rains of *Mirigsira* were not considered good and hence no sowing was done in that asterism. In *Aradara* sowing was recommended and transplanting was done for the winter (aghani) crop. This went on into *Puranhas* and *Pukh* if the rains were late. In *Hathiya* rains were very important both for the winter crops and for the sowing of rabi. Timely rains were crucial for the success of any agricultural operation and many of the local proverbs hinged on this. One proverb noted: 'If *Aradara* does not rain at the commencement and *Hathiya* at its end, saith dak here o Bhillari the cultivator is crushed.' Another noted: 'If a single shower comes in *Swati* it enriches the people so much that even kurmi women get gold earrings to wear.'[39]

The normal practice adopted after harvest ensured that a greater part of each crop was retained for subsistence until the next crop. However, for large sections of the peasantry, who were tied to the market by various rent and debt obligations, so little was left of their own crop for subsistence needs that grain had to be bought on the open market. Where money rents prevailed a considerable portion of the rent of the land was paid from the sale of food grain. The largest revenue instalment, known as *quist*, was fixed by the time of the ripening of aghani rice which took place in mid-winter. At the time when the rents were due, that is, at the periods of the chief harvests, the retail prices of foodgrains were at their lowest.[40] The peasant often

[38] There were 27 of these in each year and consequently 2 ¼ in each month. Each asterism was not of equal length. The longest was *Hathiya*, which was 16 lunar days. The others ranged from 13 lunar days in the dry months to 15 lunar days in the wet.

[39] John Christian, *Bihar Proverbs*, London, 1891, pp. 211, 220. For a paper that tests the scientific value of traditional sayings, see V.K. Okhri, 'How Custom-bound is the Indian Farmer?', *Indian Economic and Social History Review*, 18, 1981, pp. 159-88.

[40] Settlement Reports and *Report on Food Grain Supply and Famine Relief in Bihar and Bengal, 1874*, p. 9.

sold below this price to traders, ensuring profits to the latter. In fact, a large portion of the crop would be hypothecated to traders from year to year. They made financial advances against it and the crop was then removed by these middlemen as soon as it was reaped. Grain sold for rent was sold at stated times and was cheap, while the grain sold for other necessities which were not so pressing was sold later. If the harvest fell short the cultivator could not hope to hold on to his meagre stocks but was forced to sell the grain at the expected time to meet the rent charges, which were fixed and constant. It was noted by the Kisan Enquiry Committee Report in 1936 that if tenants tried to delay rent payment to the end of the year they were charged a 14 annas fine for every failure of the quist. In the case of poor harvests the *bania* benefited enormously, for he was able to play the market by stockpiling the grain.[41] In this way the bania retained nearly 6 months supply of grain. In such a situation, when the peasants had to buy their food on the market, famines and scarcity were the common lot of the rural poor.[42]

An examination of the market nexus is relevant here. At the base of the local trade were found the *grihastha beparis*, who occupied a predominant position in the complex of agrarian relations which existed at the village level. These were often rich peasants who possessed *golas* (or stores) where grain could be kept for 5-6 months. They purchased the produce of their neighbours at harvest by means of advances, even before the crop standing in the fields was ripe.[43] However, they generally lacked the ability to move grain from the village to the market which the *baladiyas* possessed. They generally sold the grain they collected to the baladiyas or hired oxen from the latter to carry it to the market.[44] The baladiyas were responsible for forging links emanating from the muffussil towns and *gunges* (markets) towards the countryside. In Shahabad the baladiya beparis clustered around towns such as Beloli, Dumraon, Sasaram and Tilothu. In turn the merchants of these towns used the beparis both to buy up the produce from the countryside, and to push into rural areas

[41] As there were few organized and regulated mandis the cultivators were forced to sell their produce to the village bania. Very little of the produce was brought by the peasants themselves to the secondary markets.

[42] See Reports on the 1866, 1874, 1896 famines and the Kisan Enquiry Committee Report 1936.

[43] Ibid.

[44] Buchanan and Hamilton, *Shahabad*, p. 431.

commodities that they had imported from outside. Such merchants were usually men of substance and were variously known as *goldar mahajan, bhoosah mahajan,* or *bepari mahajan.* They very often carried on moneylending activities alongside other trading activities.[45] Where the peasants tried to sell their grain in the market independently of the landlord or the trader, they were taxed. Furthermore, a whole variety of taxes were imposed on village *hats* (markets) by landlords. Ultimately these taxes implied a further reduction in price for the tenants.[46]

An impression of the extent of grain price fluctuations can be gained from the report of one district officer in his recording of the following facts. The harvesting of each of the three crops ushered in a distinct fluctuation in prices. Grain was cheap in September immediately after bhadoi, but less cheap in May when the rabi had all been gathered home. It was cheap again in November when the rice reached the market. Conversely, grain was dearest immediately before each of the three harvests was reaped, that is, in October, February and July. The advent of rice on the market in November did not usually have a dramatic effect on prices, even though it was the largest and most important crop in the province.[47] This was because it was mainly exported and was never cheap. Rice was rarely, therefore, the staple food of the poor in many districts but was rather the staple of the well-to-do. In Patna division the majority of the population lived on bannocks made of flour prepared from wheat and barley or from many kinds of coarse grain and pulses. These cakes were eaten with vegetables, salt and a few condiments. Maize was an important part of the diet of the poor and was consumed as thick pottage or flour. The other coarse grains consumed north of the river Ganges were gram, marua and common rice. It was for this reason that the bhadoi harvest (which was mainly coarse grain and millets) had the greatest influence on local prices and the availability of foodgrain. The most important indicator prices in these districts, for the purpose of understanding the dynamics of subsistence, were those of coarse grains: marua, maize, and millets. It was noted in Darbhanga that in normal years very little rice was consumed by the lower classes and that, during famine years, maize and marua disappeared from the market long before stocks of

[45] Ibid., p. 684.
[46] Kisan Enquiry Committee Report.
[47] *Report on Food Grain Supply and Famine Relief in Bihar and Bengal, 1874.*

rice did.[48] While, in terms of the overall crop percentage, the bhadoi harvest rarely exceeded one quarter of the crops cultivated, it did provide the main foodgrain for the lower classes so that the loss of bhadoi was serious in districts such as Champaran, Muzaffarpur and Darbhanga. In south Bihar, rice cultivation actually declined in the twentieth century. One reason was the decay of the traditional irrigation systems. By 1914-15 crop composition had changed in favour of cheaper gram and pulses in Patna, Gaya and Shahabad.[49] Even in normal times, as the poor peasant usually sold his harvest on the market to meet various rent and debt obligations (as well as buying his foodgrain on the market), the prices of these coarse grains affected him deeply.

Bihar was hit by severe famines in the last decades of the nineteenth century. The British generally followed a policy of interfering as little as possible with normal trade in response to such extremity. This non-intervention only worsened the situation.[50] British famine policy throughout the nineteenth century remained stubbornly subordinate to the development ethos. In the 1880s the Famine Commission had formed a new policy of famine relief but it proved ineffective in dealing with any large-scale food crisis.[51] The people were therefore often left to their own devices as in Saran district. In the course of the 1896 famine, the collector of Saran was led to remark extensively on the precarious condition of the peasants in his district. His comments

[48] *Appendices to the Resolution of the Government of Bengal on the Famine of 1896-1897*, p. 32.

[49] Nirmal Sen Gupta, 'The Indigenous Irrigation Organisation in South Bihar', *Indian Economic and Social History Review*, vol. xvii, April-June 1980.

[50] It was noted that great famines hit the country at intervals of 50-100 years. In the eighteenth century Bihar had a severe famine in 1770. This was a famine which was historically important in popular memory. A new era and a new population seemed to reckon from this date. In the nineteenth century the important famines were those in 1866, 1874 and 1896. For British policy on the 1874 famine, see *Minute by Sir Richard Temple Lieutenant Governor of Bengal, 1874*.

[51] As I. Klein notes, 'the new codified policy was tested in the blitzkrieg famines at the end of the century. It was found most wanting, least able to save lives where the code was applied in orthodox terms. People perished in large numbers where local officials were most suspicious about distress and most restrictive in offering relief or villagers were least able to take relief at large distant works'. See I. Klein, 'When the Rains Failed, Famine, Relief and Mortality in British India', *Indian Economic and Social History Review*, 21, 2 (1984), p. 200.

are instructive. One fourth of the villagers were agricultural labourers, some of whom owned meagre plots of land. Half of them were poor peasants cultivating up to three acres. Nearly 95 per cent of the peasants were in debt to the moneylender for various rent and cultivating obligations. Moneylending terms were usually very harsh and the interest was commonly 25 per cent. Grain advances were made by the moneylender at bazaar rates and refunds were made in grain at the time of harvesting. During the slack season hundreds tended to migrate to Bengal and Assam to tide over the inter-harvest period, returning for the bhadoi harvest in May. The 1891 census recorded that 180,000 men and 97,000 women from Saran were residing in other districts of Bengal.[52] Migration was thus a common means of survival for the poor peasantry, both men and women, and was not confined only to famine years. Most districts recorded large numbers of the poorer peasantry migrating during the slack season. In Saran the collector recorded that nearly 10 per cent of the adult male population of Sewan and Gopalganj migrated. Most of it was for seasonal agricultural work in the contiguous districts of rural Bengal.[53] Even in times of good harvests, for the poor cultivator little or nothing remained of a good harvest after the landlord and the moneylender had been paid. All this meant that, during a poor harvest, the peasant was frequently pushed across the critical resource level and succumbed to starvation.[54]

Critical to the subsistence economy of the poor peasant in Bihar was the retention of traditional rights to forests, pasture, and irrigation. Forest rights in particular were crucial to survival in times of

[52] *Appendices to the Resolution of the Government of Bengal on the Famine of 1896-1897*, p. 197.

[53] No detailed studies exist on the nature of migration from rural areas in Bihar. For Chotanagpur, however, Prabhu Mohapatra has examined tribal migration to the coal fields. See 'Coolies and Colliers, a Study of the Agrarian Context of Labour Migration from Chotanagpur, 1880-1920', *Studies in History*, vol. 1, no. 2, July-December 1985. For a more general study, see Gail Omvedt, 'Migration in Colonial India: The Articulation of Feudalism and Capitalism by the Colonial State', *Journal of Peasant Studies*, vol. 7, no. 2, 1980, p. 188.

[54] Even in normal times the incidence of rural under-feeding was quite high. In periods between harvests the cultivator's resources were usually stretched to the limit and he depended on certain subsidiary crops known as *baisatha* and *jethia*, (which included sweet potatoes, melons and cucumbers) for survival. See K. Prasad, *The Economics of a Backward Region*, p. 252.

famine.[55] District officers in 1895 recorded that access to forests, fruits and other products during the famine was decisive in allowing some sections of the population to survive the crisis. These rights included access to wood for fuel and forest fruits, honey and other products.[56] The forests also included what were only nominally 'forest lands'. Originally these had often included the vast area of grazing grounds held by villagers as common land. This land provided valuable grazing and fodder supplies, and, to a smaller extent, also provided building timber and firewood. Other important common rights in nutritional terms included those relating to fishing and hunting.

Few studies have been conducted on the dynamics of the erosion of common rights in Bihar or elsewhere in India in the late nineteenth and twentieth centuries. However, there is no doubt that a substantial and continuous erosion of access to common property resources took place. Throughout the period under discussion landlords introduced grazing fees, contracted fishing rights away and curtailed forest privileges. Customary rights connected with grazing cattle, cutting timber and collecting forest produce (in such parts of Bihar as the Nabinagar and Aurangabad thanas), were at first recorded by the British in a special *khaitan* called *Fard Rewaz Jungle* which was then published as part of the record of rights. Later, when the Board of Revenue objected, it was decided to discontinue this record of jungle rights and to record them as special rights in the khaitan of the landlord's uncultivated land. This important development legally bolstered landlord control over common lands. In these areas it wiped out, at a stroke, traditional peasant rights over common property resources which could now only be held at the behest of the landlord.[57] A similar process of erosion took place in Purnea district. Here grazing rights were well established through tradition and custom.

55 To date, most studies of famine have rarely included a study of environmental changes. This is unfortunate as it is becoming increasingly clear that a knowledge of the ecological basis of different peasant economies is crucial to an understanding of the capacity of certain communities to withstand drought and other famine-related hazards. See V. Damodaran, 'Famine in a Forest Tract: the 1897 Famine in Chotanagpur', forthcoming paper.

56 Buchanan has recorded how in times of scarcity the poor were permitted to consume leafy plants and fruits of trees growing on zamindars' lands. But even in his time the zamindars were beginning to claim all the mango trees as their own property, Buchanan, *Eastern India*, p. 239.

57 See *Gaya Settlement Report*, p. 89.

From the end of October to mid June the great grassy plains or *ramnas* were left as open commons on which the peasant's cattle could browse. During the cultivating seasons, the Darbhanga raj would set up a bamboo in the field, a sign that occupation had been resumed and the peasants would move their cattle to the Nepal terai areas. These grazing privileges had always been free and the peasants were acutely conscious of their rights. In 1886, when an attempt was made to levy grazing fees on the cattle of graziers outside the local area, the raiyats made common cause with the outsiders arguing that all cows should be allowed to graze free of charge as they were sacred animals. By 1911, however, the raj was charging a sum of 2-8 annas as grazing fees. All through the twentieth century, not only were grazing fees introduced in Purnea, but pasture lands declined as they were brought under the plough. The peasants were thus forced to move their cattle to the hills more frequently.[58] This was a situation that was common to several districts. In Saran district, by the beginning of the twentieth century, the remaining forests and wastelands were under threat from the extension of cultivation and increasing population pressure.[59]

In the state forests, colonial forest policy further added to these problems and peasants were in many cases excluded from the forests to which they had previously had easy access.[60] The resource crisis was aggravated by the general deterioration of village and private forests, many of which were suffering from excessive grazings and unregulated fellings. The Royal Commission on Agriculture reported in 1928 that the disappearance of these village forests was imminent and that, in consequence, the demands of villagers would bear more heavily on forests specifically under the Indian Forest Department

[58] See *Purnea District Gazetteer*, p. 207-9.

[59] See A. Yang, *Limited Raj*, p. 36; see also P. Robb, 'State, Peasant and Moneylender in Late Nineteenth Century Bihar, Some Colonial Inputs' in P. Robb (ed.), *Rural India*, p. 137.

[60] British forest policy in India attempted to regulate forest exploitation by controlling the collection of forest produce by local people and forest dwellers. The forest acts of 1868 and 1878 resulted in the introduction of imprisonment and fines for what was a new category of crime designated as 'forest offences'. See R. Grove, 'Conservation and Colonialism', (unpublished Ph.D. dissertation, University of Cambridge, 1988) and R. Guha and Madhav Gadgil, 'State Forestry and Social Conflict in British India', *Past and Present*, no. 123, May 1989. See also S. Kulkarni, 'Forest Legislation and Tribals, Comments on the Forest Policy Resolution', *Economic and Political Weekly*, vol. xxii, no. 50, 12 December 1987, pp. 2143-5.

control. Between 1837 and 1937 approximately 35 per cent of forests in north Bihar had actually entirely disappeared.[61] It was reported that, as a result, the propensity of the area to flooding had vastly increased.[62] One manifestation of the resource crisis was the periodic fodder shortage in large parts of Bihar. Peasants beleaguered by dwarf holdings and wages below subsistence level responded by poaching wood and fodder on a large scale. Forest 'crimes' increased in the period as a result. In 1936 the Kisan Enquiry Committee report recorded a widespread erosion of forest rights and rights to common lands. Traditional grazing grounds and even burning ghats in some villages were recorded as being newly settled for cultivation. In south Monghyr, for example, the *ahars* were allowed to silt up and the landlord would deliberately cultivate very luxuriant rabi crops on their beds. This practice effectively destroyed these indigenous irrigation reservoirs.[63] Such measures both cut down the ability to withstand adverse climatic conditions and provoked active resistance.

In 1938 a widespread apprehension of famine developed in Dalbhum where, due to the progress of forest reservation since 1937, many people had been deprived of their pasture and forest rights.[64] In the same year the tenants of Ramgarh raj agitated extensively for the return of their traditional common rights. In the Tharuhal area peasants joined together to demand that free grazing be allowed at the rate

[61] In a recent paper Richards, *et al.* have shown that between 1890-1970 forests and woodlands in Bihar declined by 30 per cent. This deforestation was associated with the creation of 262,107 ha of semidesert scrubland. By the late nineteenth century the remaining submontane forest belt along the northern border between Bihar and Nepal was gradually cut back. The demand of the new shipbuilding industry in Calcutta, of international timber markets, and of the growing commercial markets for indigo and jute reduced the area of these forests. The colonial state was also interested in reducing the Nepal border jungles for military and security reasons. See J.F. Richards, James Hagen and Edward Haynes, 'Changing Land Use in Bihar, Punjab and Haryana 1850-1970', *Modern Asian Studies*, 19, 3 (1985), pp. 706, 710.

[62] For a recent study that notes in passing the impact of ecological degradation in one village over time, see A.N. Das, 'Changel: Three Centuries of an Indian Village', *Journal of Peasant Studies*, p. 16. He records that, as the common lands in the village began to be cultivated, the natural depression which had provided the drainage was demarcated into holdings, thus obstructing the flow of water and resulting in extensive waterlogging after monsoons.

[63] See the Kisan Enquiry Committee Report and *Survey and Settlement Report, South Monghyr*, p. 76.

[64] *Searchlight*, June 1938.

of 7 cattle per bigha, and for non-Tharuhal villages at the rate of 5 cattle per bigha. They demanded that *sabai* grass should be allocated to Tharuhal and Dhagrah villagers at the rate of one anna per *bhar* and that village *purjis* should be issued on this matter. They also demanded that salwood, which was used for building purposes, be granted at half the rate paid by the British contractor working the forest. In the Dumraon raj, in 1938, the zamindari estate started to claim half-rights in timber on the cash-paying holdings of their tenants. This was strongly resisted. Similar agitations were taking place the same year in Mashaurhi where the tenants were agitating against restrictions on the exploitation of trees such as sal for their own basic needs. In Dumri Eaka the tenants raided the Parasnath hill forests and cut forbidden timber on a large scale. In Giridih criminal cases were instituted against Santhals who had been arrested in connection with their attempts to assert their rights to the free use of forest produce. Such incursions, coupled with the erosion of irrigation rights, became major grievances of the peasantry during the period 1937-46.[65]

It had long been noted by district officers that the availability of easy irrigation was crucial for subsistence agriculture and acted as a bulwark against famine and scarcity. However, large parts of the province remained unirrigated by any modern system. While canal irrigation under the Son system watered parts of Shahabad and Patna, most of Bihar was still irrigated by surviving indigenous systems.[66] The district in which such irrigation was best perfected was Gaya which had, since precolonial times, possessed a comprehensive system of reservoirs and channels so that a high portion of precipitation, in rain and streams, was conserved and turned to account. In Gaya this meant that up to 55 per cent of the cropped area was irrigated. More than three-quarters of this was irrigated by local irrigation systems. (See Table 4)

[65] See *Searchlight*, 16 November 1937 and *Searchlight*, 12 November 1937.

[66] Canal irrigation in Bihar was not without its problems and it also became the context for popular actions. While it had increased the fertility of the soil in the areas through which it traversed, the state levied a flat water rate. When in the 1930s the Depression resulted in an abnormal fall in the prices of foodgrains the tenants found it difficult to pay the water rates and an agitation for non-payment developed in Shahabad in the context of the civil disobedience movement. See Kalyan Mukherjee and Rajendra Yadav, *Bhojpur Naxalism on the Plains of Bihar*, p. 2. For a review of canal irrigation in Bihar, see *Triennial Review of Irrigation in India*, 1918-21.

TABLE 4

IRRIGATION SYSTEMS IN THE DISTRICT OF GAYA

Method	*% of area irrigated*
Government canals	4.29
Private canals	15.96
Tanks and Ahars	26.83
Wells	5.98
Other sources	1.77

SOURCE: *Gaya Settlement Report*, p. 136.

In Gaya the irrigation from *pains* and *ahars* was the most important as these two sources between them were responsible for irrigating 43 per cent of the net cropped area. An ahar consisted of embankments, or *alang* running across the line of drainage with other smaller embankments or *bandhuas* running backward along the line of drainage. This structure formed a series of catchment basins or storage places for water. Sometimes there was a spill weir in the embankment (or at one end of the bandhua) called a *kanwa* to prevent excess water accumulating. When the water was required it was obtained either by cutting the alang or by opening a pipe made of a large tree trunk along the alang which at ordinary times was closed.[67]

A pain or irrigation channel could be of any size. The smaller ones were those found originating in ahars and carrying the water of the ahars to wherever it was required. The larger ones had their origin in rivers. Where the river was deep with a clay bed, as in the case of the Poonpoon or the Jamuna, the water had to be physically raised. This required a substantial *bandh* across the river. When the water in the river was low the level was raised by erecting an embankment across the river or by cutting a channel with a sand plough (*kurur*) into the underlying sand in which there was generally water. The larger pains divided into branches (*shakhs*), the water flowing into these branches either naturally or by being diverted at particular times by embankments in the main pain. The water was then taken from the pains to the fields by lifting with a lever and bucket (*Latha kunri*) or by cutting an opening (*mohana*) in the bank which led into distributory chan-

[67] *Final Report on the Survey and Settlement Operations in the District of Gaya (1911-1918), pp. 138-40*, p. 139.

nels.[68] Very often the water from the pain was not used to irrigate the fields directly but was used in an ahar and was distributed from that ahar when required.[69]

Where water was taken by turn from a pain (*parabandi*), the village drawing its customary allotment erected an embankment, usually near the place where the pain left the village, so as to raise the water to a sufficient level to flow into side channels. When its allocation was used up, the village next in line, having previously prepared its own embankments, had the right to cut the embankment of the village whose turn had ended. The most important customs relating to this system were those that took place in connection with parabandi or with the distribution of water by turns. The parabandi arrangements usually began in the month of *Asin* and lasted as long as necessary. These were rights maintained by custom and recorded in the ancient records for the irrigation system. In Tikari estate, for example, the rights were recorded in the *Lal bahi.*[70] Such parabandis were fairly flexible so that, as the pressure on irrigation mounted, changes could be made in parabandi arrangements by landlords. At the Tikari estate, as in others, the irrigation works needed periodic repair, which was usually undertaken by the landlord on the basis of voluntary labour exacted from the local peasants.

A complex system of long-evolved local institutions maintained these irrigation works. These had been gradually eroded as landlords started to use their control over irrigation as a means of enhancing rents. In particular, they often neglected the local system in cases where commutation from produce rent to cash had occurred. From hindsight, there was a peculiar and significant symbolism in the 'interruption' of this life-blood of the agrarian system, which had previously bound landlord and tenant together through the equity of water. Once this hydrological symbiosis was broken, it laid the way open for other forms of petty exploitation. The abuse of the 'hydrological moral economy', one of the most vital parts of the whole agrarian moral economy, was almost uniquely destabilizing.[71]

[68] It was noted by the district officers that, although the word 'mohana' denoted 'opening', it was commonly used to mean 'the channel leading from the opening'.

[69] *Gaya Survey and Settlement Report*, p. 139.

[70] Ibid., p. 144.

[71] While Nirmal Sen Gupta has summarized the information on the local irrigation system in south Bihar, he makes no attempt to relate the unrest in the countryside in the 1930s and 40s to the decay of traditional irrigation methods.

An appreciation of this agrarian structure and the nature of its indigenous hydrological and ecological basis is essential to conceptualizing and explaining the emerging agrarian unrest in the countryside. In the first place the Settlement Reports allow one to construct a picture of growing tenant disputes. High rents and illegal exactions had been a phenomenon in Bihar ever since 1793. Evictions of tenants and indiscriminate raising of rents continued even after the tenancy legislation of 1885 had laid down rights of occupancy, as landlords circumvented the legal barriers in various ways. By about 1910 these largely illicit pressures had increased even more. Landlords rarely gave rent receipts to occupancy tenants. Instead they enhanced rents arbitrarily through the exaction of *abwabs*, used their 'certificate' procedure to deprive tenants of their lands, and troubled them in other ways. Traditional rights were being taken away, including vital common access to forests and water resources. Local irrigation systems were also deliberately neglected. As they fell into disrepair the removal of these supports threatened the fragile economic base of large sections of the peasantry. It is not surprising, therefore, that when popular resistance took an active form in 1937-9, and in 1946, these grievances, emerging from an often wholesale ecological expropriation, were the subject of specific attention and illegal popular action.[72]

It has been argued that the commercialization of agriculture, the growth of cash crops like poppy and sugar-cane, increasing export demands, and better communications since the early part of the twentieth century had benefited some sections of the peasantry. I have argued earlier that in some very limited way this may have been true. However, there was a marked differentiation among the peasantry and the majority of the poorest people did not benefit from these changes. Those who did gain, the wealthier sections in rural society, included those with larger areas of land, occupancy rights, higher ritual status and some ability to corner growing demand and the benefits of rising prices for themselves. These sections included tenants of a higher caste status such as the Rajputs and Bhumihars while the middle sections included Kurmis, Koeris and Ahirs. Reports of the period mention a class of better-off cultivators who transported their own

---

See Nirmal Sen Gupta, 'The Indigenous Irrigation Organisation in South Bihar', *Indian Economic and Social History Review*, vol. xvii, April-June 1980, pp. 157-90.

[72] Kisan Enquiry Committee Report, and *Settlement Reports for Monghyr, Shahabad, and Gaya.*

crops to the market and were able to take advantage of the price inflationary tendency in the 1920s. Nevertheless, it can be said that this betterment of the middle sections of the peasantry was highly fractured. It was constantly being eroded by landlord attacks on occupancy rights, by the whittling-down of common rights to forests and irrigation and by the extra-economic coercive powers of landlords, especially those of smaller landlords who were in many ways emerging as more oppressive and exploitative than their richer peers. By 1930 the effects of the Depression served to further worsen the situation of large sections of a peasantry already weakened by a continuous series of attacks by landlords on various parts of their agrarian survival system.

GROWING PEASANT UNREST AND TENANT DISPUTES

Rents in Bihar were collected both in cash and in kind. Traditionally the better lands paid money rents while the poorer lands paid produce rents. Buchanan had noted in 1818 that:

> Except for the rich inundated land near the Ganges, and the fields and gardens around the villages, the whole rent is paid by division of crop. On the banks of the Ganges where alone all the rents are paid in money and where the rent is fixed, the tenants are by far the most industrious and the least necessitous though the rent is high (Rs 5 per acre). The fertility, together with a ready market, the fondness of Hindus for the sacred stream . . . and money rent have rendered this the most populous and thriving part of the district.[73]

A century later the relative attractiveness of money rents was changing. While landlords realized the benefits of a produce-rent in a period of rising prices peasants were beginning to agitate for cash-rents. South Bihar showed a higher incidence of produce-rent, while north Bihar had developed a more cash-rented character. In Gaya, during the 1930s, the percentage of land held under produce-rent was 67.9 per cent. In Patna it was 4.4 per cent, in south Monghyr 32 per cent and in Shahabad 21 per cent; in Champaran and Darbhanga, it was 4 per cent and 8 per cent respectively. Both cash-rents and produce-rents were usually very high. However produce-rents, unlike stable cash-rents, were much more arbitrary in their administration.

[73] Buchanan, *Eastern India*, p. 325.

There were various systems of produce-rents in operation, of which the most common were the *danabandi* and *batai*. In Gaya a combination of these two systems was found in certain places. In batai the crop was usually divided on the threshing floor after the harvest, while in danabandi the amount of crop was appraised before the harvest by the zamindar's amlas, who were often very arbitrary in their decisions. The produce-rent system greatly hampered the tenants. Under it even occupancy tenants paid away rents which aggregated nearly three-quarters of the value of their crops. Hence there developed a campaign by tenants in the late nineteenth century to commute the rent in kind to rent in cash. The price-inflationary movements since the early part of the twentieth century contributed to the attractiveness of a stable money rent.[74] In contrast, and at the same time, a growing tendency developed among landlords to collect their rent in kind, even in formerly cash-rented areas, to take advantage of the price rise.[75] Pressure by cultivators, especially the more substantial ones, to resist landlord attempts to de-commute cash-rents into produce-rents, and an increasing appeal for commutation in areas of produce-rent, resulted in a growing struggle between landlords and tenants. The arena of conflict centred around the issue of commutation, with both zamindars and raiyats trying to take advantage of inflationary price tendencies in various ways.[76]

The more substantial cultivators had already taken advantage of the occupancy rights conferred on them by the 1885 tenancy legislation. When the record of rights was being prepared in 1910, they applied for commutation.[77] These applications increased from a yearly average of 2,628 between 1911-12, to 3,288 between 1926-7. Where the commutation was carried out in south Bihar areas, the landlords sought to get their own back on these tenants by intentionally neglecting irrigation works and scaling cash rents to very high

[74] *Final Report on the Survey and Settlement in Saran, 1915-21*, p. 36; *Shahabad Settlement Report*, pp. 57, 92.

[75] See Bihar Board of Revenue, average prices of staple foodcrops from 1888 for Bihar, 1928, quoted in Rakesh Gupta, *Bihar Peasantry and the Kisan Sabha*, p. 33.

[76] In Madhupur village, Piro thana, there were large-scale disputes with the landlord claiming all land as held on produce-rent with cash-rents only for special crops, while the raiyats, to a man, claimed to hold the lands on cash-rents. See *Shahabad Settlement Report*, p. 57 and *Gaya Settlement Report*, p. 117.

[77] *Report on the Administration of Bengal, 1908-1909*, p. 10.

levels. In most cases the tenant was actually able to meet these rents, given the rising trend of prices.[78]

The main pressure the landlords attempted to exert in order to force peasants, especially in south Bihar, to forgo commutation, consisted in their hold on irrigation. The costs of maintaining the separate kinds of local irrigation systems, with their ahars, pains and wells were often difficult to estimate. Estimates given by landlords were liable to be excessive, due to the fact that the labour for the original works had traditionally been given *gratis* by the raiyats themselves.[79] Throughout our period the custom of assembling one man from each house to assist in annual repairs remained in full force. Nonetheless the official records estimated the following costs for constructing and repairing these systems. [See Table 5]

TABLE 5

COST OF CONSTRUCTING AND REPAIRING IRRIGATION WORKS IN GAYA

Class	*No examined*	*Cost (in Rs)*	*Annual Cost (in Rs)*	*Area Irrigated (in acres)*
Pains	75	3,54,389	10,045	39,798
Ahars	103	94,488	3,352	14,538
Wells	223	32,480	443	3,670

Allowing 10 per cent on the capital costs, plus the annual cost of repairs, the costs to irrigate an acre per annum were as follows,

	Rs	*a*	*p*
Pains	1	2	0
Ahars	0	14	0
Wells	1	0	0

SOURCE: *Gaya Settlement Report*, 1918.

[78] The rents ranged from Rs 15-40 an acre in Patna, Gaya, Monghyr districts, Kisan Enquiry Committee Report.

[79] Accounts from Basti district in neighbouring UP suggest that the construction of a *pucca* well was both a communal and ritual act. Zamindars sent men, women and children to collect wood for the firing and baking of bricks and the *sattu, gur* and liquor received by them were regarded more in the sense of a marriage feast than of remuneration. See Shahid Amin, 'Gandhi as Mahatma: Gorakhpur District, Eastern UP, 1921-22', in *Subaltern Studies*, vol. 3, pp. 45-6.

Taking everything into consideration irrigation never actually cost the landlord more than 8 annas an acre based on 1918 figures. Most of the repairs consisted of cleaning out the pains or restoring the embankments of ahars.[80] However, by the early years of the twentieth century, the landlords were already beginning to neglect irrigation works. Especially in areas where produce-rents had been commuted to cash, landlords threatened to neglect irrigation works unless produce-rents were reinstated.[81] Settlement officers noticed that 'peasants applying for commutation to the government have to consider that they will be unable to keep up irrigation works, so that general commutation is a bad idea'.[82]

As prices rose higher in the early twentieth century the landlord directed his efforts not only against commutation, but also against the occupancy rights of tenants. The latter had previously ensured some kind of protection against arbitrary enhancement of money rents. The landlord sought, therefore, to extend his *zirat* lands and convert the original tenant holdings into bakasht.[83] By converting occupancy holdings into bakasht cultivated by short-term tenants, landlords could secure the advantages of cultivation for themselves and make

[80] *Gaya Settlement Report*, p. 145.

[81] The decline of these traditional systems increased the propensity of the area to drought, leading in some cases to hydrological instability. Traditional tank systems formed the mechanism for increasing the recharge of ground water by increasing percolation from surface storage of rain water. The decay of these indigenous percolation tanks meant that, ultimately, a ground water drought ensued resulting in the non-availability of drinking water. As early as 1866 the famine reports had recorded that the cessation of sufficient percolation through the western districts of Bengal (i.e. Bihar) had dried up the streams and generated malaria causing a depopulation of areas close to the Hoogly and Bhagirathi. By the twentieth century the decay of these traditional systems had intensified the problem. In north Bihar severe drought near the tracts close to the Himalayas was common. The Royal Commission on Agriculture noted in 1928 that drought conditions were common in north Darbhanga. Often malaria was generated where streams had ceased to flow. See *Report of the Famine Commission in 1866* and *Report of the Royal Commission on Agriculture, 1928*.

[82] See *Gaya Settlement Report*, p. 129.

[83] Zirat lands formed part of the zamindar's property and were homestead lands of zamindars on which tenants' rights could not accrue. Bakasht lands consisted of lands in which zamindars exercised direct control over cultivation. Although more occupancy rights could accrue here in certain circumstances, the zamindars usually cultivated the land through short term tenancies or sharecropping.

better profits.[84] This process had been going on since 1910. Some detailed examples of the transition are appropriate here.

In Shahabad district the Settlement Report recorded in 1915 that the process of commutation was being resisted by the landlords through exaggerated zirat and bakasht claims. In Arrah a landlord converted 400-500 bighas in a village into his zirat. In Sasaram, the landlords tried to convert the best land in a village into bakasht and to reduce tenants to tenancies-at-will. In Dariapur one landlord, Basudev Singh, decided, unilaterally, to record 500 bighas as his zirat. His occupancy tenants were simply recorded, quite falsely, as tenants-at-will. One well-to-do tenant, Autar Singh, (who owned 29 bighas) decided to take matters into his own hands and, one day in April 1912, Basudev Singh was murdered. Autar Singh and four others were tried and sentenced to death for the crime. As the Settlement Report recorded: 'This case is typical of a large number of encroachments of the occupancy rights of raiyats in Bhabua and Sasaram.'[85] During settlement operations in Gaya district many other landlords claimed original tenant land as their bakasht. In Shahabad, in Baliara village (Arrah thana), there were riots between landlords and tenants immediately preceding attestation. The Settlement Report recorded that 'the most striking feature of the year was the large increase in rent suits due to survey and settlement operations'.[86] In Patna district the Settlement Report recorded that 'it is the aim of the ordinary petty proprietor to have as large an area as possible recognized as his *Khudkasht*, chiefly because he may then hope for a premium, making new settlements'. It is not surprising, then, that the suppression of raiyati rights by proprietors in almost every thana was recorded.[87] Consequently, by 1929, a long tradition of disputes between landlords and raiyats had developed. This level of conflict was to be further exacerbated by the Depression.

The price decline of the Depression period was so dramatic that almost all sections of the peasantry were badly affected, tied as they were to the market by rent or debt obligations. The fall in the prices of agricultural produce began in about October 1930 and by December

[84] The Kisan Enquiry Committee noted the tendency of landlords to settle land on annual *pattas* with high rents.

[85] *Shahabad Settlement Report*, pp. 52-3.

[86] *Gaya Settlement Report*, p. 60.

[87] *Patna Settlement Report*. p. 38.

1931 grain was selling at half the 1929 price.[88] The fall in prices in Gaya district is given in Table 6.

TABLE 6

HARVEST PRICES—GAYA DISTRICT

	1929-30	*1932-3*	*1937-8*	*1942-3*
Winter Rice	100	67.53	57.50	160
Autumn Rice	100	67.53	57.50	160
Gram	100	47.50	50.00	150
Barley	100	80.00	80.00	250

SOURCE : Agricultural Statistics of India 1929-43.

The fall in prices made it difficult for the tenant to pay rents and there was a remarkable increase in rent suits in this period as can be seen from Table 7.

TABLE 7

RENT SUITS IN BIHAR, 1928-35

Year	*No. of Suits*
1928	100,496
1929	123,458
1930	122,030
1931	122,116
1932	121,937
1933	149,036
1934	157,321
1935	180,001

SOURCE: Kisan Enquiry Committee Report.

Tenants generally found it difficult to pay when their rents, as frequently happened, were unilaterally commuted to cash at times of high prices. In north Bihar the burden of cash-rents actually doubled so that raiyats had to sell twice as much produce to pay them. The

[88] R.A.E. Williams, *Bihar 1931-32*, pp. 19-20. See also Kisan Enquiry Committee Report.

evidence tendered before the Kisan Enquiry Committee set up by the Congress in 1936 shows that the problem was bad in all the districts. The poor market for jute, tobacco and chillies had caused a local economic depression among all sections of the peasantry in Purnea at the time. The cattle prices collapsed as well and Ahir traders in Muzzafarpur found they could not sell any cattle. Furthermore, peasants who had traditionally migrated to beat the effects of seasonal hardship found that the Depression had hit the coal, mica, and other industries, closing outlets for migrants to Bengal and Burma.[89] Unable to pay the rents, the tenants fell into arrears. Landlords then took this opportunity to further extend their bakasht cultivation, attack occupancy rights and settle lands on a non-permanent basis.[90] The bakasht problem thus increased considerably during the depression period as the lands, not only of poor peasants but also of the more substantial cultivators, were converted into bakasht on the inability of tenants to meet rent demands. Both big and small zamindars took to large-scale evictions in this period and to settling lands instead with short-term tenants.[91] Some small zamindars took a more personal interest in the land and cultivated it themselves or with their own labour. However, only a limited number of landlords behaved in this way.

The bakasht problem affected almost all north Bihar and south Bihar districts. In evidence tendered before the Kisan Enquiry Committee the tenants listed their major grievances.[92] In Gaya, the tenants of the Amawan Raj reported that:

(1) They were charged very high rents, Rs 10 in the case of *nagdi* rents, and gross overassessment in the case of produce-rents. The tenants complained that where the actual produce of a field was only 3-9 maunds it was assessed at 16 maunds.

(2) They were forced to render *begari* or forced labour and supply vegetables and ghee free of cost.

(3) They were rarely granted rent receipts, and, in the case of smaller landlords, it was noted that rent receipts were never granted at all.

89 Kisan Enquiry Committee Report.

90 Ibid.

91 In fact, small zamindars were much more grasping and rarely gave rent receipts as the evidence before the Kisan Enquiry Committee showed.

92 Kisan Enquiry Committee Report.

(4) Original tenants were being dispossessed and in their places other tenants were being settled.

(5) Their traditional rights to forest produce were being taken away.

The evidence tendered by the tenants in Gaya was reiterated in a similar form by tenants in other districts. High rents and forcible evictions were the most common complaints. For example, in Nawadah area, nagdi rent had been enhanced to Rs 15 per bigha while in Patna it was as high as Rs 20. Where produce-rents were exacted, there was very often overassessment.[93] Imposition of begari or forced labour was a common complaint. In Khuda estate tenants were not allowed to begin cultivation of their fields until the Khas lands which the Khuda raj owned in every village had first been cultivated by means of forced labour. It was noted by the Kisan Enquiry Committee that as 'the malik has got his cultivation in every village, irrigation water was diverted from raiyats' lands to that of the maliks'. In Nawadah subdivision, labourers were paid with half seer grain. The issue of rents was further complicated by the exaction of abwabs in most districts. In Aurangabad ghee was exacted by landlords from Ahirs (poor peasants who reared buffaloes) whilst carpenters and blacksmiths were forced to render free services.

The Kisan Enquiry Committee showed that landlords were also very negligent in maintaining irrigation works. This was also the case in Aurangabad, Patna, Bihar Sharif, Niamatpur, Ramgarh, Darbhanga and other areas. In Ramgarh estate ahars and bandhs were now utterly neglected. Canal rates were also rising and water supply was used in many areas as a form of social leverage upon tenants. In Sasaram common lands were threatened; and pasture lands and even the burning ghats in some villages had been settled for cultivation by the landlords. In moral terms, the expropriation and cultivation of the burning ghats was particularly offensive. It represented both an ecological and cultural alienation.

The committee found that evictions and conversion of tenant land into bakasht and erosion of common rights seemed to be a universal problem.[94] The sale of occupancy holdings was increasing dramatically. In Aurangabad district 6,000 bighas of land were sold in Ibrahimpur. In Jitura, in Shahabad area, 500 bighas were sold in one

[93] Ibid.
[94] Ibid.

village for arrears of rent and 300 bighas were converted into zirat.[95] In one instance, in the zamindari of Nirmal Kumar, lands of former occupancy tenants were given for annual settlement at high rates. In Niamatpur 500 bighas of land were sold for non-payment of arrears, while in Darbhanga, 1,700 bighas had become estate property. This dismal picture forced the committee to conclude that in almost every village land was gradually passing out of the hands of the tenants on default of rent payments, rendering them both homeless and workless.[96]

The peasants had, however, resisted this gradual change in their lives. Sometimes this showed itself in a fragmentary manner through everyday forms of resistance or even limited group resistance. Sometimes these episodes were more forceful.[97] Resistance of a more determined kind was not new to the peasantry. It had first been exhibited in a most emphatic and comprehensive fashion in north Bihar in 1917 against the oppressive excesses of indigo planters.[98] In the 1930s, in particular, the dispossession that had occurred as a result of the widespread seizure of peasant land, as well as the ecological dislocation that resulted from landlord incursions, shocked the affected sections of the peasantry, and drove them to protest vehemently against the erosion of their traditional rights. There is no evidence to show that landlords rationalized their attacks on tenant lands and common rights in any way or attempted to justify their actions. In this sense, one may argue, they were operating in some sort of ideological vacuum and thereby laying themselves more open to agitation and protest. The intervention of the elite leadership of the Kisan Sabha in

[95] Ibid.

[96] Ibid.

[97] Scott argues that most subordinate classes throughout most of history have rarely been afforded the luxury of open and organized political activity. As he notes, 'Most forms of struggle stop well short of outright and collective defiance . . . the ordinary weapons of relatively powerless groups, (are) foot-dragging, dissimilation, desertion, false compliance, pilfering, feigned ignorance, slander, arson, sabotage and so on . . . when such strategems are abandoned in favour of more quixotic action, it is usually a sign of great desperation.' See James Scott, *Weapons of the Weak: Everyday Forms of Peasant Resistance*, Yale, 1985, p. xvi.

[98] Guha notes, 'Just as the time-honoured practice of holding the rural masses in thraldom has helped to develop codes of deference and loyalty, so has the recursive practice of insurgency helped to develop fairly well established structures of defiance over the centuries.' See R. Guha, *Elementary Aspects of Peasant Insurgency in Colonial India*, Delhi, 1983, p. 12.

the countryside further favoured protest as a viable strategy of peasant action. The coalescing of all these factors meant that by 1937-9 the rural areas in Bihar were ablaze with a widespread peasant movement.

SUPPRESSING OPPOSITION: THE CONGRESS AND THE KISAN SABHA, 1937-9

The pressures of peasant unrest during the mid 1930s provided the necessary fuel for the development of a peasant organization formed specifically to defend tenant rights. It also provided the opportunity for elite leaders to intervene and articulate peasant demands more forcibly. This is not to say that a movement would not have developed had the kisans been left to themselves. Given the economic situation outlined above and the growing dispossession and ecological marginalization of tenants it might be fair to conclude that the peasantry were desperate enough to attempt to redress their ills by any means at their disposal. As Charlesworth has pointed out, 'for peasant militancy to be effective, fierce grievances with a wide appeal are necessary'. It may be said that the bakasht issues was one such grievance, and one that held a broad appeal to large sections of the peasantry. The specific conjuncture of the political and economic forces at the time combined to create a movement that had a clearly articulated political leadership. It is the conjuncture I now turn to.

It is not necessary to argue for the primacy of any one factor. The intervention of elite leaders should be seen in historical perspective. This was a period when the politics of nationalism and socialism had begun to penetrate the countryside on a large scale and the intervention of elite leaders was to be expected. It is of course possible to argue that within a historical setting the peasants defined their own goals and targets of agitation.[99] Since 1920 minor peasant associations had sprung up in Bihar to deal with kisan demands. They were of an *ad hoc* nature, usually led by local activists and were rarely effective

[99] This was a phase of more intensified class conflict, widespread anti-imperialist struggle and generally a higher level of militancy among the masses. By recognizing this, one is not negating the participation of the individual peasant in such struggles of his volition or his own perceptions of the constricting nature of landlord exploitation. Even Guha is careful about extending his notion of subaltern autonomy to the 1920s and 30s by which time, as he points out, the politics of nationalism and socialism had penetrated the countryside on a large scale. See Guha, *Elementary Aspects of Peasant Insurgency*, p. 13

before 1928.[100] In 1928, however, Swami Sahajanand Saraswati, a local leader and Congressman, began to interest himself in kisan problems and formed the West Patna Kisan Sabha in March 1928. This organization was a forerunner of the Bihar Kisan Sabha (BPKS), which was formed in 1929. It was at this stage exceedingly moderate and advocated harmony between landlords and tenants. As Sahajanand himself recorded, he had begun the organized Kisan Sabha as a staunch class collaborator in order to try to prevent the landlord-tenant struggle from escalating and thus weakening the national movement.[101] The attitude of the Congress to these associations was still being formulated in this period.[102] When the civil disobedience movement began, the activities of the Sabha were suspended and civil disobedience activities were taken on instead. In rural areas the main issues adopted for mass mobilization were non-payment of canal rates and chowkidari tax. These were not the main grievances of the peasantry but rather irritants that the Congress took on for tactical reasons, as they involved little that was socially divisive.[103] As a result, by the end of the civil disobedience period the Congress had exhausted them as effective tools of mass mobilization and a drive had developed among some Congressmen for changes in policy. This group, organized into the Bihar Socialist party in 1931, stated that their ultimate aim was the formation of a socialist state which would free land and capital from individual and class ownership and vest them in the state.[104] They wanted the Congress to develop the day-to-day struggle of workers and peasants. Though Rajendra Prasad pointed out that the Congress had nothing to do with the Socialist party the pressure mounted for the Congress to interest itself in a more meaningful way in peasant problems.[105] With the

[100] Walter Hauser, 'The Bihar Provincial Kisan Sabha 1929-42', p. 36.

[101] Swami Sahajanand Saraswati, *Mera Jeevan Sangharsh*, Patna, 1952, p. 320.

[102] The formation of the Sabha was regarded with interest by Congressmen such as S.N. Sinha, Baldev Sahay and Mathura Prasad but not many were willing to participate in its activities. But there were others like Braj Kishore Prasad who saw the organization as dangerous and advised Congressmen against joining it.

[103] K.K. Dutta, *Freedom Movement in Bihar, vol. 2*, pp. 116-17.

[104] Constitution of the Bihar Socialist party in BSA (Home Political Special), File 3 (pt 1-A)/1935.

[105] FR (2), July/1931.

organization of their party at the national level in 1934 the Congress socialists soon became a force to be reckoned with within the Congress. This development was compounded by another event. This was the move by the larger landlords in the province, (organized into the United party under the aegis of the British in 1932, as a counter to the Congress), to secure peasant support by affecting minor changes in land tenure law.[106] Rajendra Prasad noted that this event caused anxiety among many Congressmen in the province. That such a reactionary party as the United party could aim to co-opt peasant support meant that the Congress could no longer choose to neglect peasant grievances.[107] The Kisan Sabha under Sahajanand was thus revived in March 1933 and many Congressmen became its members.[108] The Sabha was at this stage considered to be a wing of the Congress, and Congressmen in general welcomed it.

The ensuing period saw the organizational work of the Kisan Sabha being actively pursued, while at the district level bodies of the Sabha were reorganized under its new constitution. Indeed, subsidiary branches of the Kisan Sabha were formed at the level of the district, the subdivision, the thana, and the village.[109] Towards the end of 1934 efforts were made to organize these bodies on an elective basis and increase membership.[110]

As the widespread and growing grievances of the peasantry became apparent the Kisan Sabha moved in to articulate their demands politically. Many meetings and conferences were held as the politicization of the peasantry was seriously embarked upon. Between 1933 and 1935 there were over 500 Kisan Sabha meetings in ten districts of the province.[111] These meetings were addressed mainly by

[106] *Bihar Prantiya Kisan Sabha ke Report 1929-35*, pp. 4-5.

[107] Rajendra Prasad, *Autobiography*, p. 454. See also Sahajanand's meeting with S.K. Sinha where the latter suggested reviving the Kisan Sabha, Swami Sahajanand Saraswati, *Mera Jeevan Sangharsh*, p. 421.

[108] Under the initiative of genuine peasant workers like Jugal Kishore and Yadunandan Sharma, the Kisan Sabha carried on an agitation against the Agrarian Reform Bill of the United party and succeeded in exposing the bogus kisan leaders.

[109] Rakesh Gupta, *Bihar Peasantry and the Kisan Sabha*, pp. 98-9.

[110] Awadeshwar Prasad Singh, 'Kisan Movement in Bihar', *Congress Socialist*, 9 April 1939, p. 3.

[111] *Bihar Prantiya Kisan Sabha ke Report*, p. 10. For the nature of peasant grievances in 1934, see 'Documents Relating to Atrocities Committed on Peasants 1934,' Ramnandan Mishra Papers.

Sahajanand. He later wrote that, by 1934, he was disillusioned by Gandhi's reluctance to sanction kisan demonstrations after, for example, the Bihar earthquake or to pressurize zamindars to reduce rents.[112] Sahajanand had also come to realize the strength and power behind mass kisan demonstrations. This fact was brought home to him most clearly when, in 1933, a simple Kisan Sabha notice for a demonstration for rent reduction in Gaya resulted in thousands of kisans gathering outside the collector's residence.[113] In 1933 the BPKS went on to conduct enquiries into the conditions of peasants in districts like Gaya, Darbhanga and Purnea and published a report on Gaya delineating the distressing conditions of the tenantry. This, they hoped, would strengthen their hand politically. The popularity of the Kisan Sabha was gradually increasing and by 1935 the membership of the BPKS was 33,000.[114] By this stage the Kisan Sabha was fast evolving into a class organization of the peasantry, a development which was accelerated by the close association of the Congress Socialist party with the Sabha. In September 1934, a young socialist, Jayaprakash Narayan, at a meeting of the BPCC, had moved a resolution criticizing the constructive programme of the Congress and urging that the party should instead concentrate on organizing the peasants and the workers on the basis of their economic grievances.[115] Sahajanand, it appears, was strongly in favour of CSP members entering the organization.[116] CSP members therefore started to join the Sabha. The reformed Bihar Kisan Sabha executive included socialists such as Jayaprakash Narayan, Abdul Bari, Puryanandan Sharma, Shyam Nandan Sinha, Ambica Kant Sinha, Ramakrishna Benipuri, and Kishori Prasanna Sinha.[117] The association of the CSP with the Kisan Sabha helped to give the former body a stronger ideological basis while, at the same time, strengthening the links of the Kisan Sabha with the Congress. The CSP continued to recognize the Congress as the principal existing mass organization of diverse elements seeking national liberation. The Socialist party also underlined the urgent need to develop and broaden the Congress base and

[112] Swami Sahajanand Saraswati, *Mera Jeevan Sangharsh*, p. 425.
[113] Ibid., pp. 430-1.
[114] *Bihar Prantiya Kisan Sabha ka Report.*
[115] *Searchlight*, 19 September 1934.
[116] Sahajanand Saraswati, *Mera Jeevan Sangharsh*, p. 435.
[117] Awadeshwar Prasad Sinha, 'Kisan Movement in Bihar', *Congress Socialist*, 9 April 1939.

bring it under the influence of revolutionary socialism.[118] The alternative programme suggested by the CSP aimed at developing peasant and worker organizations and providing Congress committees with a programme of day-to-day work among the masses. A close link between these class organizations and the Congress committees was to be established. Such a programme, it was thought, would lead the masses mobilized in these organizations to enter the Congress and transform its whole structure and leadership. However, a note of caution was introduced in the 'Meerut thesis', which argued that since the main task of the party was to bring all anti-imperialist elements under its ideological influence, it was necessary to be tactful and not to alienate these elements.

The CSP programme for changing the balance of forces in the Congress through the development of class organizations, while at the same time endeavouring not to alienate the right-wing anti-imperialist forces, represented an awkward course to negotiate. Armed with a strategy for developing class organizations, the CSP set about building the Kisan Sabha on a solid ideological basis. The fruits of this soon became manifest in numerous district kisan conferences where the socialists moved resolutions demanding the abolition of the zamindari system. A campaign was then built up to incorporate the demand for the abolition of landlordism into the programme of the BPKS.[119] Sahajanand was wary of this resolution since, as we have noted, he had founded the Sabha when still a staunch class collaborator. He later wrote that his views on the zamindari system were as yet very naive and that he had therefore opposed the resolution at the meeting of the Bihar Provincial Kisan Council in 1934. However, since the district level leadership was demanding it and as his own views were changing, he bowed down to pressure and the BPKS adopted a resolution in 1935 which demanded the abolition of landlordism.[120]

The larger manifesto of the BPKS was also adopted in this period. It continued to define the kisans as all those who lived by cultivation,

[118] The Meerut and Faizpur thesis of the CSP, quoted in A.K. Chaudhuri, *Socialist Movement in India*, Appendix 1 and 2.

[119] Sahajanand Saraswati, *Mera Jeevan Sangharsh*, p. 444.

[120] In its third session at Hajipur, in Muzaffarpur district, Sahajanand writes that he was greatly distressed by the activities of the zamindars during the Bihar earthquake in 1934 and this forced him to realize how grasping the zamindars were. Sahajanand Saraswati, ibid., p. 443.

including petty zamindars and agricultural labourers. According to the manifesto, the poverty and misery of the tenant were due fundamentally to the nature of the land tenure, revenue and credit system and 'the merciless exploitation of imperialism'.[121] The BPKS felt, moreover, that the situation of the peasants could be altered only when changes were made in the land tenure and revenue system and when imperialism was replaced by a national state, which would represent the masses and guarantee enough land to the tillers of the soil. The long term demands of the BPKS were thus that (1) the zamindari system be abolished, (2) the agrarian debt be wiped off, (3) a system of land tenure be established which would permit the peasants the ownership of their holdings, and (4) that employment be assured to the landless. Since these demands could not be fulfilled under the existing system, the BPKS made a list of immediate demands, the important ones being as follows:

(1) Conferment on tenants of fixity of tenure and right of free transfer of holdings.
(2) Provision by legislation of free common pastures in every village and free utilization of forest products.[122]
(3) Abolition of all systems of rent in kind.
(4) Exemption of all uneconomic holdings from rent and taxes.
(5) Immunity from arrest and imprisonment; and attachment of minimum holdings in execution of civil decrees.
(6) Abolition of certificate procedure of the zamindars.
(7) Stiffening of regulations against begari and failure to grant rent receipts.
(8) Cancellation of all previous debts and fixing of a low rate of interest of not more than 6 per cent per annum.
(9) Minimum living wage for agricultural workers.
(10) Abolition of chowkidari tax.
(11) Repeal of all anti-peasant, anti-labour and anti-national laws.
(12) Reinstatement of all peasants deprived of their lands, etc. owing to their participation in movements for their economic

[121] Manifesto of the Bihar Kisan Sabha in the *Congress Socialist*, 25 July 1936.

[122] The prominence of this demand is highly significant.

and political freedom and also owing to their failure to pay revenue or rent during the economic depression.

(13) Abolition of indirect taxes, particularly duties on salt, kerosene and matches.

(14) Adoption by the government of measures for adequate drainage, cheap manure, seeds, livestock and other facilities.

In this fashion, by 1936, the character and demands of the Kisan Sabha had begun to emerge. While in a long-term perspective it visualized a national state with land handed to the tiller, its short-term programme related to the pressing grievances of the peasantry and demands for their resolution within the zamindari system. In fact, as we have seen, in its official programme the Kisan Sabha directed its attack mainly against larger zamindars, while staying silent on the exploitation of smaller zamindars.[123] This was not surprising as, for tactical reasons, the leadership did not want its demands to appear too strident or to alienate petty landlords. During the struggles, however, most of these directives were ignored and, with the intensification of the bakasht struggle, a direct attack on the zamindars, both large and small, was launched to force them to reduce rents and restore bakasht land. It has been suggested that the agitation over bakasht lands, rather than being a radical initiative, comprised an attempt to work within the zamindari system to re-establish rights that had previously existed.[124] It is true that non-payment of rent and a moratorium on debt were not called for by the Kisan Sabha leadership. Instead, at the popular level, a call for the reclaiming of bakasht lands formed the main focus of the struggle, so that the zamindari system itself was rarely directly attacked. This may have been due to the specificity of the bakasht experience and the context of the struggle.[125] Piven and Cloward have recently argued that people first experience deprivation and oppression within a concrete setting, rather than as an end product of large and abstract processes. It is the concrete experience that moulds their discontent into specific grievances against specific tar-

123 The BPKS defined the truly oppressed elements in a rather amorphous manner stating ambiguously that 'a peasant is known as a *grihasta*, a person who earns his livelihood by cultivation and agriculture, be he a petty landlord, raiyat, or labourer working on wages for ploughing fields'.

124 S. Henningham, *Peasant Movements in Colonial India*.

125 Though anti-zamindari slogans were raised in the course of the struggle in some areas the zamindari system was not often the direct focus of attack.

gets.[126] In the same way the Bihar peasant experienced an erosion of common rights, increasing land rents, oppressive landlords, ruinous interest rates and dissolution of occupancy rights rather than the 'zamindari system' in the abstract. Nevertheless the movement effectively challenged the power of the landlords in local society, in an assertion of strength and defiance by subordinate classes.

The dissatisfaction of all sections of the peasantry set the context for the struggle. It may be argued that the widespread dissatisfaction of the wealthier tenantry gave the movement its main impetus. However, the movement also depended on the poorer peasants and sharecroppers, who were losing their petty holdings, and who joined the agitation in large numbers to protest against landlord excesses.[127] Though the movement had little to offer landless agricultural labourers, and was not able to take up any of their demands in some areas, the land/tenant confrontation did provide the context for some of the landless to demand increased wages.

The growing influence of the socialists in the Kisan Sabha, with their interest in transforming the Congress, resulted in an increasingly strained relationship developing between the orthodox leadership in the Congress and the Kisan Sabha itself. Though the Kisan Sabha was seen to be a wing of the Congress a growing struggle took place, especially at the district level, to control the Congress Committees, which were the locus of nationalist political power and influence. In Gaya, for example, the Kisan Sabha succeeded in entering the executive of the Congress in 1933.[128] But differences arose between the Kisan Sabha workers and the right-wing workers, as it was alleged that the Kisan Sabhites wanted to keep the Congress organization under their control, specifically to advance the cause of class organizations. Their differences grew so acute that the activities of the District Congress Committee had to be suspended in 1935 for some time. It was reorganized again in 1936 by a consensus arrangement in which A.N. Sinha, a prominent right-wing Congressman, was made president and Jadunandan Sharma, a Kisan Sabha worker, took over as secretary of the DCC. As we shall see, this kind of struggle was to

[126] F. Piven and R. Cloward, *Poor People's Movements: Why They Succeed, How They Fail*, New York, 1977, p. 20.

[127] Sahajanand admitted in Bezawada in 1944 that the middle and big tenantry benefited most from the Kisan Sabha.

[128] *Searchlight*, 8 January 1938.

spread to other DCCs in Patna, Darbhanga, Muzaffarpur, Champaran and Monghyr, once the Congress accepted ministerial office. A related problem also remained to be faced during the selection of candidates for the elections to the legislature. During the office acceptance debate, the Kisan Sabha had sided with the CSP position and condemned the slide towards constitutionalism.[129] Sahajanand pointed out that the Congress would achieve little in the Councils, apart from cosmetic changes, while the need of the hour was to keep the revolutionary mentality.[130] However, once the decision to contest seats had been taken and the Congress election manifesto was presented, the Kisan Sabha, along with the CSP, agreed to support the Congress election campaign. However, it lobbied for the nomination of as many Kisan Sabha candidates as possible. In practice this did not work out smoothly. Sahajanand, as a member of the PCC and president of the Kisan Sabha, presented a list of candidates from the Kisan Sabha for nomination. In Darbhanga, for example, Sahajanand suggested Jamuna Karjee for one seat and Dhanraj Sharma for another. But Satyanarain, as president of the DCC, insisted on not having either and was supported in this by a majority of Congressmen. It was, therefore, decided to drop Dhanraj Sharma and a different seat was given to Karjee as Satyanarain wished.[131] These events hinted at the nature of the power struggle to follow.

For the time being, however, the intra-party strains were pushed into the background as the Kisan Sabha actively campaigned for the Congress candidates in the elections of 1936. Leaders of the Kisan Sabha believed that the Congress election manifesto was fairly progressive, although it was silent on the prolonging of zamindari.[132] Kisan workers were directed by Sahajanand to work in close co-operation with the Congress, while continuing to maintain the position of

[129] *Searchlight*, 16 September 1936.

[130] Proceedings of the Lucknow Congress in *Indian National Congress Report 1936-48*, p. 79.

[131] Rajendra Prasad to Vallabhai Patel, 17-11-1936, Rajendra Prasad Papers, File no. 1/36, col. 3. In Muzaffarpur, where there were 7 seats, trouble arose over nominations to the Hajipur seat. Sahajanand wanted Kishori Prasanna Sinha, a full time Congress worker, for the seat. However, as the president of the PCC wanted another candidate Kishori Prasanna Sinha was dropped. Sahajanand resigned on this issue from the BPCC and it was only Rajendra Prasad's placatory intervention that settled matters. See Rajendra Prasad to Sahajanand, 13-11-1936, Rajendra Prasad Papers, File no. 1/36, col. 3.

[132] *Searchlight*, 16 September 1936.

the Kisan Sabha.[133] On one occasion, while inaugurating the election campaign at Ranchi, Sahajanand explained the Congress point of view with regard to council entry. He pointed out that Swaraj was meaningless for the masses unless the burden of taxes could be brought down.[134] Jamuna Karjee, another Kisan Sabha worker, averred that the masses had been exploited by the British and that the Congress would restore the balance. Alongside this typically standardized and non-specific Congress propaganda numerous kisan conferences were held and Kisan Sabha workers toured the countryside holding public meetings, most of which obtained a good response from the peasantry.[135] Such meetings were held in Monghyr, Muzaffarpur and other places, where more specific kisan demands were reiterated.[136] For example, at Muzaffarpur a meeting of 500 kisans took place at which, straightforwardly, a reduction of *malguzari* and an ending of begari were called for. Sahajanand noted after his return from Nawadah constituency in Gaya that, even though it was harvest time, kisans assembled in thousands and seemed eager to hear the Congress message.[137] In Shahabad, for example, a group of 60 Congress and Kisan Sabha workers toured the villages of the district with tri-coloured flags, singing national songs and holding public meetings in all the big villages.[138] The apparent end result of this propaganda drive was that the Congress won a sweeping victory in the 1937 elections. Undoubtedly the Kisan Sabha had made a critical contribution to this total rural sweep by the Congress. The victory was important for, although the franchise was limited to about 10 per cent of the population, the electorate nevertheless included the more substantial rent-paying tenantry, so that the pressure to satisfy their important demands could no longer be completely ignored. The extension of the franchise and the participation in the electoral process by new groups

[133] *Searchlight*, 28 November 1936.
[134] *Searchlight*, 13 September 1936.
[135] *Searchlight*, 8 December 1936.
[136] *Nav Shakti*, 17 May 1936 and 31 May 1936.
[137] *Searchlight*, 18 December 1936.
[138] *Searchlight*, 2 December 1936. In Jahanabad area Sahajanand was so sure of his popularity that he proudly proclaimed that in that area anyone backed by the Kisan Sabha would win. Sahajanand noted that the Kisan Sabha wholeheartedly campaigned for the Congress even in those areas where the Congress had sent up zamindar candidates. See Sahajanand Saraswati, *Mera Jeevan Sangharsh*, pp. 481, 483.

were very significant developments. Large proportions of the population were embraced by the rituals of electoral campaigns and were promised a share in the rewards of the electoral system. Political attitudes thus nurtured would not erode easily.

The Congress accepted office in June 1937 and almost immediately the pressure mounted to meet the kisan demands. This situation was compounded by the fact that the Kisan Sabha and the CSP were pledged to develop the day-to-day struggle of the peasantry and to develop a closer association between the Sabha and the Congress in a bid to transform the structure and leadership of the latter. Mass activity outside the legislature thus became the main preoccupation of the left wing of the Congress during the ministry period. The first important manifestation of pressure by the Kisan Sabha surfaced in the charter of demands sent to the assembly on behalf of 50,000 kisans immediately after the Congress accepted office.[139] This charter demanded that (1) arrears of rent be wiped out, (2) malguzari be reduced in half, (3) debt be cleared, (4) interest rates be reduced, (5) illegal exactions be halted, and (6) rent decree proceedings be stalled. The charter was, however, signed by only 60 Members of the Legislative Assembly (MLAs). The Kisan Sabha then mounted an intensive campaign to force the MLAs who had not signed to do so.[140] Many meetings were held by CSP/Kisan Sabha workers such as Kishori Sinha, Awadeshwar Prasad and Dhanraj Sharma to mobilize the peasants into forcing the resignation of those Congress MLAs sent from rural constituencies who had betrayed the kisans and, instead, to send 'genuine' kisan representatives to the legislature.[141] As these meetings gathered force in 1937 it was reported that socialist slogans were being propagated openly against zamindars and 'capitalists' and

[139] Report of the procession in a letter from Sridhar Sharma to Rajendra Prasad, 27-9-1937, Rajendra Prasad Papers, File no. XII/37, col. 1.

[140] Nand Kishore to Rajendra Prasad, 17-9-1937, Rajendra Prasad Papers, File no. XII/37, col. 1.

[141] Ibid. Report of the speeches, Sridhar Sharma to Rajendra Prasad, 25-9-1937, Shyam Sunder Malviya to Rajendra Prasad, 2-10-1937, Nand Kishore to Rajendra Prasad, 17-9-1937. In one speech, Kishori Prasanna Sinha pointed out that the kisans should not trust all the Congress representatives in the assemblies, as some of them, like Ramnandan Sinha, Rameshwar Sinha, and Brajnandan Sahay, had betrayed the kisans by not signing the kisan charter. These men should be asked, he said, to sign or resign and the kisans should pressurize the Prime Minister to send to the assembly the real kisan representatives.

that the kisans were being urged to organize against their own exploitation. Many of the slogans and speeches demanded that the peasants should resist illegal demands by zamindars by 'dragging them to the magistrate'. Indeed the tone of the speeches and the polemic had rapidly become more radical. In November 1936 the Bihar Provincial Kisan Conference had pointed out that the forcible occupation of disputed land was not part of the Sabha's policy. In spite of this, by 1937, many of the Kisan Sabha activists were in the forefront of the bakasht agitation and were allegedly propagating 'objectionable' slogans such as, '*Malguzari kaise loge, lathi mera sathi hai*'.[142]

The radical tone of the kisan workers' speeches was in tune with much of the spreading tenant agitation in the period and with localized direct action against landlords aimed at reducing rents and restoring bakasht land. The simmering rural unrest stimulated by evictions and non-payment of rent in Bhagalpur, Muzaffarpur, Gaya, Patna, Monghyr and Purnea gave force to the demands of the elite leadership of the Kisan Sabha. As the unrest gained strength the elite leadership realized that mere representation of kisan demands through processions, meetings and pamphleteering was not enough. Without doing more the Kisan Sabha ran the risk of being sidelined by a developing mass movement over which it would then have little control.[143] The articulation between peasant protest and organized politics as represented by the Kisan Sabha had thus reached an important stage. Quite clearly and significantly, it was the dynamics of peasant protest that was influencing elite demands by this stage rather than vice versa.

In Monghyr, in March 1936, raiyats started to refuse to do begari in the Barahiya Tal area.[144] Under the leadership of Karyanand Sharma, a Kisan Sabha activist, they appealed to the collector for reduction of rents and restoration of bakasht land.[145] This provoked

[142] AICC Papers G 98/1937.

[143] One British officer on a railway journey through Gaya commented on the large crowds of peasants travelling to kisan meetings in 1938 who clambered on to trains defying railway authorities. He noted that anti-British sentiments were also expressed by the kisans and 'everytime I appeared on the platform the crowd cried: "*Angrezi raj nash ho*" '. See Macdonald Papers, IOL, Mss Eur E 360.

[144] Agrarian Trouble in the Barahiya Tal Area, Monghyr, Home Political Special, File 42 (III)/1938.

[145] Ibid.

the local landlords into retaliating by attempting to oust the tenants who had originally been allowed to cultivate their old lands on a batai basis, thus sharpening the conflict. In most areas, as soon as the tenants showed resistance, the landlords responded by extending their bakasht claims. By 1937 the confrontation was at its peak. It was then reported that gangs of tenants were roaming round seizing the crops of the landlords.[146] In February 1937 Karyanand was arrested, along with nineteen others. At the same time Rajendra Prasad intervened and sought to arrive at a compromise between landlords and tenants. He stipulated that the landlords should settle with the tenants at least that amount of the land which the latter had cultivated prior to the agitation of 1936.[147] This award failed to satisfy the tenants as the landlords could simply transfer the land, given to each tenant, so as to prevent the accrual of occupancy rights. Another problem arose from the fact that since landlords rarely granted rent receipts, particularly on produce-rent areas, tenants had no documentary proof of the title to the land they worked.[148] For these reasons Congress intervention entirely failed in settling the issue, and the rural unrest grew more acute during 1937-8. The agitation in Monghyr in early 1937 was thus followed by a no-rent campaign in Gaya, provoked by the refusal of the landlords to grant rent receipts.[149] In the Patna and Gaya regions Kishori Prasanna Sinha, a CSP leader, was especially active and agrarian tension correspondingly high. In Jahanabad, in April, several cases of arson were reported in which it was suspected that a landlord had set a tenant's house on fire. This was, again, followed by a series of localized peasant protests.[150]

The developing tension in most of the south Bihar districts showed few signs of abating, even after the Congress accepted office in July 1937. This was due to the continuously growing strength of the Kisan Sabha in the countryside. District Kisan Sabhas were functioning regularly in 12 out of 16 districts by April 1937. In 1939 the secretary of the BPKS, Awadeshwar Prasad Sinha, claimed that the Kisan Sabha membership had swelled from 80,000 in 1936 to 4 lakhs by 1939.[151] Congress membership figures had also shown a great in-

[146] FR (1), April/1937.
[147] BSA File no. 42 (III)/38.
[148] Ibid.
[149] FR (2), March/1937.
[150] FR (1), May/1937.
[151] Awadeshwar Prasad Sinha, 'Kisan Movement in Bihar' in *Congress*

crease, from 104,743 in 1934, to 575,139 in 1938-9. This rise was partly a consequence of determined efforts by the socialist kisan workers to win new members, expressly to help them gain control over the Congress.[152]

The bakasht and 'no rent' campaigns, once under way, could not be halted. The movement spread to Tirhut, Champaran and Purnea, while in Gaya and Patna an increase in the number of agrarian disputes was associated with the beginning of the cultivating season.[153] Backed by a developing kisan agitation, the CSP and the Kisan Sabha seemed to be on firm grounds in their effort to challenge the orthodox Congress leadership. A political struggle for control of Congress committees and delegate seats was thus set in motion. The undermining of established Congress authority also occurred in another fashion. The British noted that the activities of left-wing Congressmen had resulted in the emergence of a parallel administration in some areas.[154] In Dinapore subdivision, an Intelligence report noted that, due to kisan activity, police thana officials found their authority being threatened. The Commissioner of Patna also observed that Congress thanas in many areas were functioning for the settlement of agrarian disputes which had not been reported to the police at all.[155] While the British understood the implications of this development for the functioning of the imperial administrative machinery, such activity also undermined the authority of the established Congress leadership which was responsible for the administration in this period. In the eyes of villagers in some districts, the Congress-run state government lacked the legitimacy or the authority to deal with kisan disputes. Instead these were settled by the Kisan Sabha–dominated District Congress Committees.[156]

Socialist, 9 April 1939 and All India *Kisan Bulletin*, 31 March 1939, in Indulal Yajnik Papers, File no. 1/1939.

[152] B.R. Tomlinson, *The Indian National Congress and the Raj*, p. 86.

[153] FR (1), September/1937 and FR (2), July/1937.

[154] Linlithgow to Hallet, expressing concern, 12-9-1937, Linlithgow Papers, F 125/14.

[155] Ibid., quoting extract from an Intelligence report.

[156] In this context, Gramsci's comments on the crisis of authority that develops in a situation of conflict between the represented and representatives are instructive. He noted that 'a crisis of the ruling class's hegemony, occurs . . . because huge masses (especially of peasants and petit-bourgeois intellectuals) have passed suddenly from a state of political passivity to a certain activity, and put forward demands which, taken together, albeit not organically

The Congress leadership attempted to deal with this developing situation with its tenancy legislation of 1937-8.[157] However, the new legislative measures failed to take the wind out of the sails of the bakasht struggle. The reasons for this will be discussed later. For the moment it is important to notice that, apart from the fact that the legislation fell far short of peasant demands, the very manner in which it was presented as a Congress/Zamindar agreement reduced its credibility. It is also important to note that such legal provisions took a long time to be implemented and were rarely effective in solving the problem. The localized bakasht agitations thus continued unabated after the first tenancy bill, while the failure of the provincial government to tackle the problem started to attract serious criticism. In Gaya, a Kisan Sabha notice was circulating which pointed out that the kisans had sent the Congress to the legislature with great hopes. These had not been fulfilled, since the way in which the Congress ministers had dealt with the legislation was piecemeal and unsatisfactory.[158] Sahajanand reiterated this when, in a statement to the press, he pointed out that the Bihar Tenancy Amendment Bill fell far short of kisan demands.[159] Attacks on the Congress intensified as the impression took root that the party had shielded many zamindari elements.[160] Despite these tensions the Congress was still viewed as the foremost national organization in the country by the leadership of the Kisan Sabha and it was thought to be tactically unwise to promulgate an alternative organization. A survey of the speeches of Kishoṛi Prasanna Sinha, Dhanraj Sharma (both socialists) and Ramnandan Mishra (Kisan Sabha) shows that, while they were openly critical of the Congress ministry, they still consistently urged the kisans to change the balance of forces in the Congress by joining the organization in large numbers to strengthen the left wing. As a result no real alternative to the Congress organization was presented.[161]

---

formulated, add up to a revolution. A "crisis of authority" is (thus) spoken of: this is precisely the crisis of hegemony, or general crisis of the state'. Gramsci, *Prison Notebooks*, p. 210.

157 Ibid., Maurice Hallet to Linlithgow, 25-9-1937.

158 Gaya Kisan Sabha notice to kisans in Rajendra Prasad Papers, File no. XII/37.

159 Ibid., Sahajanand's statement to the Press.

160 As noted earlier, a mass procession of peasants was organized with the intention of persuading Congress MLAs to sign their charter of kisan demands.

161 Gist of letters of Kishori Sridhar Sharma to Rajendra Prasad, 27-9-1937,

Ramanandan Mishra's speech in October 1937 in Darbhanga noted that there were two forces in the Congress, the zamindar group and the workers' and peasants' group. The zamindars were dominant, he said, because 'of their money'. He therefore appealed to the kisans to join the Congress in large numbers and thereby help to attain a socialist majority.[162] At the same time, Dhanraj Sharma, a socialist activist, condemned the Congress for failing to reduce the membership fee of the Congress so that it could become a truly kisan organization.[163]

That attempts were actually made to carry through this policy is confirmed by the position the Kisan Sabha occupied in the Congress in December 1937. In alliance with the CSP, the Kisan Sabha controlled the Patna District Congress Committee (DCC), the Gaya Provincial Congress Committee (PCC) and were a substantial minority in the Darbhanga, Champaran, Muzaffarpur and Monghyr DCCs. In Monghyr, all the 11 members of the Monghyr kisan council were members of the DCC.[164] In the legislative assembly, however, of eight members in the Working Committee of the BPCC, Sahajanand was the only kisan representative.[165] In the AICC the socialists' strategy

---

Dhanraj Sharma's speech in Shyam Sunder Malviya to Rajendra Prasad, 2-10-1937.

[162] Ibid., Srinarain to Rajendra Prasad, 2-10-1937. The classical justification for this policy can be found in Engels' introduction to class struggles in France in which Engels writes of the successes achieved by the German party through the parliamentary vote. 'It has been discovered that the political institutions in which the domination of the bourgeoisie is organised offer a fulcrum by means of which the proletariat can combat these very political institutions . . . consequently the bourgeoisie and the government have become much more alarmed at the legal than at the illegal activities of the labour party dreading the results of elections far more than they dread the results of rebellion', quoted in F. Piven and R. Cloward (eds.), *Poor People's Movements*, p. 16.

[163] Ibid., Dhanraj Sharma's speech cited in letter from Shyam Sunder Malviya to Rajendra Prasad, 2-10-1937. In another speech, Kishori Sinha pointed out that the British along with their Indian lackeys, the zamindars, were exploiting the kisans. He therefore asked the kisans to join the Congress and send their representatives to the Congress ministry which would then act to end such exploitation.

[164] See AICC Papers, File no. G 98/1937.

[165] These were Jamuna Karjee, Shivdhari Sinha, Budhan Rai Varma, Sheel Bhadra Yajee, Biresh Dutta Sinha, Shyamnandan Singh, Ramashis Thakur and Jugal Kishore Sinha.

was more successful in that there were eight CSP/Kisan Sabha members in the body.[166]

These attacks on the Congress at the organizational level were stimulated by the build-up of popular agitations in the countryside all through 1937-9.[167] At the organizational level such assaults could not go unchallenged by the right-wing Congress leadership. Rajendra Prasad was inundated with letters from districts which claimed that Congress prestige was being lowered by the activities of the Kisan Sabha and socialists.[168] However, as the speeches I have quoted above indicate, the main attack was actually mounted against the established Congress leadership itself, while at the local level the kisans were still being urged to join the Congress. Equally important was the anxiety, expressed especially among the high command, that the organizational attack launched by the Kisan Sabha and the increasing tempo of mass agitation in the countryside constituted a threat to the anti-imperialist united front forged by the Congress and to its Gandhian programme. Rajendra Prasad noted that 'we shall be doing a great disservice to the cause we all hold dear and to the country at large if we allow the principles and programme for which we have worked to be distorted or displaced by others we know are bound to injure the best interests of the country'.[169] On a more specific matter, Rajendra Prasad also objected to the use of the red flag at the kisan conference at Gaya which, he pointed out, was given more prominence than the national flag.[170]

At the district and provincial levels, therefore, the Congress leadership immediately took measures to check the Kisan Sabha opposition. The Champaran and the Saran DCCs were the first to react and banned the entry of Sahajanand into their districts.[171] It was claimed that his

166 These were Sahajanand Saraswati, Jayaprakash Narayan, Ganga Sharan Sinha, Ram Krishna Benipuri, Ram Nandan Mishra, Dhanraj Sharma, Awadeshwar Prasad Sinha and Kishore Prasanna Sinha.

167 Ibid.

168 Letters to Rajendra Prasad in Rajendra Prasad Papers, File no. xii/37, col. 1.

169 Rajendra Prasad to S.K. Sinha, 2-12-1937, Rajendra Prasad Papers, File no. XV/37, col. 3. Elsewhere he said: 'There is insidious and open propaganda attacking its (Congress) fundamentals . . . . Gandhi's programme is the only programme which can lead to success. Any other programme will lead to ruin and disaster.' Ibid., Rajendra Prasad to Ram Dayal Babu, 7-12-1937.

170 Rajendra Prasad to Ram Dayal Babu, 7-12-1937.

171 AICC Papers, File no. G 98/1937.

speeches advocated the cult of *danda* (violence), thus contravening the principles of the Congress. This was followed by a resolution of the Monghyr DCC which banned the Monghyr Kisan Sabha for agitating amongst the tenants. It was claimed that the tone of Kisan Sabha speeches had been objectionable and that the socialist slogans of '*Malguzari kaisa loge*', '*Zamindari ka nash ho*', and '*Lathi mera sathi hai*' fostered class hatred.[172]

Approving the decision of the DCCs of Champaran, Saran and Monghyr, the working committee of the BPCC in a resolution of 18 December 1937 stated unequivocally that:

> In the opinion of this Committee, the kind of propaganda that is being carried on in this province has produced a poisonous atmosphere and attacks are being made on the principle of Ahimsa which is the cherished creed of the Congress . . . In view of the fact that the Committee has before it enough material to prove that many a Kisan Sabha worker is engaged in such activities and public meetings and Congress activities are being hampered, the Committee impresses on such Congressmen as are working in the Kisan Sabha, that not only their activities but also their passive association with them is improper and directs DCCs to keep an eye on these activities and report them to the PCC.[173]

In a letter to Nehru, Rajendra Prasad explained that, since the Kisan Sabha had recently started preaching violence and as in the mass mind the Kisan Sabha was seen to be identical with the Congress, it had become necessary to distinguish one from the other, hence the resolution.[174] Significantly, the resolution was passed within six months of the Congress accepting office. Undoubtedly it amounted to a virtual ban on Kisan Sabha activity. The Bihar Kisan Sabha, not surprisingly, did not take the resolution lying down. In a trenchant rejoinder, the council of the Sabha pointed out that the working committee resolution amounted to the suppression of the mass struggle and of extra-parliamentary activity, the latter being an integral part of the constitutional programme of the Congress.[175]

[172] Ibid., see BPCC to General Secretary, AICC, 30-12-1937.

[173] BPCC, resolution in *Bihar Administrative Report, 1937-38*, p. 10.

[174] Rajendra Prasad to Jawaharlal Nehru, 18-12-1937, AICC Papers, File no. G 98/1937.

[175] Ibid., BPKS resolution. In another petition, signed by all the socialists and Kisan Sabha members of the AICC, the resolution was criticized as being ill-conceived, while the Kisan Sabha's loyalty to the Congress was reiterated.

Sahajanand Saraswati claimed that the campaign of vilification which had emerged against him was based on utterly false premises. He pointed out that he was not opposed to the Congress and that all he had asked the kisans to do was to defend themselves with force if necessary. As regards the use of the red flag he claimed that nothing was further from his mind than insulting the national flag and that the BPKS would continue to work for the Congress. In protest against the resolution, however, he immediately resigned from the Congress working committee.[176] Jayaprakash Narayan reiterated the socialist view that the Kisan Sabha had been attempting, in line with Congress policy, to link up mass agitation with work done in the legislatures, and create a mass backing for the measures of the Congress ministry. At a council meeting in January 1938, the Kisan Sabha reiterated its loyalty to the Congress as the 'sheet anchor of the anti-imperialist movement' and confirmed that the Kisan Sabha was 'pledged to independence'.[177]

The breach was, however, now out in the open and during the elections for delegates to the Haripura Congress the right wing further closed its ranks.[178] At the national level the activities of the left had served to warn the right of the long-term dangers of socialist manoeuvring. In this connection Patel wrote a strong letter to Rajendra Prasad, emphasizing the need to organize the right-wing forces at Haripura. 'We must organize all our forces before Haripura', he wrote, for, 'unless we are strong and maintain a stable majority in the AICC, we cannot effectively work our programme.'[179]

Following up this notion, Rajendra Prasad proceeded to issue directives to the provincial high command to strengthen the forces of the orthodox Congress and ensure the return of the largest number of men of their way of thinking as delegates and members of the

[176] Ibid., Sahajanand's letter resigning membership of the working committee, quoted in the *Kisan Bulletin*, 14-1-1938.

[177] *Searchlight*, 12 January 1938.

[178] These delegate elections were very important in controlling the Congress organization as, according to the Bombay Constitution of 1934, a single annual election was held at which primary members decided the delegates to the annual Congress Session, and the same men served on the PCCs, DCCs and the local Congress committees. Thus the Congress organization was open to those who could enrol members and mobilize votes.

[179] Patel to Rajendra Prasad, 21-11-1937, Rajendra Prasad Papers, File no. II/37, col. 5.

AICC.[180] Prasad also emphasized the necessity for an organized attempt to participate in Congress elections, especially in Gaya, Patna and Shahabad, where the contest was clearly loaded against the socialists. The new policy gave rise, as in the election of delegates to Haripura, to scenes of great disorder. In Darbhanga, the socialists and Kisan Sabhites such as Dhanraj Sharma, Jamuna Karjee, Jadunandan Thakur and Ramnandan Jha uncovered evidence of clear irregularities in the elections, including the non-availability of enrolment forms, obstruction at polling itself and hooliganism by established groups in the DCCs.[181] The report of the Violence Enquiry Committee of the Congress, set up to enquire into these cases, admitted that most of the incidents of violence had been witnessed in those constituencies where Kisan Sabha leaders were opposing orthodox Congress candidates.[182] Thus, of the 36 candidates elected to the AICC from Bihar in January 1938, only 5 were Kisan Sabha men or socialists, and not one found a place in the new provincial working committee.[183]

The growing breach between the Congress and the Kisan Sabha continued to plague the politics of the years 1938-9. There is no doubt that the Kisan Sabha had been further estranged by the activities of Saran, Champaran and Monghyr DCCs and by the BPCC resolution. It was reported that Kisan Sabhites were despairing of their repeated and frequent defeats in Congress committees and that many of them wanted to concentrate on mass politicization outside the organization.[184] In north Bihar, particularly in Darbhanga, left-wing activists

[180] Ibid., Rajendra Prasad to Patel, 28-11-1937 and also Rajendra Prasad to Ramdayal Babu, 7-12-1937 and Rajendra Prasad to S.K. Sinha, 2-12-1937, in Rajendra Prasad Papers, File no. XV/37, col. 3.

[181] In Darbhanga, the Congress election propaganda urged the kisans to place their vote in the white box, and to destroy the red box of the socialists. *'Ujjala bakse mein vote do, lal bakse ka gala ghot do'*, *Janata*, 6 January 1938.

[182] Report of the Violence Enquiry Committee, AICC Papers, File no. P 6/1939.

[183] FR (2), January/1938. The Haripura Congress Session also saw the passage of a resolution which supported the resolution of the BPCC on Kisan Sabhas. This resolution, while recognizing the right of the peasantry to organize into Kisan Sabhas, warned that the Congress should not associate with any activity incompatible with the basic Congress principles and activities of such Congressmen who, as members of the Kisan Sabha, were creating an atmosphere hostile to the Congress.

[184] *Congress Socialist*, 25 September 1938.

kept trying to capture control over the local party organization; and they even set up rival and parallel DCCs in Darbhanga and Saran.[185]

In mid-1938 the secretary of the BPCC noted that all the energies of the Congress were now spent in combating elections to local bodies, while most district elections witnessed struggles between socialists and official Congressmen to capture organizational positions.[186] In August 1938 Ramnandan Mishra reported: 'in the interest of the cause . . . we must get out of party struggle at least for a year, Kisan problems require our immediate attention . . . Situation (sic) is very critical and serious.' An increasing radicalization in the attitudes of the Kisan Sabha leaders now became evident. At the Comilla session of the All-India Kisan Sabha, Sahajanand underlined the need to maintain the Kisan Sahba as an autonomous organization to campaign for the class interests of the peasantry, which he felt the Congress, being a more broad-based national organization, could not pursue.[187] He still continued to maintain that, by making the peasants and workers conscious of their class interests and then advising them to join the Congress in millions, it would be possible to transform it. However, at this period Kisan Sabha and socialist activity seemed to be concentrating on developing the bakasht agitation. With the heightening of agrarian tensions the time appeared ripe for this. The Comilla session therefore saw the passage of a resolution which declared class struggle and not class collaboration as the basis of the Kisan Sabha involvement.[188] At Waini in December 1938 the Bihar Provincial Kisan Council also officially sanctioned support for the bakasht struggle, with the expressed intention that satyagraha should be the main means of achieving peasant demands.[189] By so doing, the Kisan Sabha further distanced itself from the Congress.

It can be argued that the right-wing offensive was justified since the activities of the socialists directly threatened the united front programme of the Congress. However, it should be pointed out that the Kisan Sabha programme was in many ways only a moderate one. It confined its attacks only to large zamindari at this period and did

[185] Jayaprakash Narayan Papers, File no. 23/1937; and Stephen Henningham, *Peasant Movements in Bihar*, p. 150.

[186] AICC Papers, File no. P 24/1938.

[187] Sahajanand Saraswati's speech in *Akil Bharatiya Kisan Sabha Tritiya Varshik Adhivaishan, Sabhapati ka Bhasan, 1938*.

[188] M.A. Rasul, *A History of the All India Kisan Sabha*, Calcutta, 1974, p. 33.

[189] *Kisan Bulletin*, 30 December 1938, 13 January 1939, in Indulal Yajnik Papers, File no. 1/1939.

not have a radically different socio-economic programme from the Congress. For its part, the Congress leadership realized that the mass agitations in the countryside implicitly threatened the petty zamindars on whom it relied in the countryside. The bulk of the leadership in Bihar had emerged from their ranks and they could not be alienated. Instead, the policies of the Congress leadership under the first ministry further cemented a strong relationship with these social groupings.

THE BAKASHT AGITATION

The period 1935 to mid-1939 witnessed the waging of numerous bakasht struggles throughout the province. The peasants were eager and aggressive in their wish to settle their most pressing problem, that of the restoration of bakasht lands, while the Kisan Sabha leaders were equally willing to take up their grievances and organize a movement. The Congress government's attempt to settle the issue with a second Congress-Zamindar agreement in 1938 came in for strong attack, since it conceded yet more to the zamindar than had the first such agreement.[190] 'Anti Congress-Zamindar agreement weeks' were launched all over Bihar, while many peasants actively demonstrated their support for the Kisan Sabha.[191] The slogans raised at these kisan rallies attacked the zamindari system and called for kisan unity with such words as 'death to the zamindari system', 'death to the British raj', 'the land belongs to the tiller', 'the government of the peasant is a village-based government', and 'we must seek to establish a *kisan rajya*'.[192] The failure of the Congress ministry to live up to its expectations had given force to the agitations and to the convictions of the Kisan Sabha leadership. Scott notes, aptly for this context, that the most common form of class struggle arises from the failure of the dominant ideology to live up to the promises it makes.[193] By the

[190] To be examined in the next section.

[191] Report of such meetings in Rajendra Prasad Papers, File no. I/38, col. 1.

[192] *Janata*, 27 January 1938.

[193] Scott notes: 'The dominant ideology can be turned against its privileged beneficiaries not only because subordinate groups develop their own interpretation . . . but also because of the promises the dominant class must make to propagate it in the first place . . . . In each case, the ideology formulated by the ruling class to justify its own rule provided much of the symbolic raw material from which the most damning critique could be derived and sustained . . . . A radical critique of existing arrangements may arise in virtually any subordinate class that takes the dominant ideology to heart and, at the same time, penetrates

beginning of 1938 a strong feeling had emerged that the Congress had made many promises but was now failing to meet most of them. A considerable momentum of mass opinion had built up under the aegis of Congress agitations in the past; now the Congress ministry was faced with the force of similar mass demands articulated from below. The 'hegemonic' ideology of the party had long attempted to pass off particular interests as 'general interests' through its policy of class harmony. This ideology had claimed that the economic and social order it advocated would operate in the interests not only of Indian elites but also of all other classes. It had made implicit promises of benefits to the subordinate groups, designed to serve as their stake in the prevailing social order. However, in order to sustain this strategy the Congress had to make good at least a portion of these promises in order to gain compliance. This is where the Congress can now be seen to have failed. By 1938 this failure had led directly to a refocusing of popular attention back to the basic processes of change that had formed the bedrock of peasant grievances. Direct action became more sophisticated in its strategy.[194]

In Barahiya Tal, in Monghyr in 1939, a renewed agitation developed in response to the failure of many landlords to meet the Rajendra Prasad award.[195] Karyanand Sharma, secretary of the Kisan Sabha in Monghyr, organized a satyagraha in eight villages consisting of 400 families. His tactics were designed to prevent the agents of the landlords from forcibly ploughing disputed land. An observer noted that the tenants were mainly peaceful, while the landlord goondas

---

in daily life the realities that betray or ignore the implicit promises of that ideology.' See J.C. Scott, *Weapons of the Weak: Everyday Forms of Peasant Resistance*, Yale, 1985, pp. 338 and 339.

194 Thus 'hegemony is not a symbolic bone tossed to subordinate groups, it requires actual sacrifices and restraint by dominant groups . . . the very process of attempting to legitimate a social order by idealising it always provides the subjects with the means, the symbolic tools, the very ideas for a critique that operates within the hegemony.' Scott, *Weapons of the Weak*, pp. 337, 338. See also Gramsci, *Prison Notebooks*. Gramsci notes, 'Undoubtedly, the fact of hegemony presupposes that account be taken of the interests and the tendencies of the groups over which hegemony is to be exercised, and that a certain compromise equilibrium should be formed, in other words, that the leading group should make sacrifices of an economic and corporate kind. But there is also no doubt that such sacrifices and such a compromise cannot touch the essential; for though hegemony is ethical-political, it must also be economic.' Gramsci, *Prison Notebooks*, p. 161.

195 See above.

often used spears and sticks.[196] In Reora subdivision, a long-standing dispute over 1,000 bighas of bakasht land broke out under Kisan Sabha leader, Jadunandan Sharma.[197] Sharma advised the kisans to plough up the disputed lands. As the lands were attacked the kisans were served with notices of eviction, under section 145, but many more kisans were ready to simply take their place. A notable development during these kisan agitations was the recruitment of volunteers through 'volunteer camps'.[198] In Nawadah 2,600 court cases were instituted under the Bihar Tenancy Act in a short time and at a rate of over 200 cases a day. It was noted in this district that, because the local Congress committee was pro-kisan, the landlord/tenant disputes had become intensified. In Sasaram, the District Collector complained in desperation that agrarian troubles were rife and that no sooner had disputes been settled in one village than trouble started in another. In this instance it was noted that the zamindars held out against arbitration because they were well aware of the sounder claims of the other parties in the dispute. In Bargaon village, in Sasaram, fifty men had already been arrested in connection with the disputes and seven men were daily offering satyagraha. They were soon arrested and handcuffed but once again, there were enough volunteers to carry on the agitation. The volunteers here included many Koeri and Chamar poor peasants.[199] In Darbhanga, a bakasht struggle broke out in Pandaul thana where, every day, Jathas of 25 kisans proceeded to march to areas of disputed land to plough it up. They were consistently beaten up by zamindar's goondas and repeatedly arrested.[200] In Gaya, there were struggles in fifty different places during July 1939. In Majhiawan, 200 people were arrested but, significantly, their place was taken by women. In Amwari, in Saran, in May 1939, 500 kisans, under the leadership of Rahul Sankrityana, decided to march to Chapra to place their demands before the government. The procession included many members of the most depressed communities.[201] On this oc-

196 *Kisan Bulletin*, 30 July 1939, Indulal Yajnik Papers, File no. 1/38, col. 1.

197 Indulal's visit to Reora, Indulal Yajnik Papers, File no. 1/1939, and FR (2), August/1938.

198 It was noted in Reora that the landlords were worried and many of them had actually begun to issue rent receipts. Local presses in Jehanabad were found to be working overtime. Patna Commissioners Reports, File no. 63/39. See also FR (1), November/1939.

199 BSA, File no. 63/39.

200 *Kisan Bulletin*, 13 August 1939, in Indulal Yajnik Papers.

201 Ibid., *Kisan Bulletin*, 23 June 1939, 29 July 1939.

casion, apart from Koeri and Kurmi women, many Bhumihar women actually came out of purdah to participate in the movement. This was a very unusual development, although one that was becoming more common.

To date, the resistance of peasant women has largely been ignored in the historiography of the Indian peasantry. Even the more prominent writers on social protest have neglected to examine the role of women in protest in any meaningful way.[202] In Bihar, during the Kisan Sabha agitations, women played an important part in defending their lands against zamindars. Most of the kisan meetings recorded the presence of a large number of women. In Gaya, for example, where 1,000 peasants turned up for a Kisan Sabha meeting as many as 200 were women. Once the agitation was actually underway the women protested by lying in front of the plough to prevent the fields from being ploughed by the zamindar's men. In the reprisals following the protest women were not spared and, as in Barahiya tal agitations, many were assaulted and injured by the landlord's *lathials*. However, the extended participation of women in kisan struggles does not imply that they played an important part in the leadership or that their participation affected their subordinate position in their own households.[203]

Several agitations specifically related to common property rights broke out during this phase of protest. In Ramgarh estate the peasants agitated for free access to village forests. In the Tharuhal area they demanded that free grazing be allowed and that salwood, which was used for building purposes, be made available at half the rate paid by

202 George Rude has argued that women were 'marginal protestors' in his work on *Protest and Punishment*. See pp. 4, 5, 10. E.P. Thompson has also been criticized for his treatment of the role of women and for ascribing to women the role of moral supporters of their more active men in his 'Moral Economy of the English Crowd', *Past and Present*, no. 50, February 1971. For a study that attempts to set the record straight for Britain in the nineteenth century, see M.L. Thomas and J. Grimmett, *Women in Protest 1800-1850*, London, 1982.

203 A recent study conducted on women's participation in the land struggles in Bodh Gaya in the 1970s shows that while women actively participated in the land grab movement they continued to be beaten by their husbands at home and women's issues were secondary to the struggle. See 'Zamin kenkar? Jote onkar' in M. Kishwar and R. Vanita (eds.), *In Search of Answers, Women's Voices*, London, 1984, p. 160. For a study of women's participation in a rural agitation, see Peter Custers, *Women in the Tebhaga Uprising, Rural Poor Women and Revolutionary Leadership 1946-47*, Calcutta, 1987.

the British contractor working in the forest.[204] Large-scale thefts of forest produce took place in many estates as well as 'illegal' cutting of bamboo and other forest products. In some areas peasants now also demanded that landlords cease their long-standing neglect of irrigation channels. Other ecological grievances formed an important plank in peasant agitations. Surprisingly, this ecological dimension of peasant unrest has, until now, been passed over by historians of rural Bihar. I detailed in an earlier section the growing and complex ecological pressures on rural people at this time.[205] The Kisan Enquiry Committee had, however, noted perceptively that floods, droughts, and earthquakes had become the common lot of the Bihar peasant. These had been aggravated by deforestation, landlord neglect of irrigation embankments and colonial railway building and canal-working policies, many of which disrupted older irrigation systems. In north Bihar the building of railway embankments had blocked waterways, thus cutting off the regular spills from the Kosi and Ganges and increasing the propensity of the area to the worst floods.[206] In Padri circle, in south-east Darbhanga, serious flooding exacerbated the problems of the peasantry. In the 1930s the Padri area contained up to 40,000 people in 50 villages, the majority of them being Koeris and Goalas. Before the 1930s the area had grown lucrative cash crops but had subsequently been subject to flooding as the Darbhanga raj officials had neglected to maintain dams and bandhs. This flooding, coupled with the impact of the Depression, had long-term and deleterious effects on the peasantry. As Ramanand Mishra noted, 'Year after year (they) would cultivate in the tragic hope of getting something. But nothing would come except the notice of the sale of their holdings.'[207] As a direct result, about ten thousand bighas became bakasht lands. In 1937 the Padri peasants turned to the Kisan Sabha to redress their grievances, engaging in an intense struggle to reclaim these lands. Furthermore, even the poorest sharecroppers and short-term tenants, most of them Ahirs and Kurmis, began to claim long-established rights to bakasht lands.

It is important to note here that the bakasht movement was able to cut across many castes, right through from Bhumihar tenants to Koeri and Kurmi middle peasants and small-holders. To represent the latter,

[204] *Searchlight*, 24 November 1937.
[205] Refer page 93.
[206] *Searchlight*, 13 November 1937.
[207] Quoted in S. Henningham, *Peasant Movements in Bihar*, p. 156.

a new organization of the depressed classes, the Tribeni Sangh, also emerged in this period. It attempted initially to organize the Ahirs, Kurmis and Koeris.[208] The Sangh at this stage was an embryonic organization and was only beginning to be used as a platform for lower-caste grievances. In Bhabua the Sangh contested the District Board elections in 1938 in what became a high-caste versus low-caste election. The Brahmins and Rajputs supported Markanday Pandey, the Congress candidate, while the Kurmis, Ahirs, Chamars and Dusadhs supported Ganesh Singh, the Tribeni Sangh candidate. On this occasion it was observed that a large number of low-caste women, who normally never exercised their vote, turned out in favour of the Tribeni Sangh candidate.[209] However, the Sangh was still mainly a 'caste sabha' and in this period did not develop as a radical organization or as an alternative to the BPKS. In contrast, the Kisan Sabha, despite its moderate programme, but perhaps because of its class outlook, was able to rally wide support. This success related to the deliberately class-based, non-sectarian, outlook of the organization. Nevertheless it is unlikely that the new political participation of poor low-caste peasants succeeded in challenging the power structure in the villages. Instead, wealthier sectional interests normally prevailed. For example, in Reora the evidence indicates that the rich Bhumihar tenants resented Jadunandan Sharma's socialist distribution of disputed land among all sections of the peasantry.[210] Such a situation was not atypical, for while many sections of the peasantry may have participated in the movement, the upper-caste richer peasantry naturally felt threatened by the prospect of a more egalitarian share-out of the benefits of agitation.

The bakasht movement of 1938-9 was strongest in Reora and Manjihiawan in Gaya, Chapra in Shahabad, Barahiya Tal in Monghyr and among the Santhal bataidars in the Kosi Diara area. The struggle in Monghyr had started in 1936. This derived partly from the fact that there had been some sort of Kisan Sabha activity here as early as 1936.

[208] Kalyan Mukherjee, Rajendra Singh Yadav, *Bhojpur Naxalism on the Plains of Bihar*, p. 27.

[209] See BSA File no. 18/38, The awakening of low castes had been a phenomenon of the late nineteenth and early twentieth centuries when the Ahirs, Koeris and Kurmis participated widely in cow protection movements to improve their ritual status. See Gyan Pandey, 'Rallying around the Cow', *Subaltern Studies*, vol. 2, p. 74.

[210] FR (1), May 1939.

In that year raiyats refused to do begari in the Barahiya Tal area. On one occasion, on 6 April 1936, a thousand raiyats marched in a body, led by Karyanand Sharma, to the office of the collector of Monghyr, and filed a written petition of grievances. The raiyats in this instance were mainly from Channan pargana in Lakhisarai, while the opposing landlords were the Maharajas of Gidhaur and Goenka. The main grievances of the tenants were the high rent demands, the aggressive attitude of the zamindar's amlas and the curtailing of their traditional rights to forests and irrigation. In the course of these agitations, in 1937, Karyanand Sharma and a few other Kisan Sabha workers were arrested. This further angered the peasants and large-scale kisan agitations broke out, in consequence, in Pali, Serora, Mahram Chak, Kothwa, Bhanpur, Kamarpur and Akbarpur.[211]

The Barahiya Tal agitation in Monghyr was confined to eight villages consisting of 616 families with a population of 4,800, while the area disputed involved 15,000 bighas of bakasht land. In the bakasht lands of Nawab Ibrahim of Terighat, who had 3,200 bighas of bakasht land, the government had issued notices under section 144 of the Criminal Procedure Code on both the zamindars and the raiyats, but the former were undeterred. On 10 October the men of zamindar Nawab Ibrahim, armed with lathis and swords, began ploughing (in spite of notices served on those lands), while the raiyats staged a sit-down agitation. A row developed and the zamindar's lathials attacked the peasants, assaulting two of them, Sukar Mahton and Naghon Mahton.[212] Two Congress workers were also assaulted while trying to pacify the villagers. This incident was repeated at Kothwa in Parahiya police area. Here, Zulum Singh, the zamindar, began ploughing the disputed land and the peasants resisted by offering satyagraha. The armed lathials of the landlord then attacked eight men and two kisan workers. The kisan workers, Ganaur Mahton and Dashrata, received considerable injuries. At Jitwar (Lakhisarai police district) Nirori Mahton, another poor peasant, was assaulted by a blow that chopped his ear off. By October 1938 several men had been injured in this way in the Tal area.[213] The usual tactic of the kisan in this period was to prevent the agents of the landlord from forcibly

[211] Agrarian trouble in the Barahiya Tal area, Monghyr, BSA File 42 (111)/38.

[212] Ibid.

[213] Ibid.

ploughing the disputed land by offering satyagraha in front of the zamindar's plough. It was a strategy that often proved effective. By 15 October the satyagraha had spread to all villages in the area. In many villages the zamindars were so violent that they did not even spare women. For example, at Mahram Chak, Lachi Narain Singh reploughed four bighas of pea crop sown by Bhatto Dhanuk on 14 October 1938. Bhattoo and his brothers' wives were then seriously assaulted by Lachi Narain's men with bhalas and lathis. The involvement of women in injuries in this way can be said to represent yet another attack on a moral economy, now couched in gender terms.

By the time of the ensuing rabi harvest season in early 1939 the kisans were well prepared for a more intensive confrontation at Barahiya Tal. The zamindars had employed sixty lathials from the diara areas to guard the standing crops. On 31 December a large Kisan Sabha meeting of 1,500 kisans (including more than 100 women) from Lakhisarai and Sheikhpura was held. At the time zamindars and their agents were mounting daily attacks on peasant holdings in the whole area of the Barahiya Tal. On 26 January 1939, two more kisans, Mural Mahton and Mushi Mahton of Kawarpur, were injured by the zamindars' men and the crops sown by the kisans were looted. Kisan Sabha leaders now made further preparations to meet the escalation of the situation.[214] Twenty-five men under training at Lakhisarai were sent to the kisan satyagraha camp at Mahram Chak in the Tal area to mobilize peasants to launch a more aggressive campaign. Kisan Jathas were despatched to Monghyr to perform satyagraha that would bring to a standstill the working of the arbitration courts. The kisans forcibly boarded trains leaving for Monghyr and when they were prevented from doing so they sought to hold up trains by lying on the railway line.[215] The local district collector immediately decided to mount another attempt at compromise. Even then the zamindars refused to appear before the arbitration committee and continued to defy prohibitory orders against them. The collector's efforts were half-hearted, however, and in many cases the landlords were effectively given *carte blanche* to take away all the crops from disputed lands. In much the same fashion an award was announced which gave occupancy rights with regard to only 800 bighas out of 14,000 disputed bighas. In fact, most of these cases were settled unilaterally and Karyanand Sharma's

214 Ibid.

215 See FR (1), May 1939

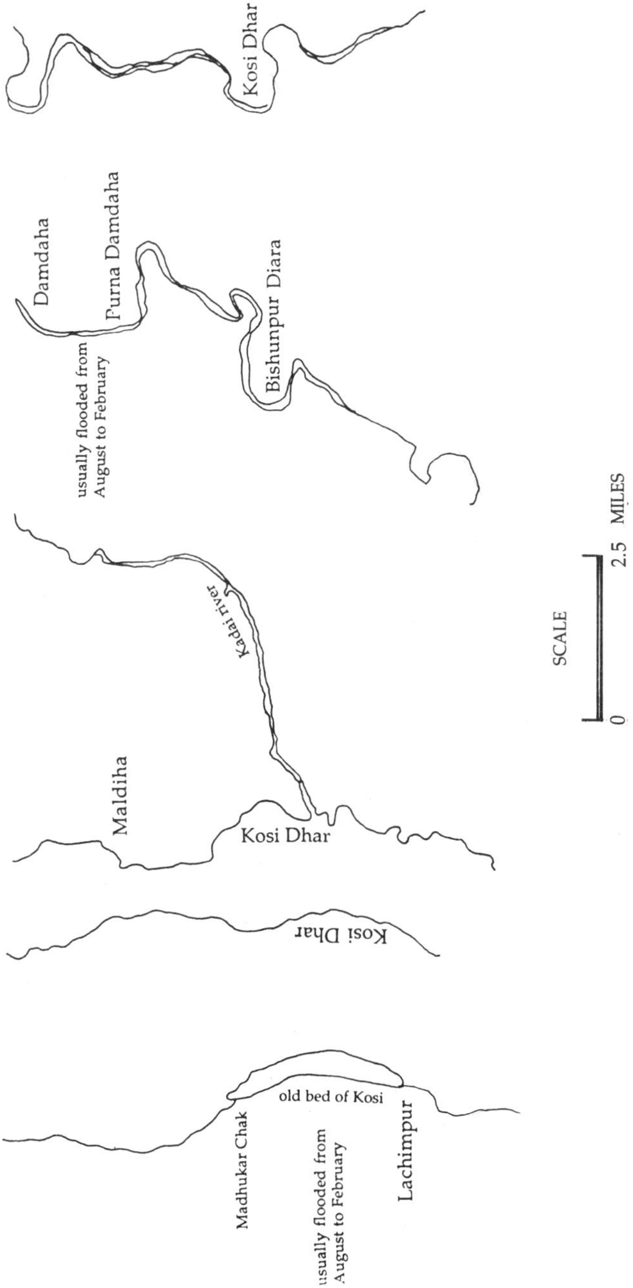

Map of Bhagalpur and Purnea Districts in 1938 Showing the Kosi Diara Area

demand that collective bargaining should be followed was not accepted. In essence, the Congress award thus failed to meet the demands of large sections of the peasantry and was thus well in line with the highly biased 'compromise' tactics so much favoured by the Congress ministry and its local institutions.

AGITATION IN THE KOSI DIARA AREA

The foregoing account of the growth of peasant unrest during the Congress ministry has already served to highlight the incidence of ecological grievances. One region where the ecological dimensions of the peasant unrest were most apparent was in the Kosi Diara. This was an area where a major dispute over rents had erupted between the bataidars and the oppressive tenure-holders of the Darbhanga raj. In this case it is quite clear that the Santhals defined their own goals and targets of agitation in a widespread movement for occupancy rights, and against landlord exactions. The Kisan Sabha was in fact only peripherally involved.

In Purnea the unusual ecological conditions of the Kosi Diara system had given rise to an oppressive batai system and concomitant peasant unrest. The Kosi river, which originated in the ranges of east Nepal, had a substantial flow with a catchment area of 23,992 square miles. It was a river that frequently changed its course, with far-reaching implications for the tenure system and peasant rights in the area. The three zillas which composed the pargana were all at one time or another subject to the tremendously destructive power of the Kosi. The area suffered from the ravages of the Kosi in 1893, 1905, and in 1935, on each occasion destroying former landmarks and putting large areas of land out of cultivation.[216]

When the river receded and the lands became stable again, the new alignments led immediately to prolonged disputes over ownership rights. The Darbhanga raj had meanwhile put into action a

[216] In 1810, Francis Buchanan found Gondwara to be a wasteland. By 1875 the situation had been reversed. Gondwara, due to a process of mineral breakdown of rotting sandgrass and milch, was now a highly fertile and productive area, while Bhawanipur, which the Kosi had recently traversed, had reverted to jungle and Birnagar was rapidly becoming sandy marsh and grazing land. This ecological shift was to occur several times over the next 40 years with far-reaching effects on the tenurial system and the peasantry. See *Purnea District Gazetteer*.

complex system of intermediary rights. These arrangements were concluded primarily with three groups of middlemen, namely *istemardars, thikadars*, and *milikdars*. Lack of accurate records and the inherently very mobile nature of the diara limited the Darbhanga raj's local control over the intermediate tenure-holders, who were for the most part independent. The Santhal raiyats were, from the start, in a weak position for, given the instability of the diara tract, it was impossible to foresee what the next year's harvest would bring. As Hill notes, 'to accommodate this volatile and shifting cropping area, a unique and adaptive tenure system had evolved in Dharmapur over the years. This tenurial plan was designed to encourage production in the diara tracts, minimize pressure on the peasant's productive capabilities and aid the Darbhanga estate by reclaiming lands from the jungle. Ultimately the system isolated the landlord and oppressed the raiyat, while entrenching the intermediaries to such an extent that they would never be pried loose from the land'.[217]

[217] Christopher Hill, 'Santhal Bataidars in Purnia District, the Ecological Evolution of a Sharecropping System', *Economic and Political Weekly*, 22 August 1987, pp. 1451-2. The cultivating rental system as it evolved had three separate tenurial components—*jotjama, halhasila*, and *birawari*. Jotjama was the first stage in resettling the deserted Kosi area on low rents. By the 1880s, jotjama had brought potential settlers from Bhagalpur, Darbhanga, Saran, Champaran and the Santhal parganas in the hopes of reclaiming the land from the Kosi jungle. As the Kosi further regressed and increasing portions of the diara land became fertile, the jotjama tenure gave way to halhasila where leases were given to cultivators for a specific period of time. Rent rates were not classified by the richness of the soil but rather by the specific crop grown. Halhasila, though more limited than jotjama, was still a very flexible tenure. Rent was determined by a measurement of the crops on the land actually cultivated rather than by the area leased to the raiyat. As the system progressed the raiyats began to anticipate when the intermediary would arrive for the annual measurement, and would purposely leave large sections of their land fallow and cultivate it only after the tenure-holder had departed. To counter this move on the part of the raiyats, the intermediaries began charging *bira bishi*, or a fee for the land that was not cultivated. This was the birawari system which was for all intents and purposes the same as halhasila, except for a fee added for the fallow land. These three systems went in cycles according to the ravages of the river. In that sense it took into account the vagaries of the river. However, it did not prevent the tenure-holders from increasing the rents indiscriminately when it suited them or to prevent the accrual of occupancy status to the tenants. The most contentious issue was the middleman's assertion that, due to the fluctuations of the Kosi, the raiyats had not cultivated the same plots for 12 consecutive years and as such they had no rights of occupancy.

Historically, each time the river flooded seriously it put large areas out of cultivation. When the water had receded and the lands became stable once again there would be a general scramble for resettlement. Soon after the completion of the first Survey and Settlement Report in 1910 the Kosi river began to retreat from Bhawanipur. Many new tenure-holders emerged, the majority of whom were relatives of the officials of the Darbhanga raj.[218] Meanwhile the old tenants, who had clung to their lands during inundation, attempted to reclaim them. This was when the largest number of disputes occurred. Intermediaries sought to increase rents by allotting the lands to new tenants, while influential tenants seized the opportunity to enlarge their holdings. The main disputes took place over lands settled to new tenants and over milik or rent-free claims. Most of the influential tenants were upper-caste—Bhumihars or Rajputs and sometimes even Goala. When the record of rights was being written up in the 1920s the main issue of contention concerned their claims to rents at old rates and to increases in land holdings. All fixed-rent claims were consequently denied by the Darbhanga raj and in several cases the tenure-holders claimed that their properties were farmed by migrants from other districts who came to the Kosi area only seasonally. There were thus as many as 1,641 disputes at the first settlement operation and 2,378 in the second, as the peasants of Dharampur pargana agitated against the false claims of the Darbhanga raj. The raj also claimed all plots lying fallow as the property of the estate and disallowed grazing on the new grazing lands, which were the re-formed beds of the Kosi. The chief anxiety of the tenants was that they should be kept as grazing grounds and not leased out for cultivation, thereby removing a necessary element of ecological flexibility.[219]

Apart from the disputes between the raj and its tenure-holders and wealthy occupancy raiyats, there were also disputes between the wealthier tenantry and some tenure-holders *vis-à-vis* their own under-raiyats. These under-raiyats were tenants on short-term leases, mainly

218 *Final Report on the Survey and Settlement of the Kosi Diara area in the district of Purnea 1923-26.*

219 The resentment against the raj for taking away grazing rights can be noted from a letter sent by the Director of Land Records to the Manager of Darbhanga raj stating that, though he was satisfied that the tenants had no right to have the land recorded as reserved for grazing, 'the raj might wish to meet the wishes of the tenants as the Hindu tenantry feel very strongly about the preservation of grazing grounds', ibid., p. 8.

sharecroppers who had no tenure rights at all. As a result they remained non-occupancy tenants, enjoying rights of cultivation only at the behest of their immediate tenure-holder, who in most cases claimed to be the occupancy tenant of the Darbhanga raj and exploited the vast mass of under-raiyats, sharecropper tenants and bataidars. The Santhals formed the majority of the latter class of bataidars in Dharampur pargana.[220] Below them were the mass of agricultural labourers, mainly the landless Musahars and Gangotias. The dependence of the bataidar on his landlord or malik for the initial lease, as well as for its renewal from time to time, left him at the total mercy of the landholder, who often made exorbitant demands on him in the form of abwabs. The chief of these included fees to be paid to the landholder for the sale of any grain, cattle, butter, honey, lac or even animal bones by the peasant.[221] Besides, this payment had to be made regularly to the *sepahi* who supervised the harvesting on the bataidar's fields, to the malik's cart driver and to the keeper of the malik's horses. The bataidars were also asked to part with grain on all social occasions at the malik's house, for example, for the birth of a son, a marriage, or the visit of a married daughter. Finally, a bataidar was subject to a series of miscellaneous demands. Each individual was required to contribute his labour and equipment for one session of ploughing at the beginning of every crop season on the khas land of the malik. Maliks who had not acquired their own draft animals called upon their bataidars to plough their khas lands every eighth day. A bataidar was also required to perform *gratis* a range of miscellaneous tasks for his malik, (also subsumed under begar) such as preparing new thatch for his houses, collecting firewood, and generally fetching and carrying. All in all, their obligations were onerous, even in good years.[222]

In the 1930s a new set of circumstances galvanized the Santhal bataidars to action. Floods hit the district every year after the great Bihar earthquake of 1934.[223] In 1934-5, 1937-8, and 1938-9 standing

[220] The gazetteer recorded that the Santhals had first migrated to the district in the 1860s to work on indigo plantations. See *Purnea District Gazetteer*, p. 111.

[221] These were termed *Sairat* collections and included *Kevali, Singhani, Karouncha, Shahad, Charsa, Haddi, Hat* and *Lahi* collections. See Survey and Settlement Report in the Kosi area.

[222] BSA (Home Political Special), 120(1) 1940.

[223] *Purnea District Gazetteer*. Deforestation of the area also increased the propensity to flooding. It was reported in 1938 that 'within recent times the

crops were destroyed by floods. In November 1937 the spill waters of the Kosi were responsible for vast areas of devastation. The paddy and the bhadoi crop was completely destroyed. Nearly 105 villages and 2 lakhs of people were affected in Supaul thana alone, and 15,000 houses collapsed. The plight of the villagers was worse in Supaul, Madhepur, Gopalpur, Chakia, Chak, Bairia, Sampal Chak, Uddami, and Jamalpur. As no crops had survived the people were forced to subsist on a kind of grass known as karmi that grew in water.[224] This distress did not, however, prevent the maliks from bringing forward rent suits to reclaim arrears or from continuing to charge abwabs. This was when the Kisan Sabha intervened. They were able to use the new legislation undertaken by the Congress ministry development to challenge the position of the maliks. The Bihar Tenancy Amendment Act of 1937 stated that under-raiyats who had been settled on the land for twelve years were, for the first time, entitled to acquire occupancy status. Swami Sahajanand visited Purnea in early February 1938 and made provocative speeches adding to the tempo of kisan activity in the province. According to the reports of the collector of the district, 'some kisan workers began to do propaganda among the Santhal bataidars. They penetrated the Santhal villages and preached to them that the land really belonged to those who cultivate the same, that they should pay no abwabs, that they should insist on rent receipts, before

---

denudation of the forest area to the north is said to have increased the intensity of floods in Purnea. Thirty years ago the country between Patthardeva on the frontier and the jungle belt was well wooded. It is now a bare cultivated plain which offers no resistance to floods'. The distress caused by floods set the stage for agrarian unrest among the Santhals. For a recent study that fails to examine the critical ecological context of unrest among the bataidars in 1938, see A. Chakravarti, 'The Unfinished Struggle of the Santhal *bataidars* in Purnea district 1938-42', *Economic and Political Weekly*, col. xxi, no. 43, 25 October 1986. Chakravarti, however, provides a detailed and useful analysis of the peasant movement in the Kosi area.

[224] The natural water drain from the west passed through the region. When the extra water from the paddy fields on the west was released by the kisans it had to pass through the village before flowing into the river. In previous years there had been a deep and broad pain to carry the water and there was a high embankment to prevent the water from running into the fields. Since then, however, zamindars neglected to repair the pains and their bandhs disappeared. Even a small volume of water from the west converted these villages into lakes ruining the crops and delaying the sowing of rabi. See Report of an Official Conference on the Causes of the Bihar Floods in *Searchlight*, 4 November–10 November 1937.

dividing the crop, that they had acquired occupancy rights under the new legislation and could not be ousted from the land'.[225]

The maliks reacted with great disfavour to these new threats to their position. Settled under-raiyats were attacked and new tenants brought in to prevent any accrual of occupancy status. In this way, the number of disputes increased markedly in the period 1938-42. In June 1938 Congressman Dhaturanand sought to draw the attention of S.K. Sinha to the problems of the bataidars by submitting a petition on the subject.[226] He also urged the kisans to commute rents into cash and to acquire occupancy rights. The petty tenure-holders became panic stricken at this and sought to wipe out evidence of all long-standing leases. The geographical area of confrontation in this dispute covered 81 villages in the Sadr subdivision of Purnea, of which 79 were in the Damdaha revenue circle. Officially it was stated that in all the affected villages there was a Santhal *tola* and that they were the prime agitators.[227] It should be pointed out here that, apart from Dhaturanand and the other kisan workers such as Dulla Tudoo, those who began to disseminate propaganda among the bataidars had never been Congressmen. Indeed the Kisan Sabha was only peripherally involved in the movement and the bataidars were left to their own initiative. By May 1939 the opposing sides had organized themselves into two hostile camps; the Santhals to keep forcible possession of lands and crops (and even to usurp possession of other lands of the maliks not in their possession), while the maliks attempted to oust the existing bataidars by settling new bataidars. The Santhals had combined in large numbers and their growing hostility threatened the maliks. The District Collector observed 'the strong combination among the Santhal bataidars, their going about in large numbers armed with bows and arrows, often using criminal intimidation whereby the small maliks dared not go on the land to claim batai, their refusal to divide the crop and in many instances removal of the entire crop from the threshing floor, and the forcible ploughing of adjoining or other lands settled with them'. The administration set up arbitration committees to deal with the growing unrest. However it was difficult for any arbiter to find a solution which could be acceptable to both sides, particularly as the bataidars wanted occupancy status before agreeing

225 BSA (political special), 120(1) 1940. See also FR (1), March 1938.
226 Ibid.
227 BSA (political special), 120(1) 1940.

to end the agitation. Commenting on the intractable attitudes of the parties the Inspector General of Police, A.E. Bion, stated that he had the impression when in Purnea 'that the Santhals will never sign the kabuliat and the maliks will never grant rent receipts to their bataidars until these kabuliats are signed'.[228]

In Paisa village (in Damdaha police area), Akbar Ali, the principal witness, stated that he had purchased 8 bighas of land in Paisa from Khela Manjhi in 1936.[229] According to a chowkidar's report on 4 December 1940, 'Kanoo Manjhi and Ram Manjhi had collected a large number of Santhals armed with bows, arrows, spears and other deadly weapons for the purpose of reploughing the standing wheat and crops of Akbar Ali and Nakchand . . . ' When police officials reached the site of the occurrence they saw 'about 200 Santhals in several batches moving around' who said that they were assembled to replough the land by force for, as Ram Manjhi firmly asserted, 'it was their land and it belonged to him'. Often the disputes arose between the Santhals and the Goalas who were the dominant caste in the district. In Jianganj village, for example, Harakchand Yadav and his kinsmen were the *sikmidars* (under-raiyats) on a tract of land held by a tenure-holder, Baidyanath Chand. Yadav's status as an under-raiyat was corroborated by rent receipts for several years.[230] The land, according to his statement, had been cultivated on batai by Nebi Gope for ten years preceding the incident. On 10 June Nebi Gope filed the following report:

This morning (Nebi Gope) went to the field to sow maize seeds, when Barka Jetha, Lerhwa, Lakhan Bara, Chhota Lakhan . . . and 20-25 other Santhals arrived at the field armed with bows and arrows and lathis. They obstructed Nebi in sowing the land saying that they cultivated the land on adhi system (adhi batai) and so why they (Nebi and others) have come. Nebi got alarmed and raised hulla hearing which Maghu Gope, Mayalal, Baldeo and Rabi Gope . . . ran to him (Nebi) and began arguing with the Santhals, but they would not listen and unyoked bullocks from the ploughs. Over this a quarrel between the two parties arose. Subsequently the Santhals accosted them with arrows and lathis. Barkha Jethi shot Rabi with an arrow which pierced his chest and he died. Nebi, Mayalal, and Meghu were assaulted with lathis.[231]

228 Ibid.
229 Special Report, no. 157/40 in BSA (political special)/121(1)/1940.
230 Special Report, no. 69/40 in BSA 121/(1)/1940.
231 Ibid.

The local officials accepted this version of the story and decided against the Santhals, arguing that the real bataidars were the Goalas. The Superintendent then concluded:

> I am of the opinion that the Goalas had the right to plough and were actually ploughing when their position was interfered with by aggressive Santhals, who possessing superior armament did not hesitate to challenge the Goalas to battle.[232]

The picture here seems clear. As the most upwardly mobile of all the lower castes the local institutions favoured the Goalas and supported them *vis-à-vis* the Santhals. In one village called Bardaha (in the Damdaha police area), a dispute along similar lines erupted in which one Punit Singh claimed to be the legal bataidar, while the Santhals were accused of being the trespassers on the land. In another village called Tarauna, Ritbaran Singh is said to have set up a Santhal to claim batai rights in order to help him dispute the claims of the real Santhal bataidar. As reported by the Superintendent of Police, the Santhal set up as bataidar 'was not a true Santhal and is not taking part in the Santhal agitation . . . the malik wanted to oust the true Santhals, who were the former bataidars, because they insisted on having a receipt for the crop . . . the malik is the ring leader in this case, and deliberately set up another party of bataidars to falsely claim the land; the malik had to admit that the accused persons were his former bataidars, and had cultivated the land in question'. These cases were typical of the many disputes that arose in the region. As had happened at Monghyr the local government finally set up an arbitration committee, but little progress could be made in the face of the opposition of the maliks.[233] Many of the disputes were settled individually and the maliks were often successful in using the courts against the Santhal claims. The maliks did not, of course, change their exploitative attitudes *vis-à-vis* the bataidars. Moreover, they made no effort to issue rent receipts and the Santhals failed to establish their legitimate rights as tenants. As such, the conflict remained unresolved and continued well beyond the tenure of the Congress ministry into 1942. Congress legislation on the rights of the under-raiyats thus remained completely unimplemented in what amounted to only a paper gesture.

[232] Ibid.

[233] Special Report, no. 85/40, BSA 120/(1)/1940.

An analysis of the agitation of the Santhal bataidars is important for it affected the lowest strata of the tenantry and highlights the attitudes of both the Kisan Sabha and the Congress *vis-à-vis* these sections of the community. The provincial leadership of the Kisan Sabha was, as I have pointed out, emerging as the foremost critic of the Congress ministry and its policies. The Congress was not concerned with the demands of the poorer tenantry and involved itself only peripherally in the movement of the Santhal bataidars. It thus clearly revealed itself to be the party of the middle and rich tenantry with little in its programme for the tenants-at-will and the landless labourers. The agitation also revealed the hollowness of the legislation of the Congress party and its lack of commitment to the problems of the poor. The inability of the Congress ministry to meet the most pressing demands of the bataidars, which their legislation had promised specifically to address, was symptomatic of the ultimate bankruptcy of almost all Congress legislation during the first ministry period. For the poorest it amounted to nothing less than a betrayal.

An analysis of the various bakasht and rent movements leads one to some interesting conclusions. It is quite clear that the movement spanned large sections of the peasantry right through from the richer occupancy tenants to the weaker Santhal bataidars. In some areas, as in Nawadah, even the landless labourers used the opportunity to demand higher wages. The Congress ministry and the legislative programme undertaken by it set the context for struggles on the basis of the mass aspirations which had been built up in the pre-ministry period. The peasants participating in the movements did not constitute a homogenous group. They consisted of a variety of different communities playing a wide variety of economic roles with frequently conflicting interests.[234] What bound them together was the strength of their grievances against the extra-economic powers of landlords in the countryside. The oppressive nature of landlordism in rural Bihar and its excessive demands on a peasantry weakened by the Depression caused a widespread outburst of peasant unrest. Upper-caste Bhumihar and Rajput occupancy tenants joined hands with Ahir and Kurmi

[234] The conflict of interests reasserted itself in Reora where all sections of the peasantry had joined hands to challenge the landlords. Once the bakasht agitation had succeeded in reclaiming the disputed lands back from the landlord, the rich Bhumihar tenants resented Jadunandan Sharma's subsequent proposal for a socialist distribution of disputed land among all sections of the tenantry. See FR (1), May 1939.

middle and poor peasants in seizing landlord lands and reasserting their rights on disputed bakasht lands. Many disparate movements also broke out under the umbrella of the Kisan Sabha movement. The bakasht agitations encompassed the agitations of the Santhal bataidars organized outside the aegis of the Kisan Sabha, as well as the movements of landless labourers. In the end, however, all were subsumed into the Kisan Sabha movement. The focusing of the agitations on the attempt to reclaim bakasht lands ensured an ultimate cohesion.

Henningham has argued that the bakasht movement was reformist rather than radical, as it did not call for a general non-payment of rent or moratorium on debt. He further argues that it attempted to redress grievances within the zamindari system and did not directly attack the system itself. This may have been true at the leadership level. In contrast, at the popular level, the peasantry often identified their own goals and targets of agitation. Thus, in Nawadah, the rumour that arrears of rent had been remitted along with an order for the reduction of current rent resulted in the complete withholding of all rent. Subsequently, over 2,600 court cases of non-payment were brought under section 112A of the Bihar Tenancy Amendment Act, at the rate of 200 a day. Clearly, in many ways the movement did indeed effectively challenge the power of the landlords in local society, at least for a short time, in what amounted to an unprecedented assertion of strength and defiance by the subordinate classes in rural Bihar. Thus in Gaya it was noted that the landlords were clearly 'intimidated' and 'worried'. In Jehanabad, the Gazetted Officer recorded that the local presses were working overtime to grant receipt books to landlords who had never issued receipts before.[235] On office acceptance, therefore, the Congress ministry was faced with a mass peasant agitation of serious magnitude.

THE CONGRESS RESPONSE

The Congress had promised many things to many people. As the foremost anti-imperialist organization in the country it had successfully challenged the colonial state and through its mass movements built up its own counter-hegemony. Congress ideology had elevated the idea of nationalist unity to an icon. 'Class harmony' and 'compromise' were the keywords in the Congress nationalist discourse.

[235] Patna Commissioner's Reports, BSA File no. 63/39.

However, it would be wrong to assume that this hegemony of ideas was in any way total or complete.[236] The subordinate classes in every case were able, on the basis of their daily material experience, to penetrate and demystify the prevailing ideology. In power, therefore, the Congress found itself face to face with a catalogue of mass demands. The attitude of the Congress to these and to the signing of the Congress-Zamindar agreement created doubts and fears among a large number of people that the Congress was not abiding by its promises. The failure to implement its promises thus allowed the left-wing opposition to gain ground. In countering this, the Congress was therefore ultimately forced to use the tactics of the colonial state to handicap the opposition.

Severe repression, both by the zamindars and the Congress, developed in tandem. The state made widespread use of its police force, in alliance with the zamindars' goondas, to repress the bakasht agitation.[237] Overweening force was used to crush the movement. In Barahiya Tal, 100 volunteers were beaten by zamindars and Karyanand Sharma was arrested by the local police thana.[238] In Darbhanga the kisans ploughing the land were beaten ruthlessly by zamindars' goondas and were arrested.[239] In Reora, Jadunandan Sharma and many others were also arrested, while in Amwari, Rahul Sankrityana was handcuffed and dragged to jail. In Shahabad, a virtual 'goonda raj' was established and the Congress ministry resorted to large-scale arrests to curb the agitation.[240] In August 1939 Sahajanand issued a statement condemning the Bihar government for arresting and jailing thousands of militant kisans, arbitrarily imposing section 144 and openly favouring the landlords.[241] Jadunandan Sharma and other

236 The concept of hegemony has been developed by Gramsci and other twentieth-century Marxists, but exactly how voluntary and complete this hegemony is likely to be is not entirely clear even on a close reading of Gramsci. Clearly, as Scott has argued, a hegemonic ideology must by definition represent an idealization which creates the contradictions that permit it to be criticized on its own terms. See Scott, *Weapons of the Weak*, p. 317.

237 As we have seen, though the Kisan Sabha continued to confine these attacks to big landlords in its official programme, large-scale attacks against petty landlords lands were common, as in Shahabad, FR (2), April 1939.

238 *Kisan Bulletin*, 25 November 1938, Indulal Yajnik Papers.

239 Ibid., *Kisan Bulletin*, 25 November 1938, Indulal Yajnik Papers.

240 FR (2), March 1939. See also Sahajanand Saraswati to Indulal Yajnik, 6-8-1939, Indulal Yajnik Papers, File no. 11/1939.

241 *Kisan Bulletin*, 28 July 1939 in Indulal Yajnik Papers, File no. 11/39.

kisan activists also criticized the Congress ministry for its ruthless suppression of the peasants. The arrests of kisan workers, Rahul Sankrityana, Karyanand Sharma and Jadunandan Sharma, were widely condemned and an agitation was launched at processions and meetings. The classifying of kisan prisoners as political prisoners was a central demand at this time.[242] In the case of Jadunandan Sharma, the kisans demanded an open trial with such slogans as '*Inquilab Zindabad*' and '*Jadunandan Sharma ko choro*'.[243] The use of coercive methods by a state which claimed to be a popular state of the people further eroded the legitimacy of the Congress and enabled the opposition to focus its attacks on the new Congress establishment far more directly than had been possible in the pre-ministry days.[244]

In order to try to restore a consensual basis for its power the Congress government attempted to intervene directly, through legislative means, in local society.[245] This was tried out initially by attempting to settle individual bakasht dispute cases through compromise and negotiation. The process was time-consuming as collective bargaining was not permitted. In one instance, in Barahiya Tal, the disputed bakasht land amounted to over 14,000 bighas in eight villages, while the arbitration committee settled individual cases leading to the restoration of only 800 bighas. In Reora only 1,500 bighas were restored. Finally, in desperation, the Congress also resorted to direct legislation to stem the rising tide of peasant unrest.

In the first few months of the Congress acceptance of office the bakasht agitation had tested this support and challenged the legitimacy of the Congress. Though the party was subsequently forced to

[242] As D. Arnold notes 'once in office, Congress ministers identified themselves wholeheartedly, and with remarkably few qualms with the old "law and order" mentality.' See D. Arnold, 'Congress and the Police,' in M. Shepperdson and C. Simmons (eds.), *The Indian National Congress and the Political Economy of India 1885-1985*, Aldershot, 1988, p. 218.

[243] Ibid., *Kisan Bulletin*, 27 January 1939.

[244] Gandhi realized that this coercive policy of the Congress ministry would quickly erode popular support for the Congress among the peasantry. In December 1938, he urged that 'if the Congress ministries find that they cannot run the state without the use of the police and the military, it is the clearest possible sign, in terms of non-violence, that the Congress should give up office and again wander in the wilderness in search of the Holy Grail'. See CWMG, vol. lxviii, p. 195.

[245] It claimed to be a popular government of the people and the speeches of Nehru, Prasad and Patel emphasized the role of the Congress as an impartial arbiter.

resort to the coercive arm of the state to suppress the agitation, it still saw itself as a party of the people attempting to build up its own hegemony. Notions of class harmony and compromise thus continued to form the ideological underpinnings of Congress interventions, both in the legislative arena and outside. Though the Congress was to claim (as the criticism of its agrarian policy mounted) that the constraint of not having a majority in the upper house necessitated legislation based on compromise, in reality the absolute majority the Congress possessed in a joint sitting was sufficient to pass contested bills, albeit with some delay.[246] In fact, the nature of Congress legislation was influenced by other considerations, the main factor being the leadership's perception of the need to maintain a united anti-imperialist front. Rajendra Prasad, the guiding force of the Bihar Congress, believed that rather than quarrelling with capitalists and landowners the Congress should try to offer them compromise solutions.[247] However, the over-eager way in which this compromise was tackled by the Congress leadership, above all by terming it the Congress-Zamindari pact, reduced much of its credibility. The terms of the compromise, as realized in the agrarian legislation that followed, conferred only marginal benefits to tenants. It favoured the landlords so that they emerged with their position *vis-à-vis* the peasant virtually intact in this period.[248]

I have partially examined three major legislative bills and the background to the Congress-Zamindar agreement against which they were framed. These are the Bihar Tenancy Amendment Bill of 1937, the Bihar Tenancy Amendment Bill of 1938 and the Restoration of Bakasht and Reduction of Arrears Amendment Bill. The first bill to be introduced in the House, in September 1937, was the Bihar Tenancy Amendment Bill. This attempted to tackle the immediate problem

[246] *Government of India Act, 1935*, Section 74.

[247] Rajendra Prasad's appeal to the Maharajadhiraj of Darbhanga, 20-4-1938, Rajendra Prasad Papers, File no. 3 B/1938, col. 1.

[248] According to some kisan leaders the failure of the Congress ministry was due to the structural weakness of the Congress movement. Jadunandan Sharma observed in 1938 that 'the masses have so far been kept in a mere state of emotional exaltation, and the Congress has always tried to reconcile the irreconcilable interests in its attempts to keep intact its national character .... It is time for the leaders to realize that the exploiters and the exploited cannot be benefited at one and the same time.' Quoted in B.B. Chaudhuri, Agrarian Movements in Bengal and Bihar 1919-1939', in A.R. Desai (ed.), *Peasant Struggles in India*, Bombay, 1979, p. 337.

of high rents. The main points the bill dealt with were the reduction of high rents, prevention of extortion of abwabs, and non-grant of receipts. While the bill sought to restrict the illegal activities of the zamindar it was also intended to smoothen the ruffled feelings of the landlord elements by hastening rent procedures.[249] Even this, it emerged, was not sufficient for the landlords. While the bill was pending at the Select Committee stage, Rajendra Prasad met the representatives of the landlords in November 1937 to hear their grievances and secure their agreement to important provisions in the bill. The discussions that followed culminated in what came to be known as the Congress-Zamindar pact.[250] The landlords placed before the Congress specific proposals regarding their suggestions for expediting the disposal of rent suits and the realization of rent decrees. In particular, they wanted more effective procedures for the judicial recovery of rent by suit to be incorporated in the bill. Their proposals aimed to hasten rent-recovery measures and to strengthen the legal powers of the landlord in this respect.[251]

The landlord's proposals were discussed at length during the talks between them and Rajendra Prasad. On the question of speedy rent disposal the Congress promised to examine the proposals carefully and, after consulting the Advocate General, make the agreed procedure part of subsequent legislation. Once this promise was secured the landlords agreed to give their consent to many of the provisions of the proposed bill and were prepared to discuss other important

[249] *Bihar Legislative Assembly Debates, vol. 1, 1937*, pp. 1830, 1836.

[250] Copy of the Congress/Zamindar agreement in Rajendra Prasad Papers, File no. III/37, col. 1.

[251] The proposals included (1) The appointment of a Separate Munsif to try rent suits in every district; (2) On the receipt of the complaint the Munsif would issue a notice to the defendant, asking him to deposit the amount claimed or file an objection in 30 days; (3) The notice was to be served by registered post and a postal *peon* would serve it; (4) If the objection was filed, the court was to deal with the decree on the day specified; (5) The disposal of suits was to be delayed for more than six months if contested and three months if uncontested; (6) After the court passed the decree, a notice was to be sent to deposit the decretal amount within one month; and if the defendant failed to pay the amount, a revenue officer would institute proceedings to recover the decreed amount according to the rules of the Bihar and Orissa Public Demands Recovery Act; (7) No fresh Vakalatnama would be permitted to be filed by the decree holder in the executing court; (8) An appeal against the rent decree would be permitted only on deposit of the decretal amount; (9) The High Court was to call for an explanation for all cases of rent suits delayed beyond six months.

intended legislation. However, as we shall see, the landlords were to retract some of these concessions in the 2nd Congress-Zamindar agreement of 1938, and also to pressurize the government to reduce the effectiveness of the reduction-of-rent legislation through subsequent executive fiats. However for the time being the consent of the landlords was secured on the following points:-

(1) The abolition of the system of danabandi or produce-rent by appraising crops before the actual harvest. The system of batai was retained but the landlord's share was not to exceed 9/20 of the produce.

(2) The commutation of produce-rent to cash-rent on the application of the tenant. The bill provided that the court accept all such applications by the tenants while it could reject any such application of the landlord. (This provision was later changed by an executive fiat, circular No. 38, which laid down that evidence tendered by zamindars must be considered and an order passed giving reasons for the acceptance or rejection of the same).[252]

(3) The suspension of the right of landlords to realize damages if the raiyat failed to pay rent and the reduction of interest charges on arrears from 6.25 to 1.5 per cent.

(4) The restriction of the landlord's right to sell the whole holding for arrears of rent. The bill provided that only such part of the holding as was enough to clear the debt could be sold, and that the price would be fixed by the court. It was, however, understood that the landlord could distrain the crops of the tenant for even a year's arrears. Even this part-sale provision was amended later.

(5) The penalization of landlords on charges of illegal exactions and abwabs, by simple imprisonment for 6 months and a fine. The bill originally made a custodial sentence mandatory but the landlords forced an amendment. It was made a non-cognizable bailable offence since, they claimed, the police would use it to harass the landlords.

(6) Reduction of all enhancements of cash-rents of occupancy

[252] Swami Sahajanand Saraswati, *Rent Reduction in Bihar, How it Works*, Patna, 1939, p. 20.

raiyats made from 1 January 1911 to 31 December 1936 (except those made on grounds of improvement by the landlord).

(7) The partial remission of rents in the case of the growing infertility of the soil and the failure of landlords to carry out irrigation arrangements. The landlords managed to obtain an amendment of this provision, entitling raiyats to remission only in cases where the landlord was contract-bound to maintain the system of irrigation. This provision was to be further qualified by a series of executive fiats later on.

(8) The immunity of the tenant from arrest for arrears of land rent.

(9) The right of occupancy raiyats or partition holdings on the same terms as a tenure-holder. However, the landlord fees or *salami* were not abolished and the raiyat had to pay a consideration fee of 4 per cent of the price he received.

(10) The right of tenants to appropriate flowers, fruits, and other plants on the land irrespective of any entry to the contrary in the record of rights.

The Congress-Zamindar agreement also touched upon other aspects of the tenancy question such as the restoration of bakasht land and the reduction of arrears.[253] These issues were to be legislated upon in a later session of the House. Regarding the issue which most agitated the tenants, the return of bakasht lands, the landlords saw to it that the provisions were hemmed in by conditions which greatly reduced the amount of land that would be returned.

All lands sold by occupancy raiyats to landlords between 1929 and 1937 were to be restored to the tenant on the following conditions:

(1) The raiyat should repay in one lump sum or in instalments payable in a period of not exceeding five years half the amount of the decree or half the costs of the execution.

(2) Failure to pay instalments for two consecutive years would entitle the landlord to recover possession of the holding.

(3) The provision of restoration would not relate to nagdi lands, the rents of which had not been commuted or enhanced since 1911.

(4) The provisions would not apply to land settled with other occupancy tenants.

[253] Congress/Zamindar agreement in Rajendra Prasad Papers.

(5) The provisions would not apply to lands in cultivating possession of petty landlords.
(6) The provision would not apply if a building, structure, or garden had been erected on the holding which was of such value that the restoration could be unfair.

Regarding the question of arrears of rent the agreement was as follows:

(1) Remission of arrears was permitted if the rent reduction officer was satisfied that the means of the raiyat were affected by the fall in prices between January 1929 to October 1936.
(2) The amount of cess paid by the landlord was refundable in proportion to the remission of arrears.
(3) The tenant was entitled to remission only if he had paid 5 per cent of the rent falling due for which arrears were claimed.
(4) If the tenant was a habitual defaulter, he would not be considered for remission.
(5) After a portion had been remitted the raiyat was bound to pay the balance in the next one and a half years.
(6) As regards realization of arrears, though the certificate power of the landlord to realize rent by retaining a holding for a period deemed by him to be sufficient to satisfy the amount due to him was abolished, the distraint provision of the act was to be amended to suit the landlord. By this means arrears of rent could be realized by the landlords through a distraint of standing crops to be enforced by a special board.

These were the terms of the 1st Congress-Zamindar agreement. Regarding rent it can be noted that, while it was agreed that the landlord's powers of rent collection would be made swifter and simpler, the concessions to tenants for the reduction of rents, remission of arrears and restoration of bakasht lands were hemmed in by conditions. These conditions were further extended by executive fiats in 1938-9 which made the whole process of rent reduction an expensive and protracted legal process. The reactivating of the distraint clause also facilitated matters greatly for the landlords. It was noted that this provision contradicted the Faizpur Congress resolution which stated that arrears of rent were to be recoverable in the same manner as civil debts. The part-sale provision, however, was intended to restrict attempts to sell holdings in lieu of arrears. The Congress was not able to compel zamindars to keep to their promise in this matter and the provision was amended in the 1938 Amendment Act. As far as bakasht was concerned, the proposals were woefully inadequate, as the subsequent legislation was to prove. In the meantime, the pending legislation before the House and subsequent on the tenancy question was drawn up on the lines set up by the Congress-Zamindar agreement.

The Bihar Tenancy Amendment Bill of 1937 was amended on the lines of the Congress-Zamindar agreement in the Select Committee stage and passed by the House in December 1937. It is important to note that the concessions made to tenants by way of rent reduction, commutation rights, and bakasht land, related only to occupancy raiyats, while non-occupancy raiyats were ignored. The more important provisions of the Act regarding the reduction of rents are itemized in the footnotes. (See below).[254] The provision in the act that sought

[254] Section 112A of the Bihar Tenancy Amendment Act stated:
(1) 'The collector may, on the application of an occupancy raiyat or a landlord in the prescribed form, or if the Governor by notification directs that a settlement of rents of the occupancy holding situated in any area . . . shall be made under this section, on an application made as aforesaid or on his own motion; (a) cancel all enhancements of the rents of occupancy holdings made . . . between the first day of January 1911, and the thirty-first day of December 1936; (b) reduce the rent of any occupancy holding, the rent of which was commuted, whether under section 40 or by an agreement between the tenant and the landlord of such holding, between the first day of January 1911 and the thirty-first day of December 1936, to such an extent that the reduced rent shall bear to the previous rent, the same proportion as the current prices bear to the average prices during the decennial period, immediately preceding the time when the rent was commuted; (c) order the partial or entire remission for such period as he considers reasonable . . . of the rent of an occupancy holding; (d) if the soil of a portion or the whole of such holding has without the fault of the raiyat become temporarily or permanently deteriorated . . . ; (e) if the landlord of such holding has failed to carry out the arrangements in respect of irrigation which he is bound to maintain: Provided that the collector may revise any order passed under sub-clause (i) at any time before the expiry of the period fixed for such order if he is satisfied that the soil of the holding has become fit for cultivation, and may at any time revise the order passed under sub-clause (ii) if he is satisfied that the landlord has restored the arrangements in respect of irrigation which he is bound to maintain: (f) reduce the rent of the occupancy holding, if there has been a fall not due to a temporary cause in the average local prices of staple food crops during the currency of the present rent to such an extent that the reduced rent shall bear to the previous rent the same proportion as the current prices bear to the prices prevailing at any time when the previous rent first became payable . . . (i) settle a fair rent in such cases, . . . as may be specified

to restrict the landlord's powers to sell his tenant holding for arrears of rent by allowing only a portion of the holding to be sold was part of the amendment of 1937, which, significantly, was subsequently withdrawn.[255]

The provision for the abolition of all increases in rent made since 1911 was an important provision, but one that was hemmed in by the condition that it would not apply if the landlord claimed to have carried out certain improvements.[256] Besides this, tenants were expected to furnish details regarding the number and date of enhancements within a specified period which could not in fact be provided without incurring heavy expenditure. Provisions for rent reduction were further qualified by certain circulars of the government in 1938-9.[257] Though it was felt that there might be instances where there might be ample scope for the application of all the clauses of sub-section 112A(1), government circular No. 39 prohibited application of clause (d) in cases where clauses (a) and (b) had been applied.[258] Clause (e) also became a dead letter and was never applied. Another government circular directed that in applying clauses (b) and (d) of sub-section 112A(1), the price of two staple crops in the area should be taken into account instead of one as decided earlier. This lowered the percentage of reduction of rents, even by government calculations, from 25 per cent to 18 per cent.[259]

---

in a notification issued for other sufficient reasons by the Governor in this behalf.
(2) No application shall be entertained under clause (a) or (b) of subsection (1) unless it is filed within a period of one year from the date on which the Bihar Tenancy (Amendment) Act came into force.'

255 The provision was as follows: 'When a decree holder makes an application, the court executing the decree may order that only such portion of the holding as may seem necessary to satisfy the decree shall be sold and the proceeds of any such sale or sufficient portion shall be paid to the party entitled under the decree to receive the same.' The landlords had agreed to this clause for the concessions secured in terms of the distraint provision and the judicial procedure for rents, to be introduced in a subsequent bill. But they were soon to withdraw their consent under the Bihar Tenancy Amendment Bill 1938. See Bihar Tenancy Amendment Act, 1937, Act VIII, in *Bihar Acts 1937*, Government of Bihar.

256 Given the notorious dishonesty of landlords, this provided a legal loophole for many of them to claim that the enhancements of rent were justified.

257 Swami Sahajanand Saraswati, *Rent Reduction in Bihar, How it Works*.

258 Sahajanand noted that this clause had not been applied, p. 5.

259 Ibid.

Another government circular, No. 42, curtailed the application of sub-clause c(i) unless it was proved that deterioration of the soil had occurred since the existing rent first became payable. It was not enough that the rent reduction officer found at the time of application that it was under water or sand deposit. If the remission was sanctioned, it would not have retrospective effect. Regarding the application of sub-clause c(ii), government circular No. 43 directed that landlords be given sufficient time to make good any irrigational defects. This sub-clause was also not given retrospective effect for a period prior to the date on which the application was filed. Earlier directions to rent reduction officers that they should ignore the promises of a landlord if his failure to maintain the irrigation system was general and habitual were withdrawn.[260] The irrigation provision was one that was very important to tenants. However, by accepting the landlord's argument that they would maintain irrigation systems only if the contract mentioned it, the effectiveness of the clause was almost entirely negated. It was also rendered easy for landlords not to bind themselves to maintaining irrigation systems which traditionally they had been honour-bound to do.[261] A further government circular directed that the order for remission would take effect from the beginning of the agricultural year from which arrears commenced, as had been decided earlier. This meant that the tenant had to file two applications separately, one for the reduction of arrears and one for remission, when instead he might have obtained the desired relief through the rent reduction application alone and be saved from heavy expenditure. This, then, was the legislation for rent reduction that followed the Congress-Zamindar agreement. The other provisions regarding tenant commutation rights, rights for distribution of holdings, and reduction of interest on arrears of rent were as agreed to in the Congress-Zamindar agreement.[262]

The following session of the Bihar Legislative Assembly of June 1938 saw the presentation of two more bills.[263] The main provisions

[260] Ibid., p. 7.

[261] The landlords could also file a revision petition to annul the remission and, if their claims were accepted, the remission would be annulled without delay.

[262] The exact provisions are available in the *Bihar Tenancy Act of 1885 (as modified up to 1st August 1940)*.

[263] *Bihar Legislative Assembly Debates, vol. II, 1938*, p. 1223 and *vol. IV*, p. 2613.

of the Bihar Tenancy Amendment Bill of 1938 dealt with the judicial procedure for rent suits and the speedy disposal of rent decrees in line with the Congress-Zamindar agreement of 1937.[264] Occupancy raiyats were given the absolute right to transfer holdings and the landlord's transfer fee was reduced to four per cent of the consideration money. For the first time too, something was done for under-raiyats and they were given the right of occupancy if they had tilled the land for twelve years. In June 1938 the bill was introduced in the House. A landlord member of the House, Chandreshwar Prasad, and two colleagues immediately stood up to claim that the new bill did not satisfy all the points of the agreement and stated: 'We cannot participate in a measure which only aims at fulfilling part of the settlement. We three represent 12 lakhs of landlords.'[265] In response to this, the House was adjourned. Rajendra Prasad and Maulana Azad then met the representatives of the landlords again on 4 July 1938, and a second agreement was arrived at.[266]

It made the following points:

(1) When an order for distraint had been passed by the civil court it was to be transferred to the revenue court.

(2) The part-sale provision of the Bihar Tenancy Act 1937 was to be amended. It was, therefore, decided that if the raiyat had defaulted in payment of rents for four consecutive years after commencement of the Act of 1937, the landlords might institute a suit for arrears. In such a suit the court may declare the raiyat to be a habitual defaulter and may direct the entire holding to be sold in execution of the decree.

(3) Regarding the reduction of arrears, the government was to reduce the amount of cess paid by the landlords in proportion to the amount of rent reduced. Section 55 of the Bihar Tenancy Act would also be amended so as to entitle landlords to appropriate any payment on account of rent towards arrears, in spite of any declaration to the contrary made by the raiyat.

(4) If after a period of six months the government found that rent

[264] It incorporated many of the suggestions of landlords on speedy judicial procedure of rent suits, which were listed in the Congress/Zamindar talks of 1937. See advocate-general's statement, ibid., *vol. IV*, pp. 3125-35.

[265] Ibid., pp. 3025 and 3034.

[266] Copy of the second agreement in Rajendra Prasad Papers, File no. 1-A/38, col. 1.

was not being realized as it should in normal circumstances, it should take such action by means of legislation or otherwise as would ensure realization of rent. The government would further issue a statement warning tenants that if rents were not paid the government would not be justified in giving special protection to tenants.

(5) It was similarly expected that landlords would do their utmost to carry out the obligations on them in terms of the agreement. A zamindar failing to do so would have himself to blame rather than the government.

(6) A board with powers to deal with all revenue matters would also be constituted.[267]

Clause 4, of course, amounted to a complete caving-in to landlord interests. Significantly, clauses 3,4,5 and 6 were concealed from the press for fear of a public outcry. The Congress made clear that all subordinate committees and members of the Congress should support the agreement and that those members or committees opposing the decision of the PCC were liable to disciplinary action. By this stage, the Congress was clearly worried about the public reaction the agreement might receive. Not surprisingly, the agreement was received with particular dissatisfaction by the Kisan Sabha. For their part, the landlords further issued a statement commending the agreement.[268] At the All-India Landholders' Conference in April 1939, the Maharaja of Darbhanga praised the Congress for its attitude of 'adjustment and peace', while Chandreshwar Singh added that, since the Bihar Congress under Rajendra Prasad had accepted the landlords' terms, the landholders as a group had little to fear.[269] Kisan Sabha Legislative Assembly members Jamuna Karjee and Shyamnandan Sinha protested against the amendment of the part-sale provisions. Nonetheless, the motion to delete the relevant clause from the bill was negatived by 73 against 7 votes, with the Kisan Sabha members, who were Congressmen, abstaining.[270] The bill was passed in July 1938. As pointed out earlier, the bill did secure some concessions for tenants.

[267] Ibid.

[268] *Indian Nation* 6-7-1938, quoted in Rajendra Prasad Papers, File no. 1-A/38, col. 1.

[269] Proceedings of the All India Landholders' Conference, April 1939 in *Indian Annual Register*, January-June 1939, pp. 396, 405.

[270] *Bihar Legislative Assembly Debates vol. IV, 1938*, pp. 3779-87.

Thus section 48A provided that any person who had for a period of 12 years continuously held land as an under-raiyat in any village, would acquire the right of occupancy.[271] Despite this the 1938 Bihar Tenancy Amendment Bill actually went further than the bill of 1937 in enabling landlord pressure *vis-à-vis* his tenant. The judicial procedure for rent suits permitted the court to pass a decree in cases of suits for cash-rent simply on an affidavit of the landlord, while the execution of the decree was transferred to revenue courts, making the process even speedier.[272] The distraint of crops provision further provided that the landlord could claim arrears of rent for a whole year by distraining the crops of his tenant.[273]

The third important piece of legislation enacted in 1938 was that dealing with the bakasht question, through the Restoration of Bakasht land and Reduction of Arrears Bill.[274] Framed along the lines of a Congress-Zamindar agreement of 1937, the bill fell far short of popular demand and provided for the return of only a small percentage of seized lands. Several loopholes were also provided for landlords to effectively circumvent the legislation. The provisions of the Act are given in the footnote.[275] The bakasht section of the bill invited much

[271] This clause was, however, never really implemented. As pointed out in an earlier section, the Santhal bataidar agitation in Purnea which was sparked off by this legislation did not succeed in its aims. The institutions of the police and the courts in most cases supported the maliks and the claims of the under-raiyats were suppressed.

[272] See Act XI of the Bihar Tenancy (Amendment) Act 1938, Sections 143-158 in *Bihar Acts 1938*, pp. 33-8.

[273] The Kisan Sabha members of the Assembly protested against this clause 121(1). See *Bihar Legislative Assembly Debates, vol. IV*, p. 3713.

[274] Copy of Act (Bihar Act IX of 1938), in *Bihar Acts 1938*.

[275] '3 (1) A *raiyat* whose holding or portion of whose holding was sold at any time between the first day of January 1929, and the 31st day of December, 1937, for execution of the decree for arrears of rent and was purchased by the landlord of such holding and is in the possession or under the control of the said landlord, may make an application to the collector for the restoration to him of such holding or portion.' The bill, however, provided that the collector could reject the application on any of the following grounds: (a) that the landlord is a petty landlord and that he has been cultivating the holding or portion by himself with his own stock and his own servants since 22nd day of March 1938; (b) that a building, structure or garden has come up on the holding which is of such value that the restoration of land in such circumstances will be unfair; (c) that the holding mentioned in the application, or any portion thereof, is in the possession of a third person and this third person is in possession of the holding under a settlement, which in the case of a holding sold before 1 January 1929 and 21

discussion.[276] It was reported that clause 'c' (which prevented lands settled by the landlord with another tenant before 19 April 1938 from being restored) had resulted in many such hasty settlements immediately prior to the passing of the bill. For example, in Bhagalpur, on hearing of the legislation, the landlords had settled every inch of restorable bakasht land with other tenants. However, despite these criticisms, the motion to amend the date to 31 July 1937 was negatived and the bill was passed.

We have seen how, while landlords succeeded in their demands for the speedy disposal of rent suits, and the power to distrain crops and sell the tenant holding in lieu of arrears, the concessions made to the tenants in respect of rent reduction and bakasht legislation were greatly hemmed in by conditions. In connection with the major grievances of the peasantry, particularly in relation to the landlord's duty to maintain irrigation rights, the bills attempted little. Landlords were legally bound to maintain irrigation rights only where they had contracted to do so. With regard to peasant rights to common property resources, rights of grazing and forest rights, the bills made no special mention at all. Individual cases were to be examined separately and no legal backing was given to these rights, apart from the rights of

---

December 1937, was made in good faith by such landlord before 19 April 1938. The land, however, could be restored on payment of a sum of money '8(1) if an application made by the *raiyat* under section 3 is not rejected under section 4 or dismissed under subsection (2) of section 6, the collector shall (a) determine the land which is liable to be restored to the *raiyat* under the provisions of this Act, (b) determine the amount which shall be payable by the *raiyat* for restoration of such land . . . .' In case the entire holding was sold, or if only a portion of the holding was sold the clause stated that the amount would be equal to 50 per cent of the aggregate of the amount mentioned in the sale proclamation for which the holding was sold and the amount of costs necessarily incurred by the landlord. The raiyat could deposit the amount in one lump sum or in instalments which should not exceed a period of 5 years. Regarding the arrears of rent, the bills provided that a tenant's application for reduction could be accepted by the collector provided it was proved that he was not a habitual defaulter, he had paid up at least 10 per cent of the arrears mentioned, and if he was satisfied that the means and circumstances of the raiyat were affected by any fall in prices during the period 1929-36 the collector could then reduce the arrears to such an extent as he thought proper. See ibid., pp. 4-8. A motion presented to amend the date to 31 July 1937, to avoid hasty settlements, was negatived. See debate on the bill, *Bihar Legislative Assembly Debates, vol. III, 1939*, pp. 1311-15.

[276] Jamuna Karjee's debate, ibid., pp. 1311-15.

peasants to fruits and flowers from their own lands. Traditional rights were thus made critically vulnerable.

The expensive and protracted legal procedure involved also reduced the effectiveness of the legislation. For example, one government circular specified that the tenant should submit separate applications for remission of rent and reduction of arrears, while the tenant could have got the desired relief from one application alone, and could have been saved the additional expenditure. This lengthened the legal procedure for tenants. As regards the reduction of rent clause, Rajendra Prasad noted that although 200 special officers had been appointed by May 1939 they had not succeeded in disposing of even 15-25 per cent of the cases.[277] Besides this, tenants were expected to furnish details regarding the number and date of various enhancements which could not in fact be provided without incurring an unrealistic level of expenditure. The official reports noted that many of the records that needed to be used to reduce rents were out of date and did not legally give sufficient information. In areas such as Chapra, Rajendra Prasad noted, officers had still not been appointed to receive applications as late as May 1939. Despite these problems the incidence of rent proceedings did alleviate pressure on some sections of the peasantry. In Patna, Gaya, Shahabad, and Monghyr districts, rent settlement operations during 1937-40 affected 25 per cent of the holdings.[278] The percentage of holdings where rents were reduced averaged from 31 per cent in Aurangabad and Jehanabad to 16 per cent in Sasaram.[279] With regard to the bakasht legislation the provisions failed to meet the most pressing demands of the peasants. The conditions for restoration greatly reduced the amount of land liable to be restored while the inability of tenants to pay their legal costs and half the auction price further hampered them. It was only through direct action and the forced settlement of disputes that some land was restored in Reora, Barahiya Tal, and Darbhanga.

In reality only the economically stronger of the occupancy raiyats could take advantage of the mechanism provided for rent reduction by the new bills. As long as the zamindar system existed, it remained an uphill task for the weaker occupancy raiyat to take advantage of

277 Rajendra Prasad to Patel, 22-6-1939, Rajendra Prasad Papers, File no. IC/39, col. 1.

278 Rent was reduced on 146,072 holdings.

279 *Annual Report on the Survey and Major Settlement Operations in Bihar 1941*.

the few measures passed in his favour. Regarding non-occupancy tenants and agricultural labourers, the bills had achieved even less. All the concessions related only to occupancy tenants. Therefore, while some sections of the richer peasantry could take advantage of the few legal benefits secured, the majority were still at the mercy of the landlords.

Perhaps the best indication of the failure of the agrarian bills to achieve real social gains was provided by the landlords' own perception of their social situation. Sahajanand noted in this regard that they did not in any way act as though their prestige and power had been impaired. At the legal level, for example, even while kisans were applying for reduction of arrears, zamindars continued to file rent suits against the very same arrear debts. Meanwhile, the government did nothing to force the courts to take note of such kisan applications and stay rent suit proceedings. Sahajanand also observed that, even though the Congress had made exactions of abwabs illegal, forced labour had still not been abolished. Furthermore, no machinery had been created that could affect the abolition of feudal dues.

Quite clearly the landlords were in fact worried by the scale of the peasant agitation. In some districts, as in Nawadah, attempts were made to improve relations with tenants by issuing rent receipts. However, this was not a common reaction. More often than not violence was simply employed to crush all opposition and landlords emerged from the confrontation with their powers virtually intact. In this the landlords were now effectively supported by the coercive machinery of the state. The realization that the Congress government was at least as willing to protect them as the colonial state had been was a factor that strengthened the psychological and practical position of the landlords immensely. It is not surprising, then, that the leader of the Landlord party, Chandreshwar Narayan Singh, felt able in April 1939 to congratulate the Congress on its policy of adjustment and peace and noted unequivocally that any danger to the landlords came not from the Congress but from 'socialism'.[280]

THE BAKASHT MOVEMENT TAPERS OFF

The last few months of the Congress ministry were politically frenzied. At the organizational level the opposition of the Kisan Sabha/

[280] Sahajanand Saraswati, *Rent Reduction in Bihar, How it Works*, p. 18.

CSP culminated in an attempt to break away from the main body of the Congress. This development was compounded by the fact that, at the national level, some left-wing groups, led by Subhash Chandra Bose, were challenging the Gandhian group. The issues over which Bose divided the Congress related to its stand on federation and the development of a 'ministerial' mentality.[281] The Kisan Sabha and the socialist delegates in Bihar supported Bose's candidature for the presidential elections at Tripuri, and the stage seemed set for a trial of strength between the two wings of the Congress both at the national and the provincial levels. The Kisan Sabha hoped that the left would emerge dominant. This crisis was forced to its logical conclusion when the Gandhian group resigned from the working committee and a dissolution of the Congress ranks seemed imminent. The Congress socialists now moved to stem the tide and preserve unity in the ranks. They remained neutral on the Pant resolution—Jayaprakash Narayan explaining their decision as the price they were prepared to pay for unity.[282] The Tripuri session thus saw the Gandhian group return to dominance with the passage of the Pant resolution. By March 1939 this socialist stance had radically changed the situation at the provincial level in Bihar. The cleavage between the Kisan Sabha and the Congress had already widened considerably all through 1938-9. While at the official level the Kisan Sabha continued to reiterate its loyalty to the Congress, it was clear that a break was envisaged, especially by Sahajanand. Indeed, such a break was openly discussed at the AIKS conference held in Gaya in April 1939 and attended by 100,000 kisans, when the main issue concerned the Kisan Sabha's relationship with the Congress.[283]

At a secret meeting of the Subjects Committee of the session two rival groups were formed. Sahajanand, who was backed by Indulal

[281] *Congress Socialist*, 19 February 1939.

[282] The resolution was moved by a right-wing group headed by a Congressman from UP, G.B. Pant. It attacked Bose over the allegations that he had made during the election campaign that the Gandhians were preparing to sell out over federation. It demanded that Bose prove these charges and proposed a vote of no-confidence in him if he failed to do so. Later the resolution was slightly modified and required Bose to follow Gandhi's wishes in appointing a new working committee. This resolution was passed by 218 votes to 135. See B.R. Tomlinson, *The Indian National Congress and the Raj*, p. 130.

[283] FR (2), April 1939; Rakesh Gupta, *Bihar Peasantry and the Kisan Sabha*, p. 151.

Yajnik and the Bengal group of kisans, advocated breaking away from the Congress and concentrating on a direct appeal to the masses, as he felt that the Congress had failed to meet mass demands efficiently. The Sahajanand group then attempted to pass a resolution of no confidence against the Bihar ministry. However, this notion was discouraged by Jayaprakash Narayan, Narendra Dev and the socialists, who all reiterated the CSP's interest in keeping the Kisan Sabha within the Congress fold. Ultimately, the issue was put on ice. The kisan conference did, however, adopt a resolution strongly criticizing the attitude and policy of the Bihar Congress ministry towards kisan agitators and in particular its free use of the notorious section 144 of the Criminal Procedure Code against the kisans.[284]

Sahajanand and a few other Kisan Sabha members had therefore to content themselves with joining Bose's Forward Bloc. Without the socialists, however, this faction did not make much headway. A further attempt was made to rope in the socialists by creating the Left Consolidation Committee; but the socialist involvement in this structure was half-hearted.[285] In July 1939 Sahajanand, along with Bose, called for an all-India protest day against the All-India Congress Committee decision that banned Congressmen from starting any satyagraha.[286] The Bihar Provincial Congress Committee promptly banned Sahajanand and Kishori Sinha from holding elective posts in the Congress for three years.[287] Consequently, by the end of 1939, Sahajanand had come to realize the futility of attempting to create a viable left alternative under the Congress. The passage of events from March to September 1939 had effectively shattered the unity of the left-wing organizations to such an extent that he and other dissident Congressmen did not feel strong enough to launch an agitation on their own. Meanwhile, the war had broken out in September 1939 and Sahajanand realized that the changed situation called for a united strategy. In a statement to the press he stated that the Kisan Sabha had forgotten its internal conflicts and was awaiting the clarion call of the

[284] *Kisan Bulletin*, 17 April 1939 in Indulal Yajnik Papers, File no. 1/1939.

[285] Yajnik to Sahajanand, 18-9-1939, Indulal Yajnik Papers, File no. 11/1938.

[286] Ibid. Sahajanand's statement to the press. Sahajanand noted that the ban paved the way to constitutionalism: 'it indicates clearly the depth of the terrible drift towards reformism. It points out the complete absence of the revolutionary approach to our national problems.'

[287] Ibid., Sahajanand's statement to the press after the ban.

leaders.[288] By the end of 1939, therefore, the right-wing Congress leadership had effectively re-established control in Bihar.

The bakasht movement was also on the decline. It was losing momentum and peasant agitations died away. The changing economic context contributed to the emergence of this new situation. Following the outbreak of the war, prices had picked up strongly and the worst effects of the Depression were over. The upper echelons of the peasantry soon began to benefit by the rise in prices. The Congress agrarian legislation, (although wanting in many respects), had also eased the situation, especially as the stronger sections among the occupancy raiyats took advantage of the mechanism for rent reduction and secured some concessions. The new compromising attitude taken by Sahajanand and the leadership after July 1939 also discouraged direct action and agitation. The government's fortnightly reports noted that the kisan movement continued to slacken, as all the principal leaders were preoccupied in giving attention to anti-war propaganda.[289] As the right-wing Congressmen in the province declared that the war effort should be undertaken by a properly constituted independent government, the left wing under Sahajanand and Jayaprakash Narayan laid down a five-point programme of opposition to the war, while carefully refraining from any direct criticism of the Congress.[290]

Despite the conciliatory attitude adopted by the left after war had broken out, the kisan movement had already seriously challenged the nascent Congress government in Bihar. The agitation had been fuelled at a fundamental level by an escalation in oppressive landlord actions and by the subsequent exacerbation of basic subsistence difficulties and ecological pressures upon an already hard-pressed peasantry. It was this very peasantry which had voted a popular Congress government to power in the specific hope that the party would redress their

288 Sahajanand's statement, 17-9-1939, in Indulal Yajnik Papers, File no. 11/1939. One can see the change in Sahajanand's stand from these letters.

289 FR (1), September 1939.

290 One may, in the light of current trends in history writing, tend to dismiss the attitude of the elite leadership as an unsatisfactory explanation for the changing course of the movement. It is, of course, true that other factors were at work which contributed to the declining vigour of the struggle. However, the attitude of the leadership was also important and it was the specific configuration of all the different factors that gave the movement its force and was ultimately responsible for its slowing down.

most immediate grievances. Once in power, the Congress proved too conciliatory to the landlords. The resulting legislative compromise led directly to unrest among the majority of the rural population so that the party found itself resorting more and more to the coercive methods of the imperial state. Equally, the Congress government in power succeeded in antagonizing the kisan movement to an extraordinary degree. Surprisingly, however, the challenge posed by the Kisan Sabha leadership did not culminate in its breakaway from the Congress. Instead the movement itself declined with the outbreak of the war and the changing priorities of the leadership. Betrayed by the Congress and let down by their own movement, the poorest peasants of Bihar were thus left to face their steadily worsening economic and ecological predicament without any effective political lobby.

CHAPTER THREE

Marking Time: The Congress in a Quandary, 1939-42

The period following their withdrawal from office saw the Congress leadership left without an official policy. The experiment in government in Bihar had not been a very successful one. Mass agitations and demands had gathered force, threatening the Congress programme and forcing it to spell out a deeply conservative policy in the agrarian and labour fields. It may be possible to argue, as Walter Hauser has done, that the intervention of the war was welcome in the sense that it ended the Congress stint in office at a time when the organization was threatened by the force of mass demands and the opposition of its own left-wing leadership.[1] However, the tempo of mass agitation in rural areas had already begun to slow down and the Congress might have been able to tide over problems for a time if it had remained in office. Indeed, the absence of an alternative programme might gradually have led the party to a compromise on federation, had not the outbreak of the war prematurely ended the drive towards it.

THE POLITICS OF WITHDRAWAL FROM OFFICE

At the Haripura session in February 1938 the Congress had clearly rejected the federation scheme. The party could not, however, come up with an agreed alternative to the scheme.[2] Linlithgow believed that in the Congress inner circle the view was held that absolute rejection was a preliminary gesture and would eventually give way to some sort of acquiescence as had been the case over the issue of the acceptance of office.[3] Gandhi had given Lord Lothian a note which, though it embodied his personal view, indicated that he would accept federation

[1] Walter Hauser, 'The Bihar Provincial Kisan Sabha 1929-42'.
[2] *Indian Annual Register*, 1938, 1, 297-8.
[3] Linlithgow papers, Mss Eur F 125/5/2. Linlithgow to all Governors.

if certain modifications were made. In a letter to Agatha Harrison he had admitted that:

About federation the position is absolutely clear so far as I am concerned. It is perfectly true that if my conditions were fulfilled, I would accept federation and so would the Congress, I feel sure.[4]

Thus before the war intervened in September 1939 there was no general expectation that the Congress would withdraw from ministerial office, at least for some time. The Secretary of State, Lord Zetland, confidently informed the British cabinet that provincial autonomy had been working successfully and that, as the Congress had worked provincial autonomy, it would also work federation.[5] The provincial Congress leadership in Bihar viewed the peasant agitations, the labour strikes and the communal riots in the period 1937-9 as part of the everyday tribulation of office. In fact, the provincial leadership actually prided itself on its success in defusing the peasant agitation by 1939. Moreover, the ministry experience was seen as a training ground for a future national government. The Congress had gained real control of many of the institutions of the state and had effectively deployed the forces of law and order to check opposition to the Congress government. It had shown itself skilful in using state power to suppress mass agitation while continuing its populist rhetoric. Nehru, who had at first vehemently criticized office acceptance in 1937, now changed his tune. In a speech in Madras in July he claimed that 'acceptance of office has done us a great deal of good. We have achieved much by way of mass awakening, and ministers have gained experience in dealing with numerous problems of administration'.[6] More importantly, the Congress had managed to secure the support of the subordinate bureaucracy to the new national state. Henceforth their services could no longer be relied on in unquestioning loyalty to the Raj. Ann Ewing records how the period of the Congress ministries severely weakened the morale of the services. Only the core of the British officers in the provincial civil service remained unaffected, and most of them reported the growing unreliability of their subordinate staff and the difficulties in dealing with Indian ministers. Indeed, many British officers alleged 'interference'

[4] Gandhi to Agatha Harrison, 23 July 1939, CWMG, vol. lxx, p. 21.
[5] Cabinet Conclusions 1939, Cab. 23, Seeley Library, Cambridge.
[6] SWJN, vol. 9, p. 591.

by ministers in administration.[7] The Bihar ministry also prided itself for releasing political prisoners, lifting censorship and removing the ban on proscribed books and other political literature.[8] It had reinstated in service those who had been discharged on political grounds and had cancelled the ban on the Indian tricolour flag, evoking a general air of freedom.

Among sections of the Muslim community, however, the Congress populist rhetoric made little or no headway. The 27 months of the Congress ministry had seen a widening of the communal rift and an increase in communal rioting in both urban and rural areas. By 1938 the splinter Muslim communal parties in Bihar had joined the Muslim League, which was gradually building up its provincial base. The Congress ministry then formed the main focus of attack as the new 'Hindu raj'. The Muslim League published a detailed report containing allegations against the Congress which claimed that the party had failed to inspire confidence in the minorities as it was a predominantly Hindu organization, and that the majority of its members, in spite of their pretensions to nationalism, were imbued with communalism.[9] While many of the Muslim League allegations were exaggerated, their perception of the pervasive influence of Hindu communal ideology, especially at the lower levels of the Congress organization, were not far off the mark. The Hindu Mahasabha in Bihar had certainly increased its activities in the province in this period, calling on Hindus to safeguard their interests, which it alleged had been neglected in the conciliatory policy followed by the Congress.[10] The partisanship of

[7] Ann Ewing papers, CSAS, Cambridge. See also the reminiscences of G. Macdonald, Macdonald collection, 1938-47, IOL, Mss Eur E 360.

[8] AICC papers, File no. PL 5/1938. See also S.K. Sinha to Maurice Hallet, 15.2.1938, in Linlithgow Papers, 125/4.

[9] See the Pirpur Report (Report of the Inquiry Committee appointed by the Council of the All India Muslim League to inquire into Muslim grievances in Congress provinces).

[10] The RSS was able to recruit several thousand young men in Bhagalpur after the riots there. In the new atmosphere of unrestrained sectarian propaganda, communal feelings were spreading rapidly in the Bihar countryside. The local zamindars actively assisted these communal forces in an attempt to diffuse the threat of agrarian unrest. It was probably for these reasons that the Maharaja of Darbhanga spearheaded a Hindu Mahasabha campaign against the alleged abduction of Hindu girls and their conversion to Islam. As a result a series of communal riots erupted in Purnea and Darbhanga in the estates of the Darbhanga raj.

certain Congressmen was revealed in their attitudes to riot situations. In Bhagalpur, after a particularly serious riot, the secretary of the DCC himself was charged with fomenting the riots through the publication of pamphlets replete with inflammatory matter and with actively participating in the meetings of the Hindu Mahasabha.[11] In this case the local institutions of the police and the judiciary had connived with the Congress administration and the Hindu rioters were allowed to go scot-free.

The main demands of the Muslim League centred around freedom of language and culture, and middle class clamourings for government jobs, and seats in councils and municipal bodies. This was probably because class issues, and worker and peasant grievances, were common to both the communities and could not be used as a basis of communal demarcation at a policy level. The League claimed that the Congress had alienated Muslims by imposing Bande Mataram as the national song, Hindi as the national language and the Congress flag as the national flag.[12] However, more important than these grievances in sparking off rioting at the popular level were the issues of cow slaughter and music-playing before mosques. The League asserted that cow slaughter was the religious right of every Muslims and could be exercised without restraint.[13] It urged Muslims to carry out the sacrifice, causing a serious worsening of relations at the local level. A study of the riots shows that both communities were particularly quick to take offence over cow sacrifice, giving rise directly to severe rioting in both urban and rural areas. Protection of the right was seen as the symbol of domination of one community over another. Muslims were, therefore, keen to emphasize this religious right, even in areas where it had not been customary, while Hindus were equally insistent

[11] In Bhagalpur town a serious communal riot occurred on 28 July 1938 over a Rath Jatra procession. Trouble arose over a procession in front of the mosque. Four Hindus and one Muslim were killed and over fifty people were injured. See FR (2), July 1938. For allegations of the Muslim League, see charges of Abdul Hamid Rasmi, Secretary Muslim Mass Contact Committee, in AICC papers, File no. G-22/1938.

[12] It was claimed that Bande Mataram was anti-Muslim as, for 50 years, the book of Bankim had been read and resented as an insult to Islam. At a meeting of the All-India Muslim League at Lucknow, a resolution had been passed condemning the attitude of the Congress in foisting Bande Mataram upon the people as the national anthem, but the Congress government had persisted with the song. Pirpur Report, pp. 17-19.

[13] Pirpur Report, p. 41.

in preventing any such thing even during Bakri Id, in areas where it was traditionally exercised. In February 1938 the celebration of Bakri Id caused minor communal riots in a village in Gaya, two places in Patna and at two locations in the Santhal Parganas, each incident necessitating the summoning of a district magistrate. In north Bhagalpur, local Muslims expressed their intention to sacrifice cows in defiance of local custom. Hindu mobs then collected to prevent the slaughter and section 144 had to be invoked by the authorities. One incident in Tilkouri village, in Hazaribagh district, caused particular agitation among the Muslim public. As a result of the rumour that the beef of a missing cow was being prepared for a marriage feast in the house of a Muslim labourer, Hindus in the village gathered to assault him and more serious communal trouble was averted only by strict supervision by the police.[14] The police were still, however, accused of siding with the Hindus.[15]

Congress policy in reaction to these riots was *ad hoc* and clearly ineffective. In March 1937 the president of the AICC instructed all PCCs to take in hand the work of increasing Congress contact with the rural and urban 'masses'. A specific plan to approach the Muslim population was sketched out by the working committee on minority rights at the Calcutta session in October 1937. The objectives of this initiative were to protect the religious, linguistic and other rights of the minorities and to ensure their widest participation in government. However, in Bihar the policy got off to a slow start. Most of the attempts to mobilize Muslims for the Congress cause were bogged down with difficulties as the contact programme was more or less a one-man affair troubled by lack of funds and was ill-planned and ill-organized. Dr Ashraf, while reviewing the work of the mass contact movement in 1938, pointed out its organizational and administrative difficulties.[16] The programme, he said, had taken precious time to initiate in most provinces and communication between provincial offices and head offices was tardy. He noted that 'as far as I can gather, Muslim contact work done by the Congress in various provinces is totally unorganised. Except for the enrolment campaign, no substan-

[14] FR (1), February 1938. For details of the riots, see V. Damodaran, 'Congress and the Muslim Communal Problem 1937-39', forthcoming paper.

[15] FR (1), May 1938.

[16] Dr Ashraf was a Congress Muslim who had been put in charge of the movement at the national level.

tial effort is being made to come into direct contact with the masses'.[17] The Muslim mass-contact leader in Bihar claimed that the programme had utterly failed in the province. However, despite its failure to secure the support of the Muslim community, the Congress in Bihar was not unduly perturbed. As a credible political advocate for the Muslim community the Muslim League had yet to gain the stature it acquired after the 1940s. It was still building up a reputation in the provincial arena and the Congress saw no reason to be particularly alarmed.[18]

To summarize, then, while the difficulties the Congress faced in office were formidable, they were not sufficient to provoke a Congress decision to withdraw from office. Instead the decision to withdraw from office in 1939 was based less on considerations of difficulties at a provincial level, and more on the Congress leadership's perception of the international situation and the larger predicament of an anti-imperialist struggle. The outbreak of war had changed the situation dramatically and brought the whole constitutional exercise under stress. The Congress and the Muslim League reacted to war with conflicting demands and created a new crisis in imperial policy, so that the war years found British policy-makers trying to grapple unsuccessfully with the Indian problem in sudden contrast to the pre-war era. Ever since the early 1920s the Congress had begun to take an increasing interest in foreign affairs and had developed a foreign policy which was said to be directed at achieving 'the elimination of political and economic subjection everywhere and the cooperation of free nations'. This was articulated with the demand for Indian independence. In 1936 the Congress had condemned the German and Italian wars of aggression and warned the British government that the fascist powers were grouping themselves together with the intention of dominating the world and crushing political freedom.[19] The Congress condemned the British appeasement of fascist aggression and in March 1939, when Hitler seized Czechoslovakia, it recorded its disapproval of the Munich pact, the Anglo-Indian agreement and the

[17] AICC papers, File no. G22/1938.

[18] It was the Cripps offer which implicitly recognized the potential for a separate Muslim state and made Gandhi and the Congress leadership realize that the Muslim League was now an actual contender for power. CWMG, vol. lxxvi, p. 28.

[19] *Indian Annual Register*, 1936, ii, p. 202.

recognition of rebel Spain.[20] The Congress further urged that India must direct its own foreign policy, 'thereby keeping aloof from both imperialism and fascism and pursuing her path of peace and freedom'. The despatch of Indian troops to Aden, Egypt, and Singapore in August 1939, as part of general defence preparations, aroused the anger of the Congress against the British government.[21] Accusing the British of flouting the Central Legislative Assembly and defying public opinion by sending Indian troops abroad, the party called upon members of the central assembly to refrain from attending its next session. The Congress provincial governments were also asked to refrain from assisting in the war effort.[22]

The imperial policy-makers had been alerted by the repeated resolutions that a declaration of war on behalf of India without Congress consent might complicate the political situation. It was against this background that Zetland and Linlithgow started to discuss their policy in the event of war. Linlithgow proposed that the federal negotiations should be put in cold storage and urged that, for the duration of the war, the government should refuse any constitutional change.[23] He had been encouraged in this view by the governors of Madras and Bombay, who had assured him that the Congress ministers wished to continue in office and would be extremely hesitant to resign even if asked to do so by the Congress working committee.[24] However, Linlithgow had also prepared contingency plans to deal with the situation likely to arise were Congress ministers actually to resign from office. At his instance the British parliament had passed the Government of India (Amendment) Act in April 1939, whereby a new section was inserted in the 1935 Act, authorizing the central government, during the operation of the proclamation of an emergen-

[20] *Indian Annual Register*, 1939, i, pp. 341-2.

[21] India's defence questions had been debated as early as 1938 by the Chatfield Commission, which advised a dramatic increase in expenditure on Indian defence to be shared between the British government and the Indian government. To prevent opposition from Indian nationalist leaders it was decided at a cabinet meeting not to publish the report of the Chatfield Commission whole. Instead a carefully worded despatch from the Secretary of State to the Viceroy would contain as much of the Chatfield report as it was decided should be made public. See Cabinet Conclusions, 1939, Cab. 66.

[22] *Indian Annual Register*, 1939, ii, pp. 214-15.

[23] Tarachand, *History of the Freedom Movement*, vol. iv, p. 280.

[24] On 11 September 1939, the Viceroy announced that the federal scheme was suspended for the duration of the war.

cy resulting from war or threat of war, to give directions to the provincial governments about the way in which executive authority was to be exercised, enabling the central legislature to make laws in the provincial field and conferring executive authority on the central government. At the same time Linlithgow was prepared to negotiate, but only on a basis which involved no substantial transfer of authority during the war.

The war in Poland broke out on 1 September 1939 and two days later Chamberlain announced to the nation that Britain was at war with Germany. Constitutionally speaking, Britain's declaration of war on Germany automatically turned India into a belligerent and, in the event, no attempt was made to secure India's consent. Linlithgow and the British policy-makers had thus presented India with a *fait accompli*. In doing so they had committed a tactical error from which further misunderstandings then arose in a more specific way. The Congress complained that when the war came Eire was allowed to remain neutral, while in Australia and New Zealand the declaration was confirmed by parliament. Only British India was made a belligerent without her consent.[25] However, it is important to note that, since India was not a dominion, Linlithgow's action in declaring India at war was constitutionally valid. But it might be suggested that Linlithgow could easily have avoided an immediate deadlock by giving India an opportunity of going to war by a political act of her own performed by her existing representatives. The Congress may have been an unofficial organization, but there were central and provincial assemblies. It would have been tactical and wise to have provided these bodies with an opportunity to declare by resolution the determination of the people of India to cast their lot with the allies. It was not to be.

When the war broke out the Congress leaders were in a quandary. There was no doubt that the Indian leaders were anti-Nazi. As the European crisis was deepening Gandhi had, in a letter addressed to Hitler, (which was not forwarded by the Government of India), appealed to him to prevent 'a war which may reduce humanity to the savage state' and, when the war broke out, he was filled with 'horror'.[26] At a meeting with the Viceroy on 5 September 1939 he told Linlithgow clearly that his sympathies were with England and France

[25] *The Statesman*, 28 October 1939.
[26] CWMG, vol. lxx, p. 21.

from the purely humanitarian standpoint. 'And as I was picturing before him the Houses of Parliament and the Westminster Abbey and their possible destruction, I broke down . . . I am not therefore just now thinking of India's deliverance. It will come, but what will it be worth if England and France fall or if they come out victorious over Germany ruined and humbled.'[27] He further underlined the importance of the part India could play in preventing a catastrophe. But Gandhi's sympathies for the British position were not well received by Nehru and other Congressmen of his mentality. At a working committee meeting before the war broke out on 11 August 1939, the committee had passed a resolution reminding provincial governments to assist in no way with the war preparations of the British government. If this necessitated a resignation from office, they must be prepared for such a contingency.[28] On 3 September 1939 the Viceroy promulgated the Defence of India Ordinance. This had the effect of substantially curtailing civil liberties. On the same day an amendment to the Government of India Act of 1935 was rushed through the Westminster parliament enabling the central government to nullify provincial autonomy.[29] On 10 September the working committee met again and passed a resolution on the war. Though the resolution condemned Nazi aggression and openly sympathized with Britain it stated unequivocally that Britain must end imperialism in her own possessions, and establish full democracy in India. It also stated that the Indian people must have the right of self-determination, the right to frame their own constitution through a constituent assembly without external interference, and that they must guide their own policy. Only a free democratic India would 'gladly associate herself with other free nations for mutual defence against aggression and for economic cooperation'. The working committee therefore called upon the British government to declare in unequivocal terms what their war aims were with regard to democracy and imperialism and the new world order envisaged and, further, how such aims were to be applied to India.[30]

In a statement to the press Gandhi pointed out that he was alone in thinking that whatever support was to be given to the British should

27 CWMG, vol. lxx, p. 162.

28 Working committee resolution on the war, CWMG, vol. lxx, Appendix vii, p. 406.

29 SWJN, vol. 10, p. 123.

30 Working committee resolution, 11 September, vol. 10, pp. 122-38.

be given unconditionally.[31] He stated that the committee felt that the nation had not 'imbibed the non-violent spirit requisite for the possession of the strength which disdains to take advantage of the difficulty of the opponent'. But in stating the reasons for its views the committee desired to show the greatest consideration for the British. At the same time that the committee passed the statement it had also appointed a board, (with Nehru as chairman and Abul Kalam Azad and Vallabhai Patel as members) to deal with the situation as it developed from time to time. Gandhi stated that he had gracefully abided by the decisions of the working committee and hoped that the resolution would receive the unanimous support of all the parties among Congressmen.

On 20 September 1939 a draft resolution for provincial legislatures was prepared which demanded the declaration of war aims and stated that no war measure or other activity should be undertaken in the provinces except through the medium of the provincial government. Patel sent a letter instructing the Congress premiers that they should, as early as possible, pass the resolution in their respective legislatures. However Gandhi felt that a resolution of this nature would create an impression that the Congress had given up the intention to co-operate in the war. He therefore intervened and instructions were issued postponing action on the resolution for a week.[32] An analysis of the resolution and the succeeding debate reveals both a sympathy for the professed objective in which Britain was engaged and a suspicion of British intentions towards Indian aspirations. The Congress had posed two basic questions: what were the ideals for which the allies were fighting and would India be given a foretaste of that freedom for which she was being called upon to fight? To many British statesman such a demand was cynically opportunistic, since far-reaching constitutional changes could not practically be attempted in the midst of a war. However it is possible to suggest that, with some imagination, the British could have reversed the gathering development of mistrust which subsequently made agreement impossible. It was clearly foolhardy of Linlithgow and the imperial policy-makers to think that they could have the total support of India without conceding anything. Given the initially sympathetic response of Gandhi and the lack of any conceived future strategy in the Congress, it seems probable that

31 CWMG, vol. lxx, p. 175.
32 CWMG, vol. lxx, pp. 189-90.

the Congress might not have raised too many difficult objections if British policy-makers had addressed themselves in some way to the main Congress demands.[33] The deadlock could have been averted.

The situation was, however, critically complicated by the attitude of the Muslims, the third point of the political triangle and already an increasingly dominant force to be reckoned with. Immediately after the publication of the Congress resolution, the Muslim League passed a resolution expressing satisfaction at the suspension of the federal scheme but urging the British government to review and revise the entire problem of India's future constitution. The resolution declared its opposition to any system of government which would result in majority rule, on the ground that such a constitution was totally unsuited to India. Being composed of various nationalities, it was asserted, India did not constitute a nation state. After cataloguing the Muslim grievances against Congress rule in the province, the resolution came to the war crisis.[34] It condemned Nazi aggression and concluded by saying that if Muslim co-operation were to be obtained the government must secure justice and fair play for Muslims. The triangle of forces that was to endure until partition in 1947 was thus already in evidence. It was at this stage too that British policy-makers reached the tempting realization that by playing one group off against the other they could stave off Indian demands.

There thus developed an increasingly nebulous situation in which none of the contending forces had any pre-determined strategy. Events moved slowly, without any really clear or discernible pattern. Thus on 30 September 1939 Gandhi noted that he had no concrete plans. 'More will be revealed to me', he wrote, 'from day to day, as all my plans always have been.'[35] For his part, Nehru wrote to Agatha Harrison, stating it as his opinion that the British should respond immediately to the working committee's statement, since not to do so would mean that unpredictable events would unfold at headlong speed.[36] On 7 October 1939 Gandhi noted that Zetland was mistaken in complaining that the Congress was wrong at this juncture to ask for

[33] These included (1) clarification of war aims (2) assurance as to India's future (3) immediate steps to ensure Congress participation at the centre.

[34] Resolution of the All India Muslim League, 17-18 September 1939, M. Gwyer and A. Appadorai (eds.), *Speeches and Documents on the Indian Constitution 1921-47*, Oxford, 1957, ii, pp. 488-90.

[35] Gandhi in the *Harijan*, 30-9-1939 in CWMG, vol. lxx, p. 205.

[36] SWJN, vol. 10, p. 143.

a clear declaration of British interests. If the British were fighting for the freedom of all, he said, then their representatives would have to state in the clearest possible terms that the freedom of India was necessarily included in the war aims. But he clearly ruled out civil disobedience; there being 'no question of civil disobedience in the aggressive sense ... launched in 1930 and 1932'.[37]

It was in the light of these opinions that Linlithgow continued to believe that the Congress, despite its initial refusal to co-operate in the war effort save on its own terms, would eventually agree to co-operate, provided it was given an excuse to do so. But he was resolutely opposed to a definition of British war aims. Instead Linlithgow felt that the wise course would be to take refuge in generalities and he proposed that he would commence his declaration with an exposition of the opposing claims of the Congress and the League and emphasize their incompatibility. In fact Linlithgow led Zetland to conclude that the Congress would be satisfied if the 14 September resolution was taken at a good deal less than its face value.[38] He was mistaken, however. There was little doubt that the Congress had no clear plans to enforce its demands, but it also showed few signs of being prepared to compromise them. On 3 October 1939 Nehru and Prasad had an interview with the Viceroy at which they pointed out that unless the declaration 'went the whole hog, it would not meet the situation'.[39] In private Nehru wrote to Krishna Menon that, though the Congress had no desire to precipitate a crisis, the matter had to be cleared up. He pointed out that Gandhi was in complete support of the working committee's statement and was 'with the Congress'. His only disagreement with the Congress was with regard to the militarization of India. In the event of external aggression Gandhi was now insistent that the Congress could only act non-violently.[40]

On 10 October 1939 the AICC met, and endorsed the working committee's statement. However, no final decision was taken or even hinted at. This was intended specifically to give every chance to the Viceroy and the British government to negotiate. The committee's resolution stated that 'India must be declared an independent nation and present application must be given to its status to the largest

[37] CWMG, vol. lxx, p. 246.
[38] Nehru to V.K. Krishna Menon, 6 October 1939, in SWJN, vol. 10, p. 175.
[39] Ibid.
[40] Ibid.

possible extent'.[41] In view of these developments it should be noted that at this time the Congress was apparently not asking for complete independence immediately, but only for as much of it as might be possible. The voting on this principal resolution was 188 to 58 and followed intense debate as the socialists, communists, and the Forward Bloc argued for immediate resistance to the war.[42] Subhash Chandra Bose stated disparagingly that 'this resolution, though finely worded, represents a policy of inaction, which I can neither appreciate nor endorse'.[43] Sardar Sardul Singh Caveesher, vice-president of the Forward Bloc, had stated that it was the right and duty of every Congressman to resist the war and that they intended to do so by all means available to them. It is clear from this that, while the Congress high command was unwilling to compromise unless on some positive action by the British, they advocated a policy of caution. The impetuous left wing, by contrast, ardently advocated direct action and immediate resistance to the war. Nehru noted that the period was one of 'wait and see' since 'we have had our say and now we have to wait for a reply'.

The answer when it came was bitterly disappointing. Having completed his talks with the leaders, Linlithgow published an official statement on 17 October in which the war aims were stated in vague terms.[44] After commenting on the differences of opinion which had emerged in the discussions with various leaders, the statement contained two points. First that, at the end of the war, the British government would be able to enter into consultations with different groups with a view to securing their aid and co-operation in framing such modifications as might be desirable. Second, that a consultative group consisting of selected Indians could be called in to help with the war effort. In the statement the war aims were not clearly explained, while the policy on India's political future remained as undefined as before. The 1935 Act might be revised, but nothing would be done before the end of the war. For any modification of the Act, however, a 'general agreement' amongst the various groups and communities would be necessary. The only concession to the Congress was the agreement to establish a consultative group, and even this was not to be nominated

[41] Congress woking committee resolution, in SWJN, vol. 10, pp. 122-38.

[42] For a debate on the resolutions see Nehru to Krishna Menon, 11 October 1939, SWJN, vol. 10, pp. 182-3.

[43] Ibid.

[44] Linlithgow, *Speeches and Statements 1936-43*, pp. 204-11.

by the political parties. Nehru noted ironically that 'in order to comfort us further we are told that the pledge given in the preamble of the Act of 1919 still holds'.[45] In a joint statement Maulana Azad and Nehru called the Viceroy's declarations a complete repudiation of all that 'India stands for nationally and internationally . . . What our next step should be, it would be premature and improper for us to say at this stage'.[46] Gandhi called the declaration 'profoundly disappointing and remarked that 'the Indian declaration shows clearly that there can be no democracy for India if Britain can prevent it'.[47]

As the Viceroy's reply was felt to be an implicit refusal of Congress demands there was little choice left open for the Congress. It was with this realization that the working committee rejected the viceregal statement on 23 October 1939 and called on the Congress ministries to resign. In a cable to Krishna Menon, Nehru wrote that 'conflict seems to be inevitable . . . no common ground is left after the Viceroy's statement and Zetland's speech'.[48]

In each of the Congress provinces the assembly passed a resolution similar to the 14 September resolution. Thereafter all the Congress-ruled provincial governments resigned one by one. Throughout the ministry period the British policy-makers had hoped that the working of offices would serve to co-opt the provincial politicians and effectively undermine the Congress as a mass nationalist movement. The British had hoped that it would be difficult to persuade the Congress ministries to resign. However, they were proved wrong, since the Congress organization was in no sense a coterie of provincial parties, while the working committee had the last word in major decisions.[49]

It can be argued that the Congress decision to resign from office was unwise and set the organization on a path of confrontation for which it was ill-prepared, and on a course which Gandhi was loathe to take in the circumstances. But he was gradually forced to take it, given the absence of alternatives. The resignations pushed the three

[45] Ibid.

[46] Joint statement issued by Maulana Azad and Jawaharlal Nehru, 17 October 1939, ibid., pp. 193-4.

[47] Ibid., pp. 193-4.

[48] Cable to Krishna Menon, 25 October 1939, SWJN, vol. 10. p. 207.

[49] Soon after the Congress withdrawal from office, the cabinet on 1 June 1940 approved the Viceroy's scheme for a revolutionary movement ordinance in the event of civil disobedience. See J. Barnes and D. Nicholson (eds.), *The Empire at Bay, the Leo Amery Diaries 1929-45*, p. 606.

points of the political triangle further apart, while Congress opposition to the war effort and the Leagues's *de facto* support for it convinced many British officials that more active support should be given to the Muslim League. Thus Jinnah called for all Muslims to observe 22 December as a day of deliverance from *Hindu raj*.[50] As a result the attitudes of the Muslim League and the Congress became even more confrontational and the British did not fail to take advantage of this. As the crisis slowly deepened in 1940, Gandhi was forced to concede the agitational programme while remaining aware that, given the opposition of the Muslim League and the intransigence of the British, the programme would be faced with formidable obstacles. However, by late 1940 he had few alternatives left and his sanctioning of individual satyagraha in October 1940 formed the basis of the approach that was to culminate in the Quit India movement of 1942.

INDIVIDUAL SATYAGRAHA IN BIHAR

The new Congress programme evolved relatively slowly. Gandhi had noted in December 1939 that 'the declaration of British policy about India is a purely moral issue, for freedom-loving India has neither the will nor the capacity to resort to armed revolt'.[51] The six months following the withdrawal from office thus passed slowly as Congress was clearly bereft of any official policy. The provinces continued to be ruled under their governors and it was reported that the expectation in Congress circles of an early return to office gradually died down.[52] In Bihar, Congress leaders began to tour rural areas in a desultory fashion, to campaign for the 'constructive programme'. The lack of any concrete plans and the professed caution of the Congress leaders left the way open for Bose and the socialists to stage their final show of strength. This took place at the Ramgarh Congress in March 1940 at which the leftists under Bose demanded an end to compromise with the British on any issue and urged direct action instead. At this stage Gandhi revealed nothing of his intentions to launch a mass movement in the near future and only criticized the Congress for its lack of discipline.[53] In effect, Gandhi hinted, the Congress organization was

50 *Searchlight*, 20 December and 27 December 1939.

51 CWMG, vol. lxxi, p. 7.

52 Quarterly reports of the Secretary of State, IOR File no. L/P&J/7/1816.

53 He noted, 'There is no discipline in the Congress. There are a number of

not ready for the launching of another civil disobedience campaign. This would have to await its proper time.[54] Gandhi still hoped, it seems, for a decisive response to the Congress from the British. He also genuinely believed in avoiding embarrassing the British in their time of crisis. What the Congress wanted, he thought, could still be achieved without direct confrontation. Gandhi pointed out that 'compromise is part and parcel of my nature. I will go to the Viceroy fifty times, if I feel like it. But my compromise will not be at the cost ... of the country'.[55] He was also clearly aware that the time was not right to launch a mass movement.

At the Congress session in Ramgarh (Bihar) the divergence of views in the Congress over future policy came to a head. The Congress was due to meet at Ramgarh on 9 March and the working committee could not afford to face this meeting without a definite programme. In early March, therefore, the committee met in Patna and clearly established that a civil disobedience campaign should be the objective of the Congress. The resolution was worded cautiously, and stressed that the Congress would unhesitatingly begin a programme of civil disobedience only when the party organization was ready for it, while full powers to begin and control the campaign were given to Gandhi.[56] This resolution was also passed by the Ramgarh Congress and instructions were sent out to the Congress committees to prepare themselves for a conversion to satyagraha committees and war councils.

The left wing in the meantime staged an anti-compromise conference at Ramgarh. To them it still appeared that the Congress was leaning towards a compromise with the British government. Sahajanand Saraswati himself was chairman of the reception committee, and he moved the main resolution affirming 'the inalienable and indefeasible right of the Indian people to complete sovereign independence'.[57] The resolution also laid down that 6 August, the begin-

---

groups and there are quarrels and squabbles. We seem not to believe in non-violence as regards our internal organisation.' CWMG, vol. lxxi, p. 306.

[54] *Indian Annual Register* 1939, vol. ii, p. 237. See also CWMG, vol. lxxi, p. 117.

[55] Mahatma Gandhi's address to the AICC at the Ramgarh Congress, CWMG, vol. lxxi, p. 353.

[56] Quarterly reports, IOR, File no. L/P&J/7/1816(A). See also R. Coupland, *Indian Politics 1936-42*, pp. 247-8.

[57] K.K. Datta, *Freedom Movement in Bihar*, vol. 2, pp. 372-3. See also *Searchlight*, 28 February 1940.

ning of the National Week celebrations, should signal the intensification of local struggles and the commencement of a new struggle on the all-India front. It added that this movement should symbolize the resolve of the Indian people to withdraw from participation in the war and make final efforts for the achievement of India's independence. The resolution also supported the immediate demands of the kisans, the most important of these being a 50 per cent reduction in rent, revenue and canal rates and the abolition of the zamindari system without compensation. However, this leftist initiative was rapidly losing its impetus. As Tomlinson points out, 'the events of March to September 1939 had shattered the unity and limited the appeal of the left-wing organisations'.[58] In particular, the CSP decision to withdraw support from Bose at Tripuri had left it in shambles. The leadership had split and, of the original leadership, only Narendra Dev and Jayaprakash Narayan were left. Furthermore the communists had resigned from it. In Bihar the CSP was more inclined to support the right-wing Congressmen on all issues. As such, then, a viable left alternative receded even further into the background.

The Ramgarh Conference clearly led the Congress nowhere. Civil disobedience was to be the programme of the Congress but the manner in which it was to be carried out had not been laid down. The left wing in the party had also done badly in the Ramgarh elections and felt further alienated. Only 646 delegates out of 3,111 elected to Ramgarh were members of left-wing organizations, while only 99 out of 376 members of AICC elected from the provinces were also members of any left-wing parties. Indeed, in Bihar, 34 out of 41 members were from the right wing. Five were CSP members, one was Forward Bloc and one Kisan Sabha.[59] The Congress and the dominant Gandhian group within it were biding their time. They were also worried about the impact of a proposed movement on the Muslim League. Gandhi, as well as Azad and other Muslim Congress leaders, feared an adverse League response to any civil disobedience movement aimed at forcing the British to concede more influence to the Congress.

However, before the all-India leaders could decide on a definite programme, the situation changed again. The collapse of France caused some nervousness among the monied classes all over India.

[58] B.R. Tomlinson, *The Indian National Congress and the Raj, 1929-1942*, p. 145.

[59] (NAI) Home Political File no. 4/4/40.

Panic withdrawals from savings banks started and a growing demand developed for coins instead of notes.[60] The success of the German advance in Western Europe forced the Congress to define its attitude to the possibility of the imminent fall of Britain to Hitler and even to the invasion of India. Linlithgow noted that the progress of the German advance since 10 May 1940 'has brought home to India the gravity of the war'.[61] Gandhi remained insistent that the Congress would only act non-violently in such a situation but the other working committee members found the policy impractical. Leo Amery, the new Secretary of State for India, was by now working on a plan to make a new offer to the Indian politicians. The war in Europe was going very badly for the British and Linlithgow began to broadcast appeals for unity. At the end of June he had talks with Jinnah and Gandhi on the probable reactions of the Congress to a further statement clarifying British intentions in India and an offer to participate on the Viceroy's executive council. Both Gandhi and Jinnah were wary of the implications of the offer. Jinnah refused to accept any statement that might compromise Pakistan while Gandhi stuck to the Ramgarh Congress resolution for independence. Some members of the Congress working committee were more eager to respond to a British initiative. Early in July Patel and Rajagopalachari steered a resolution through the committee which declared a Congress willingness to help the British 'if a national government' was established at the centre. This, it was asserted, would have the confidence of all the elected elements in the central legislature and would be able to work with the provincial government. Gandhi was opposed to this plan and condemned the resolution as showing that the Congress had lost faith in non-violence and had succumbed to the irresistible temptation that Congressmen could again become Cabinet ministers.[62]

In fact, when the British offer did come it was perceived as totally inadequate. The Viceroy's terms, published on 8 August, included an offer of places on the executive council and on the war advisory committee to representative Indians. It made little attempt to go beyond a consultative group of representative Indians on grounds of a lack of prior agreement between the major communities such as

[60] IOR File no. PY 160/1941.

[61] Ibid.

[62] Quarterly Surveys, 1 February 1939–31 July 1940, IOR L/P&S/13/207, File PY 1190/38.

would suffice to make the scheme for the harmonious working of the centre practicable. It stressed the rights of minorities and assured them that their opinions would be respected in any revisions of the 1935 Act after the war and hinted at the establishment of a Constituent Assembly of all parties of Indian opinion to devise the framework of a new constitution leading to dominion status.[63] The Congress working committee condemned the offer. 'It widens the gulf', observed Mahatma Gandhi, 'between India represented by the Congress and England.' In August 1940, in an article called 'The Parting of Ways', Nehru expressed the Congress point of view, forcefully pointing out what he saw as the intentions of the British government to tighten the hold of imperialism on India. He wrote:

> This is the goal of India, a united, free, democratic country closely associated in a world federation with other free nations. We want independence and not dominion or any other status. Every thinking person knows that the concept of dominion status belongs to the past, it has no future. It cannot survive this war, whatever the results of this war. But whether it survives or not we want none of it.

The Muslim League also rejected the offer, on the grounds that it did not concede Pakistan, and demanded that the League should be taken into full and equal partnership with the British government in running the country. By the end of September it was clear to the imperial policy-makers that both the parties had rejected the offer and Amery reluctantly assented to the Viceroy's view that the offer be put in cold storage. The Secretary of State had lobbied hard in the Cabinet for a sincere effort to secure agreement but had found Churchill and Linlithgow inflexible and unwilling to make any declaration of British intentions.[64]

The rejection of the August declaration produced a crisis in Congress policy. At the AICC session in September in Bombay it was resolved to offer an ultimatum to the Viceroy; either Indians were to be given freedom of speech for the duration of the war or the Congress would begin satyagraha. Linlithgow rejected Gandhi's demand and, on 13 October, the working committee accepted Gandhi's plan of

[63] Amery records how Churchill managed to secure the insertion of various limitations. The Congress told the Viceroy that they would not enter the executive without a prior declaration of India's right to frame her own constitution after the war.

[64] *Leo Amery Diaries*, pp. 608-9.

campaign and he was promised the fullest co-operation.[65] As a moral protest against Britain's policy regarding India and her possible entry into the war, Gandhi now advised individual civil disobedience for chosen individuals. He did not intend the movement at this stage to be a mass-based one and strict instructions were issued to all PCCs to ensure that the instructions of Gandhi and the working committee were followed. Gandhi instructed that:

> It is advisable not to hold public meetings for the purpose in cities. Meetings may be held in villages. The best and easiest way is to repeat the following slogan to passers-by as the resister walks on in a particular direction until he is arrested. My preference is for this last method. It is harmless, economical and effective. No argument is necessary. It rivets attention on the single issue of war. The idea is to prevent the movement from lapsing into mass civil-disobedience. This is the slogan: 'it is wrong to help the British war effort with men or money. The only worthy effort is to resist all war with non-violent resistance.' . . . Civil-disobedience is to be offered singly. It need not be simultaneous. The whole programme should, if possible, be finished in one month.[66]

Gandhi's decision to launch individual civil disobedience was thus based on several considerations. Britain's declining fortunes in the war had made her position in India more unpredictable. There was now a real danger that the war might reach India and Gandhi was beginning to grasp the implications of such a development. The sham of the Viceroy's August offer also made it necessary for the Congress to make some protest against it. However, Gandhi was aware that any protest at this stage should be muted and that he should follow a policy of wait and see. Given the intractable attitude of the Muslim League it seemed unwise to launch another mass movement. The individual civil disobedience campaign was thus designed to keep alive popular interest and to provide a focus for local Congress organizations while keeping the options of the leadership still open.

The satyagraha was started on 17 October 1940 by Vinobha Bhave in Paunaur village at Wardha. Bhave was arrested on 21 October and

[65] Nehru, who wanted to launch an immediate mass movement, was outmanoeuvred within the working committee by Gandhi himself, who feared that a mass movement at this stage would lead to a crisis with the League. Confidential appreciation of the political situation in India, IOR L/P&S/13/195, File PY 160/1941.

[66] CWMG, vol. lxxiii, p. 157.

sentenced to three months imprisonment.[67] In November the satyagraha campaign was extended to members of the working committee, the Congress members of the central and provincial legislatures and the members of the AICC. Nearly 600 Congressmen were arrested at this stage, including Congress president, A.K. Azad, and ex-members of the Congress ministries. The campaign was gradually extended. In January 1941 satyagrahis were accepted from lists drawn up by the PCCs. A 'working faith in non-violence and a living faith in constructive work' were the requirements. Each satyagrahi was to inform the District Magistrate of the time, place and manner of the intended breach of law. Certain discretion was allowed as to the form of civil disobedience but Gandhi recommended that it be 'harmless, economical, and effective'. The method involved the use of a single slogan ostensibly to prevent the movement from lapsing into mass civil-disobedience. The Secretary of State noted that Gandhi would direct the movement as long as he was free. In the event of his arrest Gandhi expected the movement to become 'self-acting if the people have imbibed non-violence'. Gandhi's reference to his arrest was accurately seen as enclosing a veiled threat that disorder would follow were the colonial administrators to attempt such a step.[68] In January 1941, nearly 2,000 arrests were made, but after that the numbers declined. Finally, in April, ordinary 'four-anna' members of the Congress were allowed to offer satyagraha.[69]

In Bihar an individual satyagraha was started by ex-premier Shrikrishna Sinha with an anti-war speech before a large crowd in Patna. When he was arrested and was being taken to the jail, the crowd, disregarding instructions, followed him to the jail gate. The situation then became very tense and the crowd were lathi-charged. Rajendra Prasad was forced to issue a statement condemning the demonstration and laying down instructions for future guidance along more peaceful lines. It was clear that the Congress wanted to avoid public disturbance.[70] By August 1941, 4,500 Congressmen had been convicted. In line with the working committee instructions all the District Congress committees were then converted into satyagraha committees. In some districts the progress of the enrolment of satyagrahis was slow, as in

[67] CWMG, vol. lxxiii, p. 115. See also *Indian Annual Register*, 1940, vol. ii, p. 36.

[68] IOR File no. L/P&J/7/1816(A).

[69] IOR File no. L/P&J/7/1816(A).

[70] BSA File no. 65/1940.

Monghyr, where until May 1940 the total number of satyagrahis came to only 166. In Purnea the number of satyagrahis increased even more slowly. Patna district, on the other hand, was the most active district, and officers there reported that the satyagrahis showed great zeal by starting four satyagraha camps and carrying on the constructive programme of the Congress successfully. The pattern was to repeatedly offer satyagraha. Many women took part in the process. In Saran one woman, Nagina Devi, offered satyagraha at Dighwara again and again. After her second release it was not thought proper to arrest her though she continued to offer satyagraha.[71] Many other women offering satyagraha were simply ignored, as the district administration preferred not to exacerbate the situation. The Chief Secretary, in assessing the situation in Bihar, commented that Gandhi was staging his usual publicity stunt, but that they were at a loss to know how to deal with it:

> By selecting comparatively obscure individuals and villages . . . and by professing to demonstrate his anxiety to embarrass government as little as possible by confining the practice of civil disobedience to single persons, he hopes to place us in an awkward predicament. Either we must hold our hands, in which case he steals ground from us that will be difficult to recover . . . if we are provoked to repressive measures which will excite popular feeling against the government.[72]

It was believed, and rightly, that Gandhi and the Congress high command were merely biding their time and that it would be a great mistake for the British to rely on their professed anxiety not to embarrass the government or to suppose that the clash could be avoided by dealing gently with the initial moves. The tactic the British hoped to follow in order to defuse the situation therefore consisted in a go-slow policy. As the Chief Secretary made clear:

> In taking measures against individual civil-disobedience we do not at present propose to go beyond prosecution of actual offenders in the ordinary course. . . . Gandhi's object of exciting sympathy and working up an atmosphere of martyrdom will not be served by routine prosecutions and light sentences and he was hoping for action by government which would increase excitement; such as dispersal of meetings and arrests of offenders on the spot. In a gradually developing situation such as the present one our object

[71] Bihar Intelligence Files 107-S/1941.
[72] BSA File no. 65/1940.

must be to prevent the spread of anti-war propaganda without making more of it than circumstances warrant. It would be premature at present to take more drastic action against Congress as a whole or its leaders until it has come more into the open. . . [73]

Soon after this a blanket order was issued asking all district officers not to interfere with right-wing Congress meetings or peaceful anti-war propaganda. In the case of left-wing meetings the colonial bureaucracy was merely asked to stay vigilant.

By October 1941, however, the campaign was losing its impetus. The limited satyagraha campaign did not affect the bulk of Congressmen in the provinces and localities. They were still allowed to contest elections to local boards. Gradually complaints arose that even the process of selection of satyagrahis by provincial and district Congress Committees was being manipulated by some executives to serve factional ends. For example, the Bihar Congress executive directed that preference be given to Muslim satyagrahis in line with the Congress Muslim Mass Contact policy. However this directive was not adhered to and few Muslims were encouraged to participate. In August 1941, therefore, to keep the satyagrahis free from the prevailing politics of the local bodies, it was decided that all Congressmen in Bihar should resign from the District Boards and municipalities. As a disciplinary measure the Provincial Congress Committee expelled any recalcitrant members.[74] In October only 5,600 satyagrahis were in jail all over India and the number of arrests were falling. Congress all-India membership figures also declined.[75] At this stage many Congressmen were also beginning to voice demands that the campaign should be called off. By the end of 1941 the imperial authorities were able to congratulate themselves on the fact that, despite Gandhi's campaign, the war effort remained unhindered. It was noted that 'in all the provinces government proceeds without unusual strain or difficulty and throughout India the momentum of the war effort grows massively and naturally and there is no tangible sign of obstruction'.[76] The reasons for this were neatly and perhaps not inaccurately summed up by the Secretary of State who wrote that:

[73] Ibid.

[74] Rajendra Prasad, *Autobiography*, p. 565.

[75] IOR File no. L/P&J/7/1816(A).

[76] Confidential appreciation of the Secretary of State. IOR File no. L/P& J/7/1816(A).

the Congress would claim for Mahatma Gandhi the credit for the peace and order that prevail in India, in that he has deliberately subdued the discontent of the country to the plane of non-violence and has limited revolt to a sort of ceremonial display. This is not the whole story. The mildness of the *satyagraha* is due to the indifferent response from Congressmen and the restraint shown by the Congress leadership itself due to the fear of internal disorder, as much as to the reluctance to embarrass the British government.[77]

In December 1941 individual satyagraha was suspended by the Congress after it had continued for over fourteen months. The war, by then, had reached the borders of India. There had been a succession of Japanese air raids in Assam, Bengal and Vizagapatnam and in the light of this situation Congress policy had to be reformulated.

Despite the ups and downs in the movement the nature and the political implications of the individual satyagraha campaign were important. In the mass mind the anti-imperialist struggle was kept alive. Every arrest of a satyagraha stoked up anger and resentment against an unpopular war. Gandhi was aware of the political implications of the repetition of anti-war slogans and was keen that the freedom of speech and the right to speak out against the war be constantly asserted by the satyagrahis. By selecting 'obscure individuals' from villages to offer satyagraha Gandhi sought to take the movement to the villages and make it a popular one. He was successful in that the war came to be seen as an imperialist war with little in it for India, at a time when the economic pressures of the conflict were also being increasingly felt.[78] Moreover, in many ways the individual satyagraha campaign kept the mass tempo up at a time when the Congress was unwilling to launch a mass anti-imperialist struggle. Gandhi had reasoned that little was to be gained by a mass campaign in 1940 or 1941 and he was worried, too, about the potential effect on the Muslim League. Were such a campaign to get underway in the absence of an agreement with Muslim politicians, he envisaged that the gulf between the League and the Congress would only widen. Until the Cripps mission in February 1942 Gandhi had hoped for a reconciliation between the Congress and the League and believed strongly that the divide between the two parties was actually bridgeable. Gandhi also felt, since the war had not yet reached India's

[77] Ibid.

[78] In December 1940 prices rose suddenly and provincial governments were faced with the intricate problem of price control.

borders, that the time was not right to embarrass Britain by launching a mass campaign. All this was to change dramatically with the Japanese advance into India by 1942. The war situation and changing Congress perceptions then gave rise to a violent mass struggle in August 1942. It is this crisis which is considered in the next chapter.

CHAPTER FOUR

Popular Perceptions and Faces in the Crowd: The Quit India Movement of 1942 in Bihar

Once out of office the flagging Congress organization needed a vital new programme to rejuvenate it. The period in office had revealed serious internal strains as the left wing mounted a steady series of attacks on the orthodox Gandhian group within the party. Having resigned from government, these threats had faded into the background. Despite this, a new programme was urgently required to activate the Congress organization and restructure its links with its mass base. Gandhi was, however, wary in 1940 of rushing headlong into a mass movement. At the local level mass energies that had after the mid 1930s been turned against landlords in the Bihar countryside could not suddenly be channelled against the imperial state. By 1942, the situation had changed yet again. The effects of an imperial war were being felt in India and, as the economic situation worsened, a Japanese invasion seemed imminent. In popular perceptions the weakening colonial state was held responsible for the developing crisis. In turn, such perceptions provided the context for a widespread mass movement against the state.

In August 1942 a mass movement of considerable dimensions broke out all over Bihar. The stage for this was set by the arrest of the chief Congress leaders, as part of an attempt by the colonial state to close in on its adversaries. On 9 August 1942 William Johnston, the police inspector at Patna, sent his superiors a telegraphic message about the arrests which included the information that, 'of the 15 persons on the A list, no. 1 Phulan Prasad Verma and no. 3 Rajendra Prasad have been . . . arrested . . . no. 4 Sarandher Singh is ill and I have been informed by the District Magistrate that his arrest has been postponed'.[1] These general arrests of Congress leaders had occurred several times before in the long national campaign waged by the Indian National Congress. However, this time the mass upsurge of protest that followed the arrests exceeded all expectations. The outburst began at the Patna Secretariat on 11 August 1942. After two days of widespread hartals and processions, the Sub-Divisional Magistrate noted that a large crowd had begun moving towards the Patna Secretariat. Some of them were carrying Congress flags which they made quite clear that they intended to hoist over the Assembly building. Soon the crowd resorted to throwing brickbats. At this point the Sub-Divisional Magistrate decided that firing was unavoidable and he accordingly sent the Sub-Divisional Officer and district magistrate to formally warn the crowd. The warning went unheeded and the crowd still came on, gathering around the closed gates and clambering over them. 'The fanaticism of the mob', it was reported later, 'was exemplified by one individual who approached to within 3-4 yards of the firing squad and bared his breast.'[2] The district magistrate then gave orders to fire into the right flank of the crowd. The final tally of casualties included seven killed and twenty-five injured. As the report on the riot related, in tones of some understatement: 'throughout the afternoon, the crowd continued to infiltrate into the Secretariat compound. The attitude of some of those who claimed to be government employees left much to be desired.'

This rioting in Patna was to provide the spark that ignited the flames of anti-British protest all over Bihar in the following weeks. Already weakened by the conditions of the war the colonial state faced the onslaught of a widespread mass upheaval.[3] The urban rioting which occurred throughout Patna city, and other large cities, from 8-10 August 1942 eventually provoked reverberations far into the interior of Bihar, so that disturbances became common in every district. Indeed, before long the main focus of disturbances shifted to the countryside.[4] Over the next fortnight most villages rejected the authority of the government and coerced village officials, where

[1] Freedom Movement in Bihar Papers, hereafter FMBP, file no. 70/42.

[2] Ibid., Report of the District Magistrate in Patna. In many ways this firing triggered off the movement in several parts in Bihar.

[3] For a recent study on the crisis of the colonial state in 1942, see Indivar Kamtekar, 'The End of the Colonial State 1942-47', (unpublished Ph.D. dissertation, Cambridge, 1988).

[4] District-wise summary of the disturbances in FMBP, file no. 27/42.

necessary, to join the new movement. Large crowds of armed villagers converged on the isolated administrative centres in the localities, and on the police posts and the *cutcherries* at the district and *tehsil* level. Attempts were made to storm these buildings after efforts had been made to make police and officials surrender. Simultaneously there was much looting of shops, godowns and European residences.

The summary of incidents for the week ending 21 August gives an unrelieved picture of widespread attacks in every district in the province.[5] In Hajipur, a crowd estimated to number about 4,000, composed mainly of Goalas, set fire to and looted the Hajipur railway station and goods shed. In Muzzafarpur, at Belsand, the police station staff was ousted by a mob of 5,000. The subsequent report on the incident from the Superintendent of Police concluded that 'the situation throughout the district is grave, roads and bridges all over have been badly damaged. . . .There has not been any place that has not been affected by this deadly unlawful movement . . .attendance of rural police has gone very low due to threats by the villagers . . .'. This story was repeated in Gaya, Saran and Champaran. In Sasaram, reports by the Superintendent of Police concluded that 'the disturbance has now been created by large crowds of grown-up people, instead of students as in the beginning. Reports of violence by mobs who are now usually armed with lathis, spears, swords and daggers are more frequent and loot appears to be the principal motive now'.[6] Until about the end of September the records paint an uninterrupted picture of rioting, loot and arson of the properties of the colonial state, with the police station, the treasury and the railway station (all symbolizing state control) being the main targets of attack.

To the historian, the picture presented is one of fascinating complexity. Most of the primary sources talk of 'the fanaticism of the mobs that roamed the countryside'. Who were these mobs? We have been constrained in recent years by the pioneering studies of George Rude and E.P. Thompson to avoid a loose usage of terms such as 'mob' or 'riot'.[7] With regard to the secondary works on the subject

[5] Ibid.

[6] Ibid., in Purnea it was noted that crowds of villagers surrounded the treasury in the hope that money would be given to them and disbursed freely.

[7] George Rude, *The Crowd in History*, London, 1981; E.P. Thompson, 'Moral Economy of the English Crowd in the Eighteenth Century' in *Past and Present*, no. 50, 1971, p. 79. Rude attempted to identify the crowd that stormed the Bastille during the French Revolution. He notes, 'We should attempt from the

which have attempted to add to our knowledge of such popular action, I shall attempt to limit my discussion here to the observations of Max Harcourt and Stephen Henningham, who have both in some measure attempted to characterize the nature of popular action in Bihar in this period. Unfortunately their analyses suffer from several shortcomings. Max Harcourt has been justly criticized for the limitations of his 'middle peasant thesis' in connection with the Quit India agitation. He attributes the motive force of this agitation to a traditional smallholder peasant sector, squeezed by the high prices of 1942. Such an analysis, however, ignores the real complexity of the struggle and fails to carefully examine the 'faces in the crowd'.[8]

Stephen Henningham, on the other hand, is able to understand the more complex nature of the struggle. Unfortunately his exploration of the nature of popular action leads him to characterize it too neatly as a dual revolt, with the two streams relatively independent of each other, one of rich peasants and small landlords, and the other, a subaltern rebellion of poor low castes. While it is, of course, possible to distinguish distinct levels of participation and motivation, the social breadth of the insurrection, founded both on an ideology of nationalism and on a popular consensus demanding that the colonial state be challenged, makes Henningham's over-simplified conceptualization of two relatively independent streams of agitation somewhat untenable.[9]

---

start to place the event in which the crowd participates in its proper historical context, for without this, how can we hope to get beyond the stereotypes and probe into the crowd's outlook, objects and behaviour? Next, how large was the crowd concerned, how did it act and who if any were its promoters, who composed and who led it? Such questions are important as it will help us to determine not only the general nature of the crowd and its behaviour but also its components . . . by picking out what Asa Briggs has called "the faces in the crowd" '. Rude, p. 10.

[8] Max Harcourt, 'Kisan Populism and Revolution' in D.A. Low (ed.), *The Congress and the Raj*, p. 323. He notes, 'the overall picture, then, is of a kisan revolt led and loosely coordinated by a modernist elite, partly drawn from the village population and partly drawn from urban areas'. He also concludes that the crowd was composed of middle or high castes, a conclusion not borne out by my evidence which demonstrates the extended participation of low and middle castes of Dusadhs, Ahirs/Goalas and Kurmis in the movement in large numbers.

[9] Henningham's notion of 'subaltern autonomy' is a notion shared by many other contributors to the Subaltern project. Rosalind O'Hanlon, in a recent review of the project, has noted that the notion has validity where it examines

Henningham's dual revolt can also be questioned on the grounds that he tends to underestimate the political character of the struggle and the context within which the struggle was waged. He notes that the revolt of the 'subaltern' stream was motivated mainly by the general economic crisis and the need to loot, while the elite movement had a conscious nationalist ideology. Most historians of popular movements (and this holds true of Henningham) have assumed that when rebels talk of the political, they mean the economic and social. It can be argued, however, that both streams within the movement which Henningham attempts to distinguish had definite political motives. Their actions cannot satisfactorily be defined in terms merely of the anger and disgruntlement of disaffected social groups. Instead, there was a shared consensus among different social groups about the challenge that was being made against the colonial state. The Quit India struggle was thus a political movement, with a shared conviction among most participants that articulated both a political solution to distress and a political diagnosis of its causes.[10]

---

the dominance of the elite over the subaltern, and attempts to conceptualize this domination as a relation of power. However, there seems to be a confusion surrounding this dichotomy among some of the contributors, a tendency she describes as one that 'takes the argument of subaltern autonomy quite literally, rather than making a point about power', so that the elite and the subaltern are made to appear as distinct social entities. 'The result is that the argument degenerates into an unhelpful set of assertions to the effect that the subaltern groups generated their own traditions and pursued their political projects quite independently of anyone else, and especially of the Indian National Congress.' See Rosalind O'Hanlon, 'Recovering the Subject, Subaltern Studies and the Histories of Resistance in Colonial South Asia', *Modern Asian Studies*, 22, 1, (1988) pp. 189-224.

[10] Henningham, 'The Quit India Movement in Bihar and the Eastern U.P., The Dual Revolt' in R. Guha (ed.), *Subaltern Studies*, vol. 2. He feels that the duality of the Quit India movement is epitomized by the report of a sub-divisional officer in Madhubani who noted that most of the Hindus who took part in the movement in the beginning were from the upper castes and the lower castes joined in later when they wanted to loot. It is quite clear that such a conclusion is not tenable even from the evidence Henningham himself discloses in the next paragraph, quoting from the sub-divisional officer in Begusrai, that the looting did not remain confined to any particular section and anybody who had the opportunity utilized it. 'In most villages Bhumihars and poor labourers took a prominent part in the looting.' Loot was an attractive proposition but, by ascribing the lower-caste peasant actions to loot alone, he ignores the nature of the struggle. There were many other peasant actions involving both upper and lower caste peasants in this period not involving loot, which also tend to conflict

In this context the ideological vocabulary of the elite groups in the movement should not be seen as distanced from the ideas and visions of the common people. The objectives of the movement were interpreted in more than one way by different social groups, while its parameters were defined by certain common premises, such as the fact that the colonial state was conducting an illegitimate war, that it was responsible for the severe economic and social crisis in the country and that its time had run out. As Gareth Stedman Jones has noted in his reappraisal of the Chartist movement, 'to be successful, that is to embed itself in the assumptions of the masses of people, a particular vocabulary must convey a practical hope of a general alternative and a believable means of realising it, such that potential recruits can think within its terms'.[11] Only this type of notion can help explain why the anger and frustrations of these disaffected social groups should have taken the form they did, or why the political vocabulary of 1942 should have continued to express their changing fears and aspirations.[12] What I intend to examine in probing the popular actions of that period in Bihar is the nature of the legitimizing notion that informed the crowds at the time. Essentially, the men and women in the crowds were informed by the belief that the war was illegitimate and did not serve Indian interests and that in conducting such an illegitimate war the colonial state had reached a crisis and had lost control of the political and economic situation. In a phase of rising prices and war requisitioning, the mass of the people rose to challenge

with his conclusion. In most cases the peasants seem to have concentrated on attacking the symbols of colonial rule, the police thanas, the court. In Pasraha and Fatwah, when they killed the RAF pilots, they were attacking the physical presence of colonial rule. Henningham, however, does not even record these important cases. Towards the end of his paper he is prepared to concede the notion of a shared ideology. 'The element binding the two insurgencies together was the ideology of nationalism' but he does not probe this important aspect further for the sake of preserving his argument of two separate and individual streams in the movement.

[11] Gareth Stedman Jones, 'Rethinking Chartism' in *Languages of Class*, Cambridge, 1983, p. 96.

[12] It was not simply the experience but rather a particular linguistic ordering of experience which would lead the masses to believe that their exclusion from political power was the cause of their social problems. As Gareth Stedman Jones notes, 'Consciousness cannot be related to experience except through the interposition of a particular language which organizes the understanding of experience and it is important to realise that more than one language is capable of articulating the same set of experiences', ibid.

state authority in the belief that they were supported by the wider consensus of the community. This consensus was so strong that it overcame engrained motives of fear and deference to colonial authority. It thus presented an outright challenge to the state.

In examining the events of 1942 an attempt is made at the outset to place them in their proper historical context, that is in the political and economic situation of the 1940s.

THE WAR SITUATION AND CONGRESS POLICY

Already, by the beginning of the year 1942, the war situation had taken an ominous turn, threatening India at her very doors as well as vitally affecting British interests. The Japanese were advancing on the Indian Ocean at considerable speed. As General Slim noted in his memoirs, '1942 had been a series of Allied disasters on land, sea and in the air'.[13] Hong Kong surrendered as early as 25 December 1941, Borneo on the 31st and Manila on 3 January 1942. The heaviest blow to Western prestige came with the fall of Singapore on 15 February, which shocked even Hitler.[14] Singapore, the most powerful British bulwark in the Far East, surrendered after only a brief resistance to Japanese troops who were inferior in number, but superior in tactics and morale. The morale of the British imperial forces suffered a great deal. Significantly, the psychological impact was probably greatest on the Indian troops who formed more than half of the units captured in Malaya. Over 60,000 Indian troops, the majority of the British forces, had been captured. G.W. Seabridge, a journalist of the *Straits Times*, sent out the first reports on the disaster. He noted that many Indian troops had fought magnificently although some were disappointing and others gave themselves to the enemy readily. It was here that the Japanese propagandists were at their most proficient. By means of radio and pamphlets dropped from aircraft, they rammed home the point that the British were putting Asiatic troops in the front line 'as cannon fodder while the white soldiers skulked in the background'. Moreover, they promised that any Asiatic soldier who gave himself up would go unharmed. Captured Indian soldiers were deprived of their arms and uniforms, given a singlet and shorts, a handful of rice and allowed to go free 'unless you want to come with us and

[13] The Slim Papers, file 2/4. The Slim Collection on the war in the Eastern Front is voluminous and interesting.

[14] Milan Hauner, *India in Axis Strategy*, Stuttgart, 1987, p. 394.

fight the white man'. Singapore fell so rapidly because 'the fight for it was less than half hearted . . . from the moment the garrison of at least 62,000 failed to repel an invading force of 5,000 the battle of Singapore was lost'.[15] Indian officers claimed that they were victims of a discrimination which in turn affected their morale. As one Indian Lieutenant-Colonel wrote, the 'biased treatment of Indians by the British', and 'the general discontent of Indian troops of all ranks was universal in Malaya'.[16]

The Indian defence chiefs had underestimated the dangers of an invasion by the Japanese. Following the collapse of Singapore, Java was overrun on 5 March. Two weeks later Sumatra fell, while in the Bay of Bengal the Andaman and Nicobar islands were occupied. The Japanese were also advancing through Burma, the capital, Rangoon, falling on 7 March. This was the biggest blow to British hopes and was to have far-reaching effects on Indian public opinion.[17] After the onslaught on Singapore in February 1942 the Indian command realized that, strategically, the most vulnerable points lay in eastern India where the most important industries were also situated. India's most important munitions works, located in Calcutta, would be liable to long-range bombing attack were Burma to be occupied, and would have no fighter aircraft, anti-aircraft gun defence or other air cover.[18]

[15] Notes prepared by G.W. Seabridge on 28-2-1942 and forwarded to the Dept of Naval Intelligence; Milan Hauner, *India in Axis Strategy*, p. 424.

[16] M.K. Durrani, *The Sixth Column*, London, 1955, pp. 2-3.

[17] British defence measures to forestall the Japanese advance were marred not so much by the lack of troops as by basic shortages of aircraft and ships. On the eve of the Japanese attack on India the best-trained and equipped Indian troops were in the Middle East, while only one British and six Indian divisions were available for the defence of the whole of India and Ceylon. In a confidential file to P.G. Grigg, former Finance Member in the Viceroy's Council, on India's military problems written as early as October 1941 it was noted that:

> While India is thus making this very large contribution to the imperial war effort both in men and ammunitions, she is receiving a very small supply of the essential equipment for her own defence. There is not a single aircraft in India capable of taking the air against modern German or Japanese machines; India has not at present a single modern tank or armoured car; has only eighteen light and twenty heavy anti-tank guns; and only twenty 2-pdr. anti-tank guns. The fourth and fifth divisions, which have now been in the field for two years and one year respectively, have not yet been equipped with anti-tank regiments.

See P.G. Grigg Papers, file no. 8/3, (Churchill archives, Cambridge).

[18] Ibid. The absolute minimum of anti-aircraft guns required for even a moderate defence of India was noted to be 300 heavy and 200 light anti-aircraft

The Indian battalions were also poorly equipped and insufficiently trained.[19] Despite this knowledge the High Command could not find the resources to prevent the Japanese overrunning Burma. To defend Burma, a land the size of Germany, the British had two divisions, so that even when reinforcements arrived they were no match for the Japanese.[20] It was against this background that the Japanese began their first attacks on Burma in December 1941. The attack intensified in March-April 1942, so that by 20 May 1942 the whole of Burma had passed to the Japanese. Significantly, up to 400,000 Indian refugees had fled before the advancing enemy.[21]

The defeat of the British in Burma finally brought the war to India, creating widespread fears that the British would buckle against the advancing Japanese in India as they had done elsewhere. Public morale had already suffered badly as a result of the withdrawal from Burma and the influx into India of thousands of refugees. The threat of air raids was also a great cause for alarm. As previously noted, the Indian command considered some areas especially vulnerable, in particular the coastline from Chittagong to Calcutta and the fleet bases at Trincomalee and Colombo. At dawn on 5 April, all fears of a Japanese attack came true when Colombo harbour and its installations were struck by a large force of Japanese carrier-borne planes. Off Ceylon, two heavy cruisers, the Dorsetshire and the Cornwall, and the light aircraft carrier Hermes were located by the Japanese and sunk. In addition, some two dozen merchant ships operating in the Bay of Bengal were sunk. Apart from Trincomalee, two Indian coastal towns, Vizagapatanam and Coconada, were bombed. In retrospect Wavell was to say that this was India's most dangerous hour: 'Our

---

guns. See also Churchill's correspondence with Roosevelt, *Churchill iv*, pp. 159-63.

[19] P.G. Grigg Papers, note on India's military problems, file no. 8/3.

[20] The reinforcements numbered 50,000 troops, made up of 7,000 British, 20,000 Indian and the rest raised in Burma among the hill tribes. The seventeenth Indian division, which had met the first full force of the Japanese invasion, had been trained for open warfare in the African desert; such transport as the division possessed was wheeled and unsuited to jungle tracks. The fighter aircraft available in India amounted to eight antiquated Mohawks stationed in Calcutta. Scarcely a single bomber squadron was available for operations, while no heavy bases existed for them along the India-Burma road.

[21] William Slim Papers, file 2/4; The Burma Campaign, file 2/4. The figures of the numbers are subject to controversy. Slim gives the figure as 400,000, while Gandhi refers to 800,000 refugees.

eastern fleet was powerless to protect Ceylon or Eastern India, our air strength was negligible.'[22]

With Japan's relentless drive westward the political pressure on Britain to make a resolute offer of concessions to India built up steadily. Under joint Chinese and American pressure the War Cabinet in India felt compelled to act and proceeded to work on a draft declaration. As Leo Amery, the Secretary of State, was to admit a few weeks later, 'The pressure outside upon Winston from Roosevelt, and upon Attlee and Co. from their own party, plus the admission of Cripps to the War Cabinet suddenly opened the sluice gates and things moved in a rush.'[23] Churchill and Linlithgow, however, continued to have misgivings about making any concessions. Linlithgow throughout the period had played a cat-and-mouse game with the Congress and deeply mistrusted their motives. He had no reason to change his views now, and strongly opposed any conciliatory proposals. In a letter to Churchill he wrote in February 1942: 'However we must regard the morale and fighting value of the army as more important than U.S. opinion or that of any minorities in the U.K. . . . I am quite clear that an important element in your scheme would damage gravely our power of resistance to Japanese invasion. That element is your proposal to give the Defence council the function of constitution-making as well as the duty of advising about the war.'[24] Churchill himself had serious misgivings about the likely effects that the proposed declaration would have on the conduct of the war in the Far East, in terms of its effects on the morale of the army. His fears were heightened by military memoranda showing the likely effects on the army of concessions to the Congress.[25] Lockhart, too, Military Secre-

22 Wavell, Operations, paragraphs 5-6, in Milan Hauner, *India in Axis Strategy*, p. 449.

23 Amery to Linlithgow, 10-3-1942, TOP, vol. i, no. 304.

24 TOP, vol. i, no. 124, Linlithgow to Churchill, 14-2-1942.

25 The memoranda referred to the class composition of the Indian army. According to the revised list, the total numbers of combatants, as at 1 January 1942, had reached 690,000 excluding the British element, of which 450,000 were in India and 210,000 were serving overseas. Twelve months earlier the total strength of Indian combatants had been under 420,000. The Punjab contributed 48 per cent of the total, of which 96,000 were Muslims and 51,000 were Sikhs. Nepal contributed 11 per cent. Of the total of 690,000 almost 35 per cent were Muslims, 41 per cent were Hindus, 10 per cent Sikh, 8 per cent were Gurkhas and the rest belonged to other groups. See Lockhart, Military Secretary of the India Office, memo on the class composition of the Indian army, 25-2-1942, TOP, 1, no. 180.

tary at the India Office, had expressed his doubts on the likely effects of a Congress mass movement, if it developed, on the morale of the army. Since the present expansion was still drawn from the pre-war martial classes, he believed that there was no reason to suppose that their feelings were any different from that of the old army and that there was no evidence to show that concessions to Congress would have any beneficial effect on recruiting or on the fighting spirit of the army. On the other hand he warned that 'any concessions to Congress would produce a violent reaction from the Muslim League . . . and the effect upon the army might be disastrous'.[26]

At this stage, however, it was felt that a declaration with concessions to the Congress could not be postponed indefinitely as fears of internal unrest were gaining ground in several quarters. On 15 February, the Director of Intelligence (DOI), India Command, cabled his assessment of the internal situation to the war office, in view of the pending Japanese invasion. General fear was widespread throughout India, he observed, but open expression of sympathy with Japan was confined to Bose's Forward Bloc elements. In Bengal left-wing speakers had professed their readiness to support Japanese invaders and the DoI warned that active fifth-columnists were undoubtedly in existence. A heavy exodus from Indian industrial areas might lead to rioting, and communal and tribal disturbances. This would increase the burden on field troops, diverting them to internal security duties and thereby further weakening India's defences against Japanese attack. It was also realized that communications could be endangered by sabotage. The DOI concluded that recruitment might be adversely affected by anti-war defeatist rumours fostered by enemy agents or at least by anti-British elements.[27] In April 1942, Thomas Stewart, the Governor of Bihar, wrote to Linlithgow warning that 'the possibility

[26] Ibid. He further warned that outside British India the Nepal government might require the British government to give an assurance that the Gurkha troops would not serve under an unpredictable government of independent India. Afghanistan also caused considerable apprehension since at this time she might be expected to require direct access to the sea at the expanse of British India. An even greater risk could develop in the North West Frontier Province where, despite its pro-Congress majority, the War Cabinet decision could be interpreted as having prompted the disintegration of the Raj. See also Linlithgow to Amery for similar views, 3-3-1942, TOP, vol. i, no. 224.

[27] Lieutenant Colonel Cawthorn to Major Mackenzie, 15-2-1942. He noted that 'some effect (of this) is becoming noticeable in the recruiting of technical personnel but funk may be a factor in this'. TOP, vol. i, no. 128.

of internal disorder in the event of invasion or an air attack looms large in the public eye and a great interest is being shown by all political groups in the forming of village defence parties'.[28] At first, the Viceroy himself was at pains to mitigate the effects of what he regarded as 'the darker side of the picture'. He noted that antipathy in India towards the Japanese had strengthened after the visit of Chiang-Kai-Shek and that, apart from certain elements linked to the Forward Bloc, there was no clear indication of an obvious fifth column. However, Stewart was forced to admit that recruitment had in fact slowed down and that the fall of Singapore had contributed to the widespread apprehension that the Japanese attack on India from air, sea and land could not be long delayed. 'The fear of internal disturbances was gaining hold on the public mind.'[29]

These growing official fears of a general mass upheaval contributed to the external pressure on the British to make concessions, already being exercised by China and the USA. Both countries believed that concessions would be necessary in order to maximize support for the war effort among the civilian population. At long last the War Cabinet decided to announce that Sir Stafford Cripps would visit India, bringing specific proposals with him. With this announcement even Linlithgow was forced to concede the need to make some new moves. On 10 March, on the eve of the official announcement that Cripps would be visiting India with the War Cabinet proposals, the Viceroy decided to launch an appeal to all Indians, regardless of their political or religious convictions, inviting them to join the national war-front and to close ranks against the aggressor.[30]

The details of the draft declaration on India had been hammered

[28] Stewart to Linlithgow, FR (1), April 1942.

[29] 'In several parts of the country fear of internal disturbances, possibly communal, is gaining a hold on the public mind. The press has bitterly criticised neglect of the defences of the eastern outposts of the empire. Contributory causes of defeatism are axis broadcasts, alarmist rumours, and the irresponsible utterances of political speakers.' Linlithgow to Amery, 26-2-1942, TOP, vol. i, no. 186.

[30] Linlithgow's message to India on 10 March 1942, see TOP, vol. i, no. 310 note 1. Even then he was doubtful about the merits of the scheme. 'I have considerable doubts about the merits of the scheme, which has been reinforced by Governors especially Hallet and Stewart. I confess that I am even now not confident that it will have any very considerable success in the country, but at least it can do no harm and it may well serve to rally and hearten the many people of goodwill, who at this time are looking for some such move.'

out in London when, on 26 February 1942, the War Cabinet Committee on India was specially constituted. Two months later the new committee produced a draft declaration.[31] This formed the basis of the proposals taken by Cripps to India on 23 March and made public at a press conference in Delhi on 29 March. On close scrutiny these proposals did not go substantially beyond the proposal of August 1940. Both plans had promised the Indians the right to frame their own constitution after the war in return for their immediate co-operation in the war effort. The Cripps offer also went further by stipulating that, after achieving full dominion status immediately after the cessation of hostilities, the new Indian union would have the right to separate itself from the British Commonwealth if it so decided. A further very important provision recognized 'the right of any province of British India that is not prepared to accept the new constitution to retain for the time being its present constitutional position, provision being made for subsequent accession. With such non-acceding provinces, should they so desire, His Majesty's Government will be prepared to agree upon a new constitution following the lines laid down above'.[32] On the other hand no alteration to India's constitutional position was foreseen during the war except for the extension of the Viceroy's executive council, then under expansion, which was to comprise an Indian majority, while preserving the key posts of finance and defence in European hands. Further it was declared that 'during the critical period which now faces India, and until the new constitution can be framed, H.M.G. must inevitably bear the full responsibility for Indian defence'.

The attitude of the Congress to this proposal proved politically decisive. The motives for a rejection of the proposals were made apparent in Gandhi's statement in the *Harijan* on 19 April 1942.

> It is a thousand pities that the British government should have sent a proposal for dissolving the political deadlock which, on the face of it, was too ridiculous to find acceptance anywhere . . . . But he (Cripps) should have known that at least the Congress would not look at Dominion Status even though it carried the right of secession the very moment it was taken. He knew also that the proposal contemplated the splitting up of India into three

[31] The meeting was chaired by Attlee, and comprised Amery, Cripps and Anderson from the War Cabinet, Simon, the Secretary of State for War, and Grigg, former Finance member of the Viceroy's council.

[32] Cripps offer in TOP, vol. i, no. 456.

parts, each having different ideas of governance. It contemplated Pakistan and yet not the Pakistan of the Muslim League's conception. And last of all it gave no real control over defence to responsible ministers.[33]

The first phase of the negotiations ended on 4 April with the Congress rejecting the Cripps proposals. On the following day Ceylon was attacked by the Japanese. At this critical juncture the Americans made a determined effort to mediate between the two sides through Roosevelt's special envoy, Colonel Louis Johnson.[34] All attempts to bring about a settlement failed when the Congress finally rejected the Cripps proposals, and demanded an immediate creation of an All-India government with full executive powers to act as a cabinet. Except for the influential but weak Liberal party and, with some reservations, the princes, all main political groupings followed the lead of the Congress in rejecting the proposals. The Muslim League, the Hindu Mahasabha, and the Sikhs all demanded sufficient safeguards for their particular interests. Cripps refused to make yet another attempt to reach a settlement since he insisted that the draft proposals were a package-deal and, as such, must be accepted or rejected as a whole. He finally left Delhi on 12 April.[35]

Much of the blame for the failure of negotiations has been laid on Linlithgow. He was convinced that his views were correct and often chose to ignore contradictory opinion. The day after Singapore fell the Viceroy wrote to Amery, 'I am carrying on here almost single-handedly, an immense responsibility . . . I do not think that I exaggerate too often that the key to success in this war is now very largely in my hands.'[36] To my mind it is quite clear that the fear of Japanese attack entirely overshadowed the Cripps negotiations, hardening attitudes on both sides. The Ceylon raid, in particular, took place at the time of the Cripps mission to India and seems to have significantly affected the mood and the final outcome of the mission. While the Congress hoped to ward off a Japanese attack by demanding full national government, the British were firm that nothing could be conceded. Wavell, for instance, even thought that the Japanese raid was made with the specific object of securing India's rejection of the settlement. Furthermore, he categorically refused to give in to Nehru's

[33] Gandhi's statement, 'That ill-fated Proposal' in CWMG, vol. lxxvi, p. 28.
[34] TOP, vol. i, no. 533.
[35] Note by G. Laithwaite, TOP, vol. i, no. 607.
[36] Linlithgow to Amery, 16-2-1942, TOP, vol. i, no. 13.

demand for an Indian Defence Member of the Council in place of the Commander-in-Chief.[37]

The Congress reaction to the breakdown of the negotiations was swift, as was only to be expected in the circumstances of despair that followed. Gandhi was upset, especially by the likely communal consequences of the Cripps offer. He realized that any constitutional offer by the British would inevitably recognize the right of the minorities to secede. He had hitherto postponed taking direct action, fearing that the Muslims might be alienated. Now, in the light of the proposals, he felt that the situation had irrevocably changed.[38] More important, however, in determining strategy was the growing fear in Congress circles of a Japanese invasion. How best could the Congress ward off an imminent attack? In tactical terms Gandhi felt that an India under British dominion invited Japanese invasion. He took the view that India could be saved were the Congress to succeed in forcing the British to devolve some measure of self-government at this critical time. It is interesting to note the change in his view on direct action which had developed since October 1941. In answering criticisms about the declining enthusiasm for individual civil disobedience and his failure to launch mass civil disobedience he had commented then that:

> the call will not come before the close of the war. There is neither warrant nor atmosphere for mass action. That would be naked embarrassment and a betrayal of non violence. What is more it can never lead to independence. Mass action at this stage without communal unity is an invitation to civil war.[39]

This attitude was to change and later, in June 1942, Gandhi was to say:

> I always thought that I would have to wait until the country was ready for a nonviolent struggle. But my attitude has undergone a change. I feel that if I continue to wait I might have to wait till doomsday. For the preparation that I have prayed and worked for may never come and in the meantime I may be enveloped and overwhelmed by the flames of violence that are spreading around.

[37] Memorandum on the Indian political situation, by the Secretary of State on 28-1-1942, W.P. (42) in TOP, vol. i, no. 43.

[38] CWMG, vol. lxxvi, p. 120.

[39] CWMG, vol. lxxv, p. 56.

His earlier insistence on communal unity he modified by saying, 'I am firmly of the opinion that there is no unity whilst the third party is there to prevent it. It created the artificial division and it keeps it up.'[40] 'Withdrawal', he said, 'is a necessary preliminary to unity.'[41] 'With the aggressor at the gates, more than ever unity is vital in India. I desire above all things to see a joint struggle against him and to achieve independence. In the very process of doing this, it is highly likely that we shall have forgotten our quarrels in the same common goal.'[42]

The failure of the Cripps Mission widened the breach and precluded further dialogue between the Congress and the British government, while events were clearly moving in the direction of a new mass movement. The first ultimatum to the British came in April. 'My firm opinion', Gandhi stated, 'is that the British should leave India now in an orderly manner.' Such a withdrawal he regarded as a military necessity for the safety of both India and Britain, saying, 'Britain cannot defend India, much less herself, on Indian soil with any strength, the best thing she can do is to leave India to her fate.' Gandhi believed that voluntary British withdrawal would turn hatred into goodwill and enable Indian leaders to arouse the people and organize resistance against the Japanese, help China and Russia more effectively and thus promote world peace. Alternatively, there was the possibility that if the British withdrew the Japanese would not attack and India would be spared the rigours of war.[43]

The bombings on the borders of India, in Vizagapatnam, and Coconada, and then in Chittagong in April, intensified the war crisis, and made Japanese landings on the Eastern coast seem imminent. Nehru and Azad advocated armed resistance and guerilla struggle, but Gandhi dissented.[44] He declined to attend the working committee meeting and sent, through Mirabehn, a draft resolution for the AICC meeting in April demanding British withdrawal from India and complete non-violent non-cooperation with the Japanese forces, should they attack India. He disapproved of the scorched-earth policy, and the introduction of foreign soldiers into India, and called upon the

[40] See CWMG, vols. lxxvi, p. 159 and lxxv, p. 120.
[41] CWMG, vol. lxxvi, p. 228.
[42] CWMG, vol. lxxvi, p. 25.
[43] CWMG, vol. lxxvi, p. vi.
[44] CWMG, vol. lxxvi, p. 52.

people to whole-heartedly take up the constructive programme, with slogans such as 'banish communal strife' and 'exorcise the demon of untouchability'.[45] This resolution was eventually superseded in favour of a more practical draft prepared by Nehru, Rajagopalachari and others. Nehru took the realistic view that if Gandhi's draft were accepted the Congress would be implicated in a passive partnership with the Axis powers. In early July a compromise version of the resolution, proposed by Nehru, was accepted by the Congress High Command at Wardha. It called for the stationing of Allied troops in India including British, Commonwealth and American armies (with the symbolically important addition of Chinese units), to ward off the Japanese. The stated essential precondition was that the British must become the ally of a free India on Congress terms. Alternatively, if the appeal for withdrawal were not met, the Congress threatened to launch a mass civil disobedience campaign.[46] The British authorities were thus, effectively, given three weeks to design counter-measures to disobedience before 7 August, the date when the Congress working committee resolution was to be submitted for ratification at the full session of the AICC in Bombay. The Congress decision was made in the context of a serious war crisis. The significance of the fact that it was taken at a time when the Japanese seemed about to overrun India should not be underestimated.

Ultimately Gandhi had decided to accept the reconstituted resolution in the face of the working committee decision, as his own attitude had undergone a change in the direction of positive action. Though he continued to assert his view that conditions in the country were far from propitious for the successful launching of a mass non-violent movement, he believed that it was precisely to overcome these conditions that a movement had become necessary. There was an ominous logic in the forecast that 'from the internecine warfare and unrestrained dacoities, a true India will rise in place of the false one we see'.[47] He was also prepared to take the risk of the struggle turning violent, stating:

I cannot remain inactive, I will certainly launch a non-violent movement.

[45] Draft resolution of the AICC in CWMG, vol. lxxvi, pp. 63-5.

[46] In the second week of July, in a draft resolution for the working committee, Gandhi clearly gave notice of his resolve to start a mass movement should the British not heed the appeal to withdraw. See CWMG, vol. lxxvi, p. 284.

[47] CWMG, vol. lxxvi, p. 105.

> But if the people do not understand it and there is violence, how can I stop it? I will prefer anarchy to the present system of administration, because this ordered anarchy is worse than real anarchy. I am sure that the anarchy created by our efforts to mitigate this dangerous anarchy will be less dangerous . . . Crores of people in this country have no weapons. Even if they indulge in violence among themselves how long can they do it? Ultimately they will have to listen to me even if some of them die in mutual violence. We have to take the risk of anarchy if God wills it.[48]

After Gandhi's 'do or die' speech the AICC on 8 August passed the Quit India resolution by an overwhelming majority, calling for the immediate transfer of power to Indian hands under threat of mobilizing a mass civil disobedience movement. It was intended that Gandhi should forward the Congress resolution to the Viceroy and ascertain his attitude before the threat of mass civil disobedience was put into operation. Linlithgow, however, was in no mood to wait. The rush of events in the previous weeks had revealed the determination of the Congress to launch a struggle and he moved quickly to nip their plans in the bud. This involved the launch of a secret plan codenamed 'Adolf' (sic) with which the state planned to strike swiftly and firmly at the Congress leadership before it could complete preparations for a mass campaign of non-cooperation.[49] The war Cabinet had approved of these decisions of 6 August, even suggesting that Gandhi and a dozen Congress working committee men should be deported from India.[50] The arrests took place on 9 August. Ironically, the arrests, meant to pre-empt a struggle, triggered off a mass movement of such dimensions that, for a few weeks, it completely destroyed the authority of the colonial state in large parts of Bihar.[51]

[48] Ibid., p. 160.

[49] See Chief Secretary, Bihar, G. Godbole to all District officers, 7 August 1942, FMBP, file no. 33/42.

[50] TOP, vol. ii, nos. 435 & 457.

[51] What is remarkable is that until the last moment Linlithgow and Stewart failed to forecast the scale of the struggle. On 22 August, after the movement had broken over their heads Stewart wrote to Linlithgow: 'The earliest actions took the course we expected, processions and picketing of courts without an indication of the violence to follow.' He continued somewhat abashed: 'having only a short time ago attempted an appreciation of the situation which was entirely wrong, I am loathe to embark on another. Nevertheless it is strongly suggested that the revolt and the form it took was carefully planned and our intelligence was at fault.'

THE ECONOMIC SITUATION AND POPULAR PERCEPTIONS IN BIHAR

The preceding discussion of the war situation and its connections with Congress policy only tells a part of the story. While attempting to set the event in its historical context it now becomes necessary to outline the economic situation of the 1940s. There is, of course, a major objection to economistic determinism in that it 'may conclude the investigation at the exact point at which it becomes of serious sociological or cultural interest'.[52] I shall attempt here to go beyond the self-evident truth that people protest when they are hungry by examining the developing pattern of protest and the way in which it was manifested. Economic statistics alone do not suffice to explain patterns of behaviour at a particular historical moment. For example, it is worth pointing out that, while it is true that prices were high in 1942, it is also true that prices rose much more between May and October 1943, without giving rise to definite political effects.[53] Instead, one needs to examine the more diverse conjuncture of forces that existed in August 1942. The resulting popular response, with its complex culturally-mediated character, cannot be accounted for by political or economic stimuli in any simplistic way.

The declaration of war had caused prices to rise and, at a stroke, ended the endemic price depression of the 1930s. By March 1941 the average price of coarse rice in Bihar had increased by 13.2 per cent over the average price in 1940 and by 32.5 per cent over the average price in 1939. Primary produce and other prices rose steadily higher throughout the war period, whilst food and other basic commodities became scarce. The prices of all major commodities including rice, wheat and mustard showed a tendency to rise, and there was also a marked increase in the case of white sugar, gur and sugar-cane.[54] The scarcity was exacerbated by the partial failure of crops owing to the lack of rainfall in many areas.[55] In Patna and Gaya, agriculture suffered as a result of droughts in 1941 and 1942 and, as the price increase continued, wild speculation resulted. Such speculation en-

52 E.P. Thompson, 'The Moral Economy of the English Crowd', p. 77.

53 See price statistics for that period.

54 *Season and Crop Reports*, see Table 8, showing the harvest prices of certain important crops.

55 *Land Revenue Administration Report 1941, 1942* and *The Annual Report of the Agricultural Department, 1941-42.*

TABLE 8 (A)

STATEMENT SHOWING THE HARVEST PRICES OF CERTAIN IMPORTANT CROPS PER MAUND OF 40 STANDARD SEERS, IN THE YEAR 1941-2

District	*Winter rice (cleaned)*						*Autumn rice (cleaned)*						*Wheat*					
	*1932-7**		*1940-1*		*1941-2*		*1932-7**		*1940-1*		*1941-2*		*1932-7**		*1940-1*		*1941-2*	
1	*2*		*3*		*4*		*5*		*6*		*7*		*8*		*9*		*10*	
	Rs.	a.	Rs.	a.	Rs.	a.	Rs.	a.	Rs.	a.	Rs.	a.	Rs.	a.	Rs.	a.	Rs.	a.
Patna	3	8	4	12	5	6	3	4	4	10	5	0	3	8	4	2	5	8
Gaya	3	7	5	0	5	10	2	5	6	12	10	0	3	5	4	3	6	0
Shahabad	3	2	4	6	4	7	2	12	4	0	5	6	3	6	4	0	5	5
Saran	3	7	4	12	5	11	3	4	5	0	5	7	3	9	4	11	5	0
Champaran	3	3	5	0	5	4	3	4	4	7	5	0	3	9	4	0	5	12
Muzaffarpur	3	8	4	13	5	3	3	4	4	7	5	6	3	13	4	6	6	2
Darbhanga	3	6	5	0	5	7	3	2	5	0	5	7	3	14	5	0	4	13
Monghyr	3	10	5	0	5	8	2	8	3	8	4	0	3	7	4	8	5	3
Bhagalpur	3	7	4	12	6	5	3	9	4	0	6	2	3	8	4	4	5	4
Purnea	2	13	4	12	6	6	2	9	4	2	5	11	2	12	4	0	6	2
Santal Parganas	3	1	4	8	5	0	2	12	4	2	4	8	3	11	4	8	5	12
Hazaribagh	3	5	5	1	5	9	2	15	4	15	5	4	3	11	4	11	6	12
Ranchi	3	2	4	0	5	4	3	1	3	13	4	7	4	7	5	0	6	6
			to 4	3					to 4	0	to 5	4						
Palamau	2	14	4	0	5	0	3	7	6	10	6	6	3	6	4	0	7	4
Manbhum	3	4	4	14	4	14	3	8	4	10	4	10	3	13	4	9	6	9
Singhbhum	2	14	4	5	4	7	3	1	4	5	4	10	3	10	5	4	5	11

TABLE 8 (B)

STATEMENT SHOWING THE HARVEST PRICES OF CERTAIN IMPORTANT CROPS PER MAUND OF 40 STANDARD SEERS, IN THE YEAR 1941-2

District	Barley						Indian corn						Gram						Sugar (raw)					
	*1932-7**		*1940-1*		*1941-3*		*1932-7**		*1940-1*		*1941-3*		*1932-7**		*1940-1*		*1941-3*		*1932-7**		*1940-1*		*1941-3*	
	11		*12*		*13*		*14*		*15*		*16*		*17*		*18*		*19*		*20*		*21*		*22*	
	Rs.	a.	Rs.	a.	Rs.	a.	Rs.	a.	Rs.	a.	Rs.	a.	Rs.	a.	Rs.	a.	Rs.	a.	Rs.	a.	Rs.	a.	Rs.	a.
Patna	2	2	3	2	3	10	2	4	3	1	3	3	2	8	2	15	3	5	3	2	1	15	3	15
Gaya	2	9	3	0	3	2	2	4	3	2	3	0	2	6	3	2	3	8	2	9	2	4	12	12
Shahabad	2	6	3	5	3	12	2	10	3	0	3	6	2	7	2	15	3	9	3	0	3	4	8	0
Saran	2	5	3	1	3	0	2	2	2	11	3	1	2	13	2	11	4	2	3	2	2	8	4	0
Champaran	1	14	2	11	3	0	2	3	3	0	3	6	2	11	2	13	3	12	2	15	1	8	4	0
																					to 2	8		
Muzaffarpur	2	8	3	2	3	6	2	4	3	4	3	10	2	15	3	1	4	0	3	9	1	6	4	0
																					to 2	0	to 5	8
Darbhanga	2	5	3	8	3	4	2	5	3	0	–		2	15	4	0	4	0	2	12	3	0	6	0
																							to 7	8
Monghyr	2	8	2	8	3	8	2	2	3	0	3	8	2	12	3	8	4	0	3	14	3	0	4	8
Bhagalpur	1	14	3	4	3	3	2	2	3	2	4	7	2	11	3	4	5	6	2	14	2	8	5	8
Purnea	–		–		–		1	14	2	9	2	11	2	14	4	0	4	7	2	15	4	9	6	4
Santal Parganas	2	6	3	8	4	0	1	15	2	12	3	0	2	12	3	8	4	8	2	15	4	4	5	8
Hazaribagh	2	3	4	10	4	10	2	3	2	13	4	1	2	15	3	12	4	10	2	15	4	10	5	15
Ranchi	–		–		–		–		–		–		3	1	3	14	5	0	–		–		–	
Palamau	2	4	2	13	3	10	2	1	2	8	4	0	2	13	–		4	11	3	1	2	0	4	0
Manbhum	–		4	4	4	4	2	6	3	10	3	12	3	5	3	14	4	11	3	14	3	14	5	0
Singhbhum	3	1	–		–		2	6	3	0	3	5	3	2	4	4	5	11	4	0	4	8	6	8

TABLE 8 (C)

STATEMENT SHOWING THE HARVEST PRICES OF CERTAIN IMPORTANT CROPS PER MAUND OF 40 STANDARD SEERS, IN THE YEAR 1941-2

District	*Cotton (cleaned)*			*Linseed*			*Rape and Mustard*		
	*1932-7**	*1940-1*	*1941-2*	*1932-7**	*1940-1*	*1941-2*	*1932-7**	*1940-1*	*1941-2*
	23	*24*	*25*	*26*	*27*	*28*	*29*	*30*	*31*
	Rs. a.	Rs. a.	Rs. a.	Rs. a.	Rs. a.	Rs. a.	Rs. a.	Rs. a.	Rs. a.
Patna	16 2	19 8	19 0	4 3	5 2	4 7	4 9	5 12	5 2
Gaya	17 8	17 8	17 8	4 2	4 4	5 8	4 12	5 8	6 0
Shahabad	–	–	–	3 11	4 13	4 15	3 15	4 6	5 6
Saran	13 9	13 5	14 10	3 3	3 5	4 8	3 3	5 4	4 6
Champaran	17 14	17 12	16 0	3 8	3 12	4 8	3 14	3 12	3 12
Muzaffarpur	20 2	16 0	18 0	3 10	5 9	5 4	4 4	4 8	4 8
Darbhanga	16 14	21 0	16 0	3 9	5 8	5 2	4 7	6 0	5 4
Monghyr	18 0	18 0	18 0	5 0	5 0	5 0	4 6	6 0	6 0
Bhagalpur	18 6	20 0	20 0	4 11	5 0	6 0	4 12	4 8	4 4
Purnau	–	–	–	3 1	5 0	5 8	3 14	3 6	3 12
Santal Parganas	18 11	22 8	30 0	4 3	5 8	5 8	4 12	6 4	6 8
Hazaribagh	18 14	18 14	18 14	3 15	4 0	5 0	5 1	7 0	6 6
Ranchi	–	–	–	–	–	–	–	–	–
Palamau	18 2	20 0	20 0	3 5	4 8	5 2	4 5	6 10	5 0
Manbhum	21 0	20 0	23 0	5 0	8 4	10 0	–	7 0	3 4
Singhbhum	20 3	20 0	20 0	4 9	5 0	4 8	5 2	5 8	5 4

TABLE 8 (D)

STATEMENT SHOWING THE HARVEST PRICES OF CERTAIN IMPORTANT CROPS PER MAUND OF 40 STANDARD SEERS, IN THE YEAR 1941-2

District	*Jute*			*Tobacco Leaf*			*Salt*		
	*1932-7**	*1940-1*	*1941-2*	*1932-7**	*1940-1*	*1941-2*	*1932-7**	*1940-1*	*1941-2*
	32	*33*	*34*	*35*	*36*	*37*	*38*	*39*	*40*
	Rs. a.	Rs. a.	Rs. a.	Rs. a.	Rs. a.	Rs. a.	Rs. a.	Rs. a.	Rs. a.
Patna	–	–	–	11 0	12 8	14 8	3 2	3 5	3 13
Gaya	–	–	–	–	–	16 0	3 2	3 12	4 3
Shahabad	–	–	–	13 2	14 0	18 0	3 5	4 4	3 9
Saran	–	–	–	10 12	11 17	–	3 0	3 10	4 0
Champaran	3 7	2 8	5 0	15 10	–	–	3 2	3 2	3 14
Muzaffarpur	4 11	4 8	5 0	14 9	10 0	14 0	3 4	3 5	3 15
Darbhanga	3 12	5 0	4 8	9 14	12 0	10 0	3 1	3 6	4 8
Monghyr	–	–	–	9 0	8 0	10 0	3 10	3 12	5 4
Bhagalpur	4 15	8 0	8 0	6 2	5 8	6 0	2 12	4 0	4 11
Purnau	4 7	3 8	3 0	4 2	3 0	2 0	3 4	3 8	4 6
		to 5 8	to 11 0		to 12 0	to 11 0			
Santal Parganas	3 5	12 0	15 0	8 4	12 0	15 0	2 12	3 2	3 12
								to 3 4	to 3 15
Hazaribagh	–	–	–	13 14	17 6	17 6	3 0	5 7	4 3
Ranchi	–	–	–	–	–	–	3 0	3 4	3 7
								to 3 12	to 4 4
Palamau	–	–	–	–	–	–	2 11	3 5	3 12
								to 4 0	to 4 0
Manbhum	–	–	–	12 10	19 0	19 0	2 11	3 2	3 2
Singhbhum	–	–	–	–	15 0	20 0	2 11	3 4	3 9

* Average of five years 1932-7.

SOURCE : Season and Crop Report for 1941-2

couraged individuals to stockpile food and other goods on account of the war and with the prospect of a rise in military needs. Table 8 details the harvest prices of certain important crops in 1942. As can be seen from the table, in particular districts like Gaya, rice prices registered a sharp increase in relation to other crops. In some districts the price rise was higher than in others due to the local marketing conditions.

To diffuse unrest the state attempted to control price levels and ensure an adequate supply of essential commodities. In June 1942 the government fixed the maximum wholesale price for rice at Rs 5-2-0 for standard coarse and Rs 6-4-0 for standard medium.[56] Until then, the central government had introduced no controls over foodgrain prices, (except for wheat), in view of the differing opinions of various provincial governments on such policies.

The high prices and the concomitant hardship were not borne equally by the population. On the contrary, for the small landlord and rich peasant who comprised the elite in the villages, high prices and scarcity provided many opportunities to profit through grain-dealing and blackmarketing. At the other end of the scale scarcity and high prices were clearly onerous for poor peasants and agricultural labourers. As prices rose the real value of cash wages for agricultural workers diminished whilst scarcity meant that those who received wages in kind found themselves under pressure to accept less. As for the other sections of the peasantry the benefits of the rise in prices were also generally more illusory. Many of them were cut off from any real benefit from the rise in prices by the activities of middlemen and their agents who bought up stocks as cheaply as they could. These were the beparis, the paikars and the gomastahs, whose activities seemed to converge in a tangle of predictable and opportunistic dealings.[57] The Royal Commission on Agriculture had noted in 1928 that these middlemen activities were so well organized that they

[56] B.N. Sarkar, *Some Landmarks in War Time Food Control in Eastern India*, Patna, 1944; and speech by the food member of the Governor General's council in the British Parliamentary Papers 1942-1943, no. Cmd. 6479, ix, 777. Neither in Bihar nor in Bengal did officials have any real knowledge of how much rice there was available in the province at any one time. See H.B. Martin Papers, CSAS, Cambridge.

[57] For details of how marketing was organized in and around Patna, see Kum Kum Banerjee, 'Grain Traders and the East India Company', in *Indian Economic and Social History Review*, 23.4.1986. For a later period, see *The Report of the Royal Commission on Agriculture*, vol. 14, p. 106.

successfully insulated a mass of unorganized producers from the market and then swallowed up the profits.[58]

An economic situation thus subsisted where scarcity and rise in prices benefited mainly rich peasants and trading communities. In the early years of the war these groups were able to considerably increase their profits through speculation and the panic induced by war. One revealing piece of evidence relates to a rumour floated by a mill owner of Saharsa, Rai Saheb Manoharlal, whose agents spread a story that the government would confiscate all paddy crops for the war effort. This caused widespread panic among the cultivators and allowed the miller's agents to seize the paddy at cheap prices.[59]

However, even these opportunist groups all too soon had reasons to be less happy as, in an attempt to control inflation and check speculation, the state began to intervene in the market. In 1942 the government attempted to control price levels and to ensure an adequate supply of essential commodities. It is important to note that these measures were introduced only as late as 1942. All through 1941 requests made by the Bihar government for centralized measures to check speculation were ignored by New Delhi so that it was only in 1942 that the provincial government succeeded in fixing the maximum wholesale price which was to come into operation on 15 July. Not surprisingly there was a rapid reaction.[60] The merchants were furious. Those involved in the grain trade had kept a vigilant eye on every turn of the season because periods of scarcity were what they thrived upon. At the slightest indication of an approaching bad season, grain which during normal seasons was disposed of at regular intervals would be hoarded, with a view to obtaining an exorbitant profit on it. It should be said that, for these reasons, earlier state interventions in the grain market had usually ended disastrously in the face of merchant hostility.[61] Price controls in 1942 resulted in precisely the same situation. Either grain stocks disappeared and hoarding intensified, or the grain stocks surfaced elsewhere where prices were

[58] *Report of the Royal Commission on Agriculture*, vol. 14, p. 106.

[59] *Searchlight*, 5 April 1942. See also 11 April 1942 which reports, 'Though prices of wheat and other food articles abnormally rose up the growers did not get any benefit from the rise in prices for they sold their crops at the time of harvest when prices were low.'

[60] B.N. Sarkar, *Landmarks in War Time Food Control.*

[61] See Kum Kum Banerjee, 'Grain Traders and the East India Company', p. 425.

higher. When district magistrates attempted to impose controls, they did so in a haphazard fashion so that very often prices in adjacent districts were different. A Congress Grain Enquiry Committee was set up in Bihar in 1942 to examine the grain situation. The Committee noted that immense distress and confusion had been caused by the grain scarcity and the vagaries of prices as they were altered by the grain-control officers. Transport of grain had also become markedly inefficient.[62] As the Congress Enquiry Committee discovered, merchants frequently complained of price control as a factor that drove goods underground. In the case of salt the Committee noted that, while the price remained fixed on paper, salt still sold at high prices in many areas in the *muffussil*. It is relevant to note here that even basic commodities like salt were hit hard by scarcity and inflation. What was feared far more than high prices was the more or less total disappearance of rice from the market. The effects of such shortages and price rises on the poorer classes was very severe at a politically critical juncture. The grain merchants of Deoghar told the Congress that large quantities of rice and dal were exported from the district and that this was directly responsible for the shortage of these commodities.[63] By April 1942 the scarcities were exacerbated by the fall of Burma. This stopped Burmese rice imports into India and gave the Japanese access to seven million tons of rice, the whole paddy harvest of the Irrawaddy valley.[64]

In Shahabad it was observed that a large group of businessmen were greatly perturbed by the rules on price control. Many of them believed that it would be difficult for Britain to defeat Hitler, and had strong Congress sympathies. They were thus reluctant to give any credit or do any moneylending business on a large scale, particularly on a medium-term or long-term basis. A collector's report noted that merchants were hoarding gold and particularly silver coins as a consequence of the increasingly prohibitive price of precious metals, which were no longer available in any significant quantity.[65]

62 Report of the Congress Grain Enquiry Committee, *Searchlight*, 15 May 1942. See also Gandhi's worries about an impending salt famine in CWMG, vol. lxxvi, pp. 134 & 242.

63 Local shortages during the war were caused by various factors. In Bengal it was recorded that all surplus stocks of rice were removed from coastal areas to prevent food from falling into enemy hands.

64 Slim Papers, file no. 2/4.

65 Collectors reports on Shahabad, FMBP, file no. 64/1942.

Coin hoarding was not a problem confined only to the merchant classes. In urban and rural areas, other sections of the middle class also revealed a growing panic in their tendency to hoard. In rural areas, in particular, the wealthier peasants were adding to their hoards, causing serious anxiety about the decline in specie. The government wished at the time to economize on silver, since its stocks had been depleted by the war effort. It also wished to avoid embarrassingly large returns on hoarded coin after the war, should the price of silver slump.[66] The treasury therefore started to withdraw from circulation standard rupee and half-rupee coins which had a high silver content. These were replaced by coins with a lower silver content and by a paper currency. Victorian rupees ceased to be legal tender on 31 March 1941. Edward VII rupees followed on 31 March 1942.[67] Through sustained propaganda the authorities then attempted to flush coinage out of the countryside by asking people to surrender currency that was not legal tender. Brokers had a field day when they bought rupee coins at less than their face value, even disseminating rumours that all silver coins were being withdrawn. The psychological effect of the demonetisation was tremendous. It served to strengthen the growing idea that the colonial state was in a real crisis.[68]

Among the urban middle classes, and among the small investors, the loss of faith in the ability of the colonial state to survive was reflected in the increased withdrawals from savings banks as well as in coin hoarding. Withdrawals from savings banks during the first three weeks of February 1942 were nearly double those in the corresponding period in the preceding year. If one examines the All-India figures, it becomes clear that the period from March 1941 to April 1942 saw the largest number of withdrawals, as 2,193,000 accounts were closed. In Bihar alone 134,000 accounts were closed in the same period. The panic among small investors had in fact been building up since the fall of France. The Viceroy noted this when he commented that the panic in 1942 was not unusual, but was rather the product of the uncertain political circumstances that had pertained since 1940. However the fears and the insecurities of the small investor had grown considerably since then.[69] Even the Viceroy was prepared to concede

[66] G.H. Baxter to G.O.I., Finance Dept, 16.2.1943, IOL, file L/F/7/568.

[67] Finance Dept minute, IOL, file L/F/7/568.

[68] Ibid.

[69] TOP, vol. i, no. 186. In February 1942, Linlithgow noted that, with respect to the figures for all the savings banks, including those for Post Office accounts

that the level of insecurity and public fear was more pronounced by April 1942 than in the preceding years, as indicated by the steady increase in the price of gold. Withdrawals from savings bank accounts and gold hoarding were not the only responses made by the frightened middle class. The Fortnightly Reports noted an increasing level of purchases of land and buildings, made as a precaution against the devaluation of currency.

The reaction of the colonial state to this growing panic among investors was to use force to secure money for war funds. In his personal memoirs Stanton Ife notes that, as a district magistrate in Patna, he withheld licences and used other pressure tactics to force the Banias and merchants to invest in war certificates.[70] This caused further resentment among these groups whose antagonism and mistrust of the colonial state was already well developed.

The preceding account has attempted to survey the fears and resentments of the wealthier urban and rural middle classes to the war. It now remains to consider the resentments and anger of the majority of the poorer participants in the Quit India struggle. The war crisis bore particularly heavily on this section of the population. As prices rose the real value of cash wages diminished, while scarcity meant that those who received wages in kind found themselves under severe pressure to accept less. The increased availability of employment due to war recruitment constituted a mere drop in the ocean. Leo Amery noted that the pre-war annual recruitment of 30,000 had increased to only 90,882 nationally; while in Bihar the increase was even less than the national average.[71] Increased war-related production and the institution of new army camps did provide some new employment, but it made no real difference to the overall employment figures or to

---

and the discharge of cash certificates, the withdrawals from these accounts were less than those in June and July 1940 after the fall of France. From 7 December 1941 to 14 February 1942 (before the fall of Singapore) the total withdrawals were 3.84 crores of savings banks accounts and Rs 1.99 crore withdrawals of cash certificates. In comparison, between 1 June and 31 July 1940, 11.3 crore savings accounts and 4.49 crore cash certificates had been withdrawn. In effect, he was making the point that many panicky account holders had already withdrawn their savings soon after the fall of France. But he agreed that the insecurity and panic had increased as was manifested in the sharp increase in the price of gold by 1942.

[70] The Stanton Ife Papers, CSAS, Cambridge and a personal interview with Stanton Ife, at Cambridge, June 1986.

[71] *Searchlight*, 8.2.1942 and 10.3.1942.

general economic deprivation. By mid-1942, in fact, the deprivation had increased. With the fall of Burma and the amassing of troops on India's borders, food requisitioning intensified and created greater scarcity. Burmese rice imports had also been disrupted, adding to the shortage. Scarcity and shortage in the mass mind were attributed to an unpopular war.

The pressures of deprivation found characteristic expression in a dramatic increase in the crime rate. Dacoities all through 1941 and 1942 were on the increase. By July 1942 dacoities and petty thefts in Bihar had reached a record figure. It was noted by one official that many prisoners were serving only short sentences for petty thefts committed for want of food.[72] In addition, gang robbery had also increased. In Bihar in April 1942, 92 dacoities were registered, recording an increase of 100 per cent on the monthly figure for the previous three years. A spate of market lootings was also reported. In January 1942 several markets were looted in Purnea district.[73] The following June there occurred what the district official noted as a 'serious epidemic' of hat looting in Champaran.[74] The first case took place in Raxaul on 17 June when three cloth shops in the bazaar were looted by a mob of 400-500 people. This was followed between 18-25 June by looting in six other hats situated in various quarters of the district. It is interesting to record the local district officer's view of the situation. He noted that while high prices and scarcity were the ostensible reasons, incitement by the local kisan socialist party was also an important factor. He further observed that at some of the hats that had been looted, rice was actually being sold at a rate lower than the controlled price. The district officer felt that the underlying causes of the reduction were political rather than economic.[75] The observation that the price-setting actions of the crowd were essentially political was perspicacious and important. In 1943 hat lootings were to increase, with women and children actively participating in carrying away the loot from bania shops.[76] In such crowd actions against

[72] FR (1), January 1942.
[73] FRs (1) and (2), July 1942.
[74] The term *hat* denotes a market, ibid.
[75] Ibid.
[76] Once again the participation of women has to be highlighted for in most historical writing it has been ignored. When food and water resources grow scarce, women are usually the worst hit. They have thus traditionally played an important role in food riots.

Banias one can perceive the emergence of responses to the distortion of a moral economy of just prices and the power of an assumption that the looters had the moral right to oblige vendors and Banias to give more value for money.[77]

What could also be seen here was a more generalized and growing resentment among the urban and rural poor against high prices and scarcity. This was to develop quite independently of any anti-imperialist sentiment. Building on this, the programmes of the nationalist groups were able to draw on a general sense of indignation that an externally imposed war, with its grain requisitioning and its need for war supplies, was the cause of the crisis. On another level, the war itself was questioned as being illegitimate and inimical to the interests of India. India and her people, it was said, had no interest in the conduct of the war, yet were being made to pay a price for it. Thus while at home the British state was able to marshal enough resources and support for a patriotic war against fascism and for liberty, in the Indian colony, the war was seen as an imperialist war, and the sarkar as ultimately responsible for the consequent economic deprivation. Only this can explain the fact that at the popular level, when vendors and, later, Banias were looted in 1942-3, and when hats were looted, much of the crowd's ire and moral indignation focused as much on government property as on private property.

In addition to the pressing economic situation the war resulted in growing tensions and feelings of insecurity about the life and property of each individual person. In Bihar and Eastern UP in particular the proximity of the war zone added to this tension. As the war threatened India's borders in 1942 attempts were made to strengthen the existing defence measures and land was requisitioned for aerodromes. Land was also requisitioned by the army for defence installations of all kinds. In Bengal this had amounted to 175,000 acres and involved 2,700 villages. In Bihar the annexations amounted to about 25,000 acres, dislocated many hundreds of villagers and caused immense distress.[78] In Jamshedpur, the concentration of military strength also led to the requisitioning of much property, houses, school buildings

[77] FR (1), June 1943. In a village close to Patna, a party of women was reported to have looted some pack bullocks which were carrying rice. In many of the dacoities in 1943 women and children often accompanied the dacoits carrying away food.

[78] Ibid.

and land.[79] Rajendra Prasad advocated passive resistance among villagers whose lands were thus confiscated. In Bengal, the government requisitioned all country river craft to prevent them being used by the Japanese in their inward march into India.[80] These boats formed the linchpin of the inland distribution system in many areas and their requisition and destruction contributed critically to the economic deprivation in these areas, culminating in the Bengal famine in 1943.

Japanese bombings were another real threat and, with the fall of Singapore, the widespread fear that Calcutta and Jamshedpur would be bombed increased. When Rangoon fell panic quickly developed and merchants began withdrawing their relations from Calcutta. In Jamshedpur, by February 1942, workers had become increasingly apprehensive of air attacks.[81] It is recorded that 40,000 people, mainly women and children, had already been evacuated by that period. Even animals were evacuated. The arrival of troops had added to the exodus of families in areas where the army set up new camps.[82] The Tata works complained of panic in their labour force and the Fortnightly Reports noted that the management would prefer to close works to escape destruction in the event of hostilities.[83] One incident in Buxar indicates the extent of the fear that had developed among ordinary people. Here, considerable confusion and disorder prevailed at Brahmpur *mela* on the 14 May after a passing aeroplane dropped a bundle of propaganda leaflets near the thana grounds. Taking it to be some kind of bomb, people began to flee in all directions, creating a general confusion that took some time to abate.[84]

The most serious panic, however, set in with the Japanese attack on Burma. The Japanese penetrated southern Burma on 23 December 1941 and sent 80 aircraft to bomb Rangoon. The bombs killed 2,000 while another 400,000 individuals fled, leaving the docks and railways deserted. Burmese guerrillas further hampered the retreat of the British. Finally, the Burma road was cut and in the last battles some British battalions were reduced to 100 men. The retreat of the British was an ignominious one and, by 20 May 1942, all of Burma had passed

[79] See Hugh Kemp Papers, District officers collection, CSAS, Cambridge.

[80] Paul Greenough, *Prosperity and Misery in Modern Bengal: The Famine of 1943-1944*, Oxford, 1982, p. 89.

[81] FRs (1), January 1942 and (1), February 1942.

[82] Ibid., molestation of women was a widespread and general fear.

[83] FMBP, file, 29/42.

[84] *Searchlight*, April 1942.

into Japanese hands. Slim notes that 400,000 Indian refugees fled before the enemy while 10,000 died on the road to India.[85] Nationalist sources, for their part, put it at double that number.[86] The first 'unfavourable impression' of the British defeat was created by the passage of trains through north Bihar, containing sick and wounded from the Burma front. This was compounded by the unexpected return of labour from the front which had only recently been recruited from Assam to build roads. These workers returned both with sores on their feet and many lurid stories to tell of the British retreat.[87]

Further panic was spread by the return of the Indian refugees from Burma. Many of them returned to their native villages in Bihar and UP with stories of British mismanagement. Contact with the evacuees was also provided by the passage of refugee trains through Patna and through the refugee camps set up in large parts of Bihar. Many of the refugees had reached Calcutta only a few days before, after covering 300 miles of an arduous route over thick jungle-clad hills.[88] One-fifth of those who took the land route had died on the way. Amongst them were those belonging not only to Bihar and UP but to such distant places as Peshawar and Rawalpindi. A complete story of their return has never been told. In his memoirs, Slim records the horror of the civilian retreat along the Prome Road in January 1942, remembering that 'refugees were streaming along the roads, barefoot men clasping a few possessions to their heads. One little boy had nothing but his terror'.[89]

What Slim does not note is that the British had treated the Indians badly during the evacuation and that they had closed one of the easier land routes, the Tammi Parallel route, to Indians. Gandhi noted this in anger in April 1942. 'Hundreds, if not thousands, perished without food and drink, and the wretched discrimination stared even these miserable people in the face. One route for whites, and the other for blacks! Provision of food and shelter for the whites, none for the blacks! . . .'[90] The Congress took up the case of the evacuees energetically. Syed Mahmud, visiting the evacuees in one of the trains that stopped at Patna, noted that their condition was terrible. The oc-

[85] Slim Papers, file 2/4.
[86] CWMG, vol. lxxvi, p. 55.
[87] FR (1), February 1942.
[88] *Searchlight*, 18.4.1942 and 19.4.1942.
[89] Slim Papers, the Burma campaign, file 2/4.
[90] CWMG, vol. lxxvi, pp. 195-6.

cupants were packed like sardines into compartments with two or three dying every day.[91]

The fact that large numbers of returning refugees brought back stories of the disastrous British retreat added to the stock of floating rumours and spread the panic and resentment further. The rumours gave the impression that the Japanese would be in India in a month's time, that the British had treated Indians badly in the evacuation, and that, in following a scorched-earth policy, Rangoon had been damaged more by the retreating British army than by the invading Japanese. This created fears that Britain would follow the same policy on Indian territory.[92] In Jamshedpur it was reported that there were rumours that factories and other installations would be blown up to prevent the Japanese from taking advantage of them.[93] These fears were not baseless for, in the lower Irrawaddy, the retreating British army had destroyed the Yenanguyuang oil fields, a precious asset, to prevent the Japanese acquiring a strategic advantage. This policy was strongly condemned by the Congress as a threat to India's assets, and a panic was generated about the security of Indian industries, were the Japanese to invade.[94]

By March 1942 there were one million men under arms in India, but this was subject to fluctuations as rumours frequently caused the pace of recruitment to slow down. One particular rumour, that Hindu recruits in the army would be forced to eat beef, caused a minor crisis.[95] Other common rumours related to the collapse of the colonial state and the mutiny of Indian soldiers and seemed to indicate, as one district officer put it, an obsession on the part of the public with the idea of internal disorder. In Bhagalpur it was rumoured that two troop trains had had to be cancelled because the sepoys refused to go to war.[96] Other rumours related to the shooting of European officers by Indian troops and the mutiny of Indian soldiers in Malaya. In Bhagalpur, as a result, motor-drivers expressed a reluctance to join the military service and by March 1942 the authorities were realizing that

[91] *Searchlight*, 19.4.1942.
[92] FR (1), April 1942.
[93] FR (2), February 1942.
[94] See Gandhi's critique of the scorched-earth policy, in CWMG, vol. lxxvi, p. 113.
[95] FR (1), May 1942.
[96] FR (2), March 1942.

recruitment of combatants had generally fallen and that a radical change in the system was needed.[97]

Enemy broadcasts and pamphlets dropped from the air were another major source of worry, adding to the existing level of rumour and encouraging panic. Japanese radio informed the listening public that Tanahashi, the victor of the Arakan, would be in India in one week, while Japanese pamphlets dropped from the air in border areas encouraged Indian soldiers to desert.[98] The effectiveness of the radio broadcasts can be questioned in terms of the number of people actually affected by them. Since 1939 Hindustani broadcasts had been used by the Axis powers for propaganda purposes. By 1941 broadcasts were being transmitted from Berlin, Rome and Tokyo. Apart from the official German foreign-language broadcasts on the short-wave bands, which included Hindustani and were mainly directed to North Africa to influence Indian troops, there was also a plethora of so-called secret stations, along with the Azad Hind radio transmitting from secret hideouts near the Indian frontier.[99] How many receivers there were tuned in day or night to these Axis broadcasts will never be established, although some authors claim that it was about 200,000 during the disturbances. These, as well as propaganda pamphlets dropped from the air, provided the main sources of Axis propaganda.

One should not overestimate the influence of these sources of Axis propaganda. The main means of popular communication in local society was through the medium of rumour. With illiteracy wide-spread and regular channels of communication closed for most people, rumours formed the fundamental mechanism for speaking out and expressing opinions. Moreover, the evidence of rumour can give the historian a deeper understanding of popular perceptions.[100] The very fact that the rumours analysed here became memorable indicates that they were utterances that people responded to because they

[97] FR (1), January 1942. The deterioration in recruitment and army desertions were fears uppermost in British minds. See also FMBP, file 27/1942.

[98] Slim Papers, file 2/4 and FR (1), June 1942. Evidence on pamphlets dropped from the air in Burma to create army desertions is based on personal interview with Martin Farrow in Cambridge, March 1988, ex-soldier of the 14th Route Army who recorded his impressions on the war propaganda of the Japanese.

[99] Milan Hauner, *India in Axis Strategy*, pp. 546-8.

[100] For an analysis of rumours in nineteenth-century Bihar, see A. Yang, 'A Conversation of Rumours, the Language of Popular Mentalities in Late Nineteenth Century Colonial India', *Journal of Social History*, Spring, 1987.

mirrored collective experience and attitudes. It was becoming eminently clear that, among large numbers of people, the Japanese advance into India was taken as inevitable, while the ability of the colonial state to survive the crisis was widely questioned. A perceptive Fortnightly Report noted that in these rumours there lay an obsession with the idea of internal disorder and the collapse of the state.[101] The rumours also addressed and questioned the legitimacy of the war, a feeling reflected in the rumours about forcible government requisitioning and consequent pauperisation. It was said, for instance, that the Maharaj Adhiraj of Darbhanga had been detained in Delhi because he refused to make over all his gold to the government.[102] In Sasaram, a panic selling of grain took place, originating in a rumour that the government might requisition it for the war. In Jamshedpur, fears of a scorched-earth policy were enhanced by the announcement of a scheme for free insurance. This was interpreted by the workers to imply that there was a real and imminent risk of death and mutilation. This further fomented existing anxieties.

The rumours described above fall into two categories. The first category relates to the collapse of the state and the mutiny of Indian soldiers while the second reveals a fear of forcible government requisitioning. The collective experience and attitudes mirrored here indicate a widespread distrust in the ability of the colonial state to survive, and a fear that, with the advance of the Japanese, the country might descend into chaos. Other rumours reflected a resentment and a fear that the people might have to surrender their gold and their crops for the cause of the war. The colonial state was clearly concerned by the spread of these ideas, for they reflected a growing public consensus about the conduct of the war and a maturing understanding of its implications for the life and property of common people. The rumours were also associated with the psychology of developing mass protest and the insecurity which, ironically, grew out of the crisis of colonial legitimacy, which was also, of course, a basic crisis of the state. All through early 1942 contemporary administrators were quick to read manifestations of protest into rumours, and efforts were made through propaganda and meetings to suppress them. Some of them were dismissed as originating from mischievous persons. However, the

[101] FR (1) and (2), February 1942.
[102] FR (2), March 1942.

state could do little to check their spread, or quell the growing panic among the people.[103]

ORGANIZATION AND LEADERSHIP

The preceding account of the events of early 1942 may serve to contextualize the nature of the growing economic and political crisis which affected all sections of the people, and which was exacerbated by the war. The situation was fast approaching a climax. As described above, the issue on which the storm finally broke was the arrest of the national and provincial level leadership on 9 August 1942. What, then, was the nature of the organizational build-up in Bihar up to August 1942? Many historians and contemporary nationalist political observers have been concerned with the question of spontaneity when discussing the Quit India struggle.[104] On the other hand, the British devoted immense administrative energy to proving Congress responsibility for the disturbances.[105] Though clearly there is no validity in the claim by Francis Hutchins that the events of 1942 occurred completely spontaneously, without any lead being given by elitist urban and intellectual elements, the contribution of these sections should not be overestimated. The Quit India movement was actually given its initial spark by the activities of students and socialist groups and it is to this element that we shall now turn.

An important political juncture had come about in April 1942 with the failure of the Cripps talks, followed by the possibility of the imminent invasion of India. This had served to shift Gandhi's perspective towards the view that the best way of averting a Japanese invasion might be to force Japan's enemy, Britain, to relinquish control over India. However, at the provincial and district levels in Bihar one can discern a gradual emergence of opinions of this kind, at a much earlier date, in pamphlets and meetings of the organized

[103] One Fortnight Report noted that some Sadhus from Patna and Gaya were spreading anti-British propaganda. See FR (1), April 1942.

[104] Francis Hutchins, *The Spontaneous Revolution*, Delhi, 1971. See also interview with Sucheta Kriplani in oral transcripts of CSAS who noted, '1942 was the result of the general outburst of anger of the people in all the states. The sabotage and the destruction were spontaneous. If we say we had organised it we would be wrong'.

[105] Statement published by the Government of India on the Congress party's responsibility for the disturbances. CMD 6430.

left. The Congress Socialist party, which was at its strongest in Bihar, had disagreed with the Congress on the issue of the war, as it was felt that the rightwing in the Congress seemed willing to compromise with Britain in the event that a suitable declaration of war aims eventually emerged.[106] By contrast the Congress Socialist party had, from the outset, wanted the Congress to declare its opposition to participation by India in any war to which the British government might be a party, to undertake forthwith the preparation of the entire Indian nation to actively resist the utilization of men, money and resources for the purposes of such a war, and to utilize the crisis for securing Swaraj. In 1939 the Congress had actually passed the 'National Demand' resolution, sponsored by the Congress Socialist party. This resolution demanded that India keep aloof from imperialism and fascism.[107] Ram Manohar Lohia, an active socialist, wanted every Indian to be a war resister. The Lohia plan of war resistance for 1940 had included the issuing of bulletins, anti-recruitment propaganda, stoppage of supplies of raw materials and enrolment of volunteer resisters disciplined and trained to resist war preparations and activities. The question of co-operation was, in fact, the single most important cause of discord between the Congress right and splinter groups that came under the umbrella of the Congress left. It is important to note here that the efforts of the Forward Bloc to organize an Anti-Compromise Conference had received the support of the socialists during September 1939, even though they later opposed the separatist policy of the Forward Bloc at the Ramgarh session of the Congress in March 1940.[108] Instead, the Congress socialists had chosen the path of a united front, while they believed that any planned struggle should be launched under the auspices of Congress. The communists in this period were in agreement with the Congress Socialist party on this tactical line and their activities, like those of the Kisan Sabha, had concentrated on the organizational preparation for the coming struggle. The communists were to change their tactical line to 'People's War' in June 1941 and abandon all opposition to the war. However,

[106] The communists were not very dominant in the politics of the left in this period and the Congress Socialist party were the most important critics of Congress policy, *Selected Writings of Jayaprakash Narayan*, Delhi, 1980, pp. 65-7.

[107] *Congress Socialist*, iv, no. 14, 2.4.1939, pp. 6-7, 11. See article by Ram Manohar Lohia, 'War Resister'.

[108] B.R. Tomlinson, *The Indian National Congress and the Raj*, p. 146.

as became clear, even the communists had found this Soviet-influenced policy tactically unviable in the context of 1942, and were forced to change their line; indeed many communists in Bihar actively participated in the 1942 struggle.[109]

From 1939 onwards the socialist leadership in Bihar was actively engaged in encouraging resistance. The district collector of Shahabad reported that the socialist, Budhan Rai Verma, had been moving to and fro in December 1939 in the interior villages of Bihar asking people not to support the war.[110] Bishwanath Mathur, another socialist, criticized the resignation of ministries as being politically insufficient and demanded a stronger form of action. As early as October 1939 he addressed a meeting in Arrah. His statement is worth considering in some detail. He asserted that:

> There is a tendency towards peace among the leaders. You people should be prepared for a movement and enlist volunteers . . . the movement will not be stopped though I am sent to jail. I am arranging for the non-payment of revenue. The government will have to spend much in oppression and the treasury will be empty. . . . We should sacrifice life and property for freedom which is called guerrilla war. Pamphlets will be distributed . . . there should be a committee in the villages one member should fight and the other should go to jail. There will be open fight (sic). The pamphlet will be found sometimes pasted on the back of a bullock, the news will be communicated very secretly with directions. We should catch the weak part of the British government and start a movement. The rail, post office and all business will be shut down that we may destroy her government. If foreigners attack India namely Russians and Japanese at that time we will form soldiers of our own then and there . . . we will not help the British government.[111]

At another meeting in November 1939, Jayaprakash Narayan exhorted the audience not to pay rent and to start a non-tax-paying campaign. One of the vehicles of Socialist party propaganda consisted in the use of *Prabhat Pheris* (early morning processions). Party members carried bugles, sang national songs and shouted slogans such as '*Angrezi raj nash ho, Khuni jang band karo, Hindu muslim*

[109] Interview with Bihar communist Sunil Mukherjee, oral transcript, courtesy Bipan Chandra, J.N.U.

[110] FMBP, file no. 18/39.

[111] FMBP, file no. 17/39.

ek ho, Monafa band karo'.[112] In many of the speeches of the Forward Bloc adherents specific efforts were made to criticize the Congress high command for its failure to take a harder line and to make political capital for the Forward Bloc.[113] For example, Sheel Bhadra Yajee, in his speeches during 1940, informed his audience that the Forward Bloc had distanced itself from the Congress when the latter failed to act on its earlier resolution. He said that:

> The Congress has decided that it will not help in the war. The Congress should act on this resolution and not allow the British to complete recruitment. We must not assist the British. We must close schools, tell our brethren to stop payment of tax and the British should not get a single pie as loan.

He was also critical of the political inactivity of the Congress and condemned the leadership at the All-India Anti-Compromise Conference for the Forward Bloc, Kisan Sabha and the Mazdoor Sabha in 1940.[114] The *Indian Nation* claimed that in all districts the Forward Bloc had formed councils of action to conduct the national struggle and that it had also enrolled volunteers for the purpose in every district.[115]

The *Janata*, the Kisan Sabha newspaper, also carried on vigorous propaganda against the war in 1940, exclaiming on one occasion, 'We shall proclaim it by beat of the drum. We shall not contribute even a single pie to this (war), not a single brother of ours shall take part in it.'[116] A whole variety of leaflets on the struggle were circulated in

[112] Ibid. 'Down with British rule, End this bloody battle, Hindus and Muslims unite, End this profit of war.'

[113] Speeches of Sheel Bhadra Yajee, FMBP, file no. 212/40.

[114] He said, 'You know that the Congress does not like to fight. The reins of the present Congress are in the hands of such leaders as are big seths and sahukars who have the wealth of the country under their control. Gandhi cannot start a satyagraha because blood will flow. But you find that there is an awakening among the kisans as also among the Mazdoors and young men. This means that power will come into the hands of those who will participate in the struggle. You know what advances the country made in 1930 but then there was Chauri Chaura, (sic) and Mahatma Gandhi stopped the fight. The Forward Bloc and the left ask Gandhi to commence the struggle but Gandhi does not listen. The resolution should be brought into practice. The Swami went to jail, the fault of the Swami was that he toured the villages explaining to the people that the British had grown weak . . . . Independence was to be wrested from them . . . . The police are our brethren . . . .' See BSA, file 212/40.

[115] Quoted in FMBP, file no. 114/40.

[116] FMBP, file no. 114/40.

different locations. One called on workers to go on strike and join the nationalist struggle. It also called on students to carry the movement ahead: 'There will be self-service *dals* and this will be the people's army . . . British imperialism is in danger. We will never get a golden opportunity like this. The bonds that bind us are going to collapse. How much are you helping in the process? Every youth should swear, "We will fall victim to bullets, we will climb the gallows to be free!" '[117] A little later at Palasa, in 1940, the All-India Kisan Sabha passed an important resolution on the national struggle and the war. This stated that:

> The political wait-and-see will allow imperialism to take the upper hand and leave our fighting strength paralysed . . . . This meeting declares that this policy of stopping the struggle would prove suicidal and misleading. The Sabha declares that Indians should not allow the continuance of the war and should launch a world wide struggle at an early date for the attainment of independence.[118]

The views of all the left-wing groups at this time thus favoured the primacy of the anti-imperialist struggle. In fact, as we have seen, they were openly critical of the Gandhian group for its policy of inactivity. By 1940 the Kisan Sabha began to concentrate on its anti-war speeches and propaganda and class issues were not highlighted. The bakasht movement in large parts of Bihar lost momentum in that year as the agrarian situation favoured the shift to other issues and as bakasht problems were temporarily seen as no longer pressing. However, even in areas where the agitation remained active, as in the case of the agitation of the Santhal bataidars in Purnea, efforts were made by the elite Kisan leadership to deliberately underplay the most pressing class issues and concentrate instead on the coming struggle with the British.[119]

[117] Ibid.

[118] M.A. Rasul, *A History of the All India Kisan Sabha*, p. 66.

[119] In the course of touring the disturbed areas in Purnea, Congressman Baidyananth Chaudhury visited Dharahara, the Santhal tola, where he emphasized that the differences between malik and bataidar should be obliterated and everyone should unite for the success of the struggle for independence. Despite this directive there was no mass participation by the Santhals of Purnea in the 1942 struggle. However, the Quit India movement temporarily eclipsed the localized struggle of the bataidars and it was only a year later that the agitation was revived. See Anand Chakravarti, 'The Unfinished Struggle of the

Among the first people to be recruited by the new socialist propaganda in 1942 were school and college students. They readily joined the volunteer dals of the Forward Bloc in preparation for the struggle and, when the leaders were arrested, were the first to organize protest processions in urban areas. However, evidence of widespread dissemination of socialist propaganda and of early student unrest should not lead one to the conclusion that an organizational build-up for the Quit India movement was complete before August 1942. In rural areas in south Bihar much more localized organizations, such as the *Azad Dastas*, sprang up in the course of the struggle, rather than in preparation for it. Such organizations involved local people and were largely independent of the more organized socialist activity which predominated in north Bihar.[120] Clearly, therefore, while the rebellion may have been given its initial impetus by the activities of students and socialist party workers, the events of the succeeding two weeks cannot be simplistically adjudged as deriving from a movement directed from above. The complex manner in which the movement manifested itself in different areas, whether in the Gope/Dusadh parallel government in Barh or in the levels of participation in the interior villages of Monghyr, indicates that no such simple conclusion is tenable.

THE AUGUST MOVEMENT AND THE ANATOMY OF CROWD ACTIONS

Reports made by district officials of the various districts on the outbreak of the Quit India movement tell their own story. In Begusarai, the Sub-Divisional Magistrate noted that:

> The subdivision of Begusarai was a pro-Congress subdivision, which had previously played an important part in previous civil disobedience movements, especially in the movement of 1930 . . . . The Congress had their organisation in every village more or less, and the young men of the villages, especially the Bhumihars, have an inherent tendency to work as volunteers under the Congress flag on any occasion. The village organisations were in their best form at the time of the Congress governments, while after the fall of the Congress government there was a marked decline but in many villages

---

Santhal *bataidars* in Purnea District, 1938-42', *Economic and Political Weekly*, vol. xxi, no. 43, 25 October 1986.

[120] Interview with Azad Dasta activist, Havildar Tripathi, in Patna, March 1987.

they maintained it and the village teachers played an important part in maintaining the organisation of the Congress. The news of the arrest of the all-India leaders caused wide agitation, especially among the Congressites, and the student's section. On 10th August there was a strike in all the schools in the subdivision. The students began to move in a procession through the town shouting slogans and attempting to disseminate the 16 point programme of the Congress. Till 10.8.42 the movement was practically limited to the students, and the activities consisted of strikes in schools, hoisting of flags on schools and processions in bazaars. From 13.8.42 young volunteers were deputed to villages to organize volunteers. In Ballia *thana* and Barbhan out-posts were attacked. On this day the attack on railway stations also began, the passenger trains were detained by students in some stations and passengers in European dress were insulted. The railway telegraph lines between Begusarai-Tilrath, Barauni-Rupnagar were cut. From the afternoon the rural mob began to take the upper hand and the student section began to lose prominence. The mob consisting of the residents of villages round about Barauni junction began doing mischief to the yard in Barauni Junction.[121]

By 14 August the situation was already completely out of the hands of the students and although, as the report reveals, the more sober section of the populace did not approve of the looting, they had no control of the mob. On 14 August, the railway station was burnt at Barauni. The Anglo-Indian staff became panic-stricken and fled on a troop train, leaving the station in charge of subordinates. The report goes on to note that 'the morale of the public and government officers had gone down . . .even the armed forces showed signs of hesitation . . .In Begusarai town the looting of the town was attempted by a party of 300 looters'. The state was able to impose some kind of control only by 24 August. A military party arrived from Samastipur and burnt about a dozen houses and shot three men, of whom two died.[122] Following this summary of events, the report then proceeded to analyse the nature of the movement in Begusarai. It noted that school students were among the first to launch the movement. They were then joined by sections of the Congress, and though some Congress leaders tried to maintain control over the agitation, they failed once the village masses joined them. This quickly led to much looting of government property.

121 FMBP, file no. 45/1942.
122 Ibid.

The vast properties, especially the food grains in the railway station, attracted them. Then the situation was no longer in the hands of the Congress. The loot did not remain confined in the hands of any particular section. The Muslims took part in the loot to a lesser degree. The menial staff of the station joined in. Of the villagers the Bhumihars and poor labourers took a prominent part in the looting. The report noted that the merchants in the outlying stations were worried, but somehow they managed to get the leaders on their side by making payment to the Congress cause and thus saved their properties.[123]

Accounts such as this indicate that there was no clear planned strategy in the campaign. The movement began with the processions and slogan-shouting of students and other local Congress activists. Later, incidents of sabotage and looting took place, while attacks on police thanas, the treasury and other symbols of state authority also occurred. In this second phase some form of popular consensus about the justice of challenging the state largely by attacking state property seems to have been at work. Both the urban and rural poor joined the movement at this stage. The type of duality in revolt suggested by Henningham, by which he characterized the upper castes as being mobilized over the issue of nationalism and the lower castes as being more interested in looting is not borne out by the evidence. The District Magistrate's report on Begusarai specifically notes that looting was not confined to any particular social group. Anybody who had an opportunity utilized it. Everyone, right through from upper-caste Bhumihars to poor labourers, took a prominent part in the looting spree.[124]

In Patna the Sub-Divisional Officer reported that the news of the arrest of Rajendra Prasad spread like wildfire. In the evening a procession, consisting mainly of college students, paraded through the principal thoroughfares and proceeded towards Government House. On 10 August, hartals were observed in schools and colleges and most of the Hindu shops were closed. A mob then proceeded towards the Secretariat and was joined by the students of the local schools. When the mob persisted in hoisting the Congress flag on the

[123] Ibid.

[124] While one can agree that looting might have been more economically advantageous to the poorer sections one cannot conclude that they participated in the movement only for these reasons. Other actions of the lower castes, not involving loot, lead one to a different conclusion.

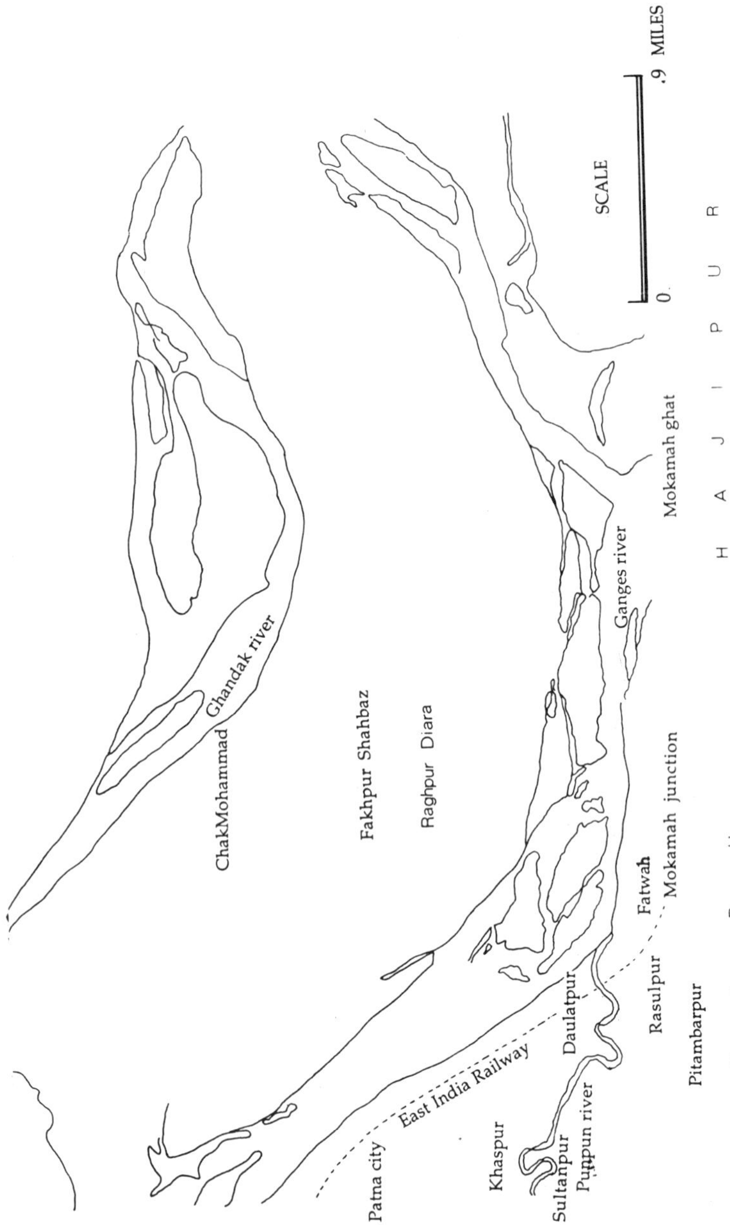

Patna Subdivision, 1935

Secretariat building, shots were fired and seven people were killed. This signalled the outbreak of large-scale protest. In the subsequent procession students, rickshaw wallas, coolies and women all participated.[125] In the evening a very crowded meeting was held at Kadamkuan when Babu Jagat Narain Lal delivered a speech. Soon afterwards the mob spread to all parts of the town, including the Sadr subdivision, cutting all means of communication and blocking all roads with fallen trees. A very large crowd, consisting of several thousand people, collected in the road in front of the jail compound at Frazer Road while prisoners were being transferred from Bankipore Jail to camp jail, and twenty-three prisoners escaped.

The timing and nature of the outbreak of the movement in Patna were much influenced by the great length of railway lines running through the district.[126] Much of the energy of the mobs was largely expended on damaging the railway lines and in burning and looting the railway stations. In Patna town, post offices and municipal offices were burnt and the police soon became powerless to deal with the situation. Throughout Bihar most of those police stations which were isolated from Headquarters were attacked. In Chandi, for example, the local police station had to be evacuated in the face of a mob of 10,000 persons. However, it seems that the crowd did not attempt to kill the police constables, even when they succeeded in capturing the police station, as they had done at Mokamah. The looting in Mokamah was particularly bad because the police station itself was overwhelmed by the crowds. As a result the looting of trains and of goods godowns at Mokamah junction went on without any police interference. Observers noted that the lower grades of the railway employees played an important part in the looting.[127] At Naubatpur the Sub-Inspector reported that Jagat Narain Lal and other Congress agitators were moving from place to place, keeping up the spirit of the public against the government. Daffadar Pirthi of Chachaul reported that on the 21st he was surrounded by several armed Congressmen who enquired from him about the number of armed police in Naubatpur. He also saw 80 Ahirs and Koeris with spears and

[125] District-wise summary of the events for the week ending 21 August 1942 in FMBP, file no. 27/42.

[126] The sabotage of the East Indian Railway during the rioting was said to have pushed back the invasion of Burma by British forces for some time. See Flack Papers, Mss Eur F 180, IOL, Cambridge.

[127] FMBP, file no. 27/42.

garasas on the move. It was also reported that many chowkidars were pressured to leave their posts, while their relatives were harassed.[128]

In Shahabad the Collector reported that up until the violent outbreak of the movement the authorities had received no indication of any abnormal Congress activity. The magnitude of the insurrection was the more remarkable because it was mainly conducted by leaders of secondary importance with the aid of students and later of 'hooligans'. The report continues:

the arrest of the Congress leaders was followed from 9th August by seditious meetings and processions consisting largely of students. At first these gatherings were not particularly violent except at Arrah where a small police party accompanied by the Superintendent of Police and the Sub-Divisional Officer was heavily stoned when attempting to disperse a meeting on a maidan. As the insurrection gathered momentum attacks were made on post offices, police stations and canal *tehsil* offices. Railway lines were removed and telegraph wires cut. A mob of students at Arrah raided the courts and the *zillah* school and the Superintendent of Police was forced to fire on students uprooting signs at the station.[129]

The situation in Arrah deteriorated markedly when students, joined by a crowd, commandeered a train on the Arrah Sasaram light railway and proceeded to Sasaram, looting stations en route. At Sasaram the constables made it clear that they were unwilling to disperse the mob by force and a message for help was sent to Arrah. This was the first major sign that the lower levels of the police were withdrawing their loyalties from the state. Henceforth, the Raj could not rely on its police functionaries to quell the insurrection. The disloyalty is well shown up in an account appearing in the memoirs of a district officer, H.B. Martin. He records that:

It was then that I got the information that a large mob had collected and commandeered a train from Arrah and were coming to Sasaram along the light railway destroying all the stations as they went along and looting . . . there was some trouble in the civil court compound where the boys from the school wanted to hoist a flag . . . . . I told the police to clear them away whereupon the Hindu constables became engrossed in contemplation . . . . In the afternoon the more violent chaps from the muffussil arrived and burned the two railway stations. I suggested that the police might be asked

[128] Ibid., S.I. to D.I.G., CID.
[129] FMBP, file no. 42/42.

to come out and at least try to stop them but the *havildar* explained that he had long experience of such matters and he knew that if we attempted to open fire on the mob we would be killed. He added that he had discussed the matter with his friends and they were all of the same opinion. Comforted by such an open expression of loyalty, I decided to concentrate on guarding the gaol and the treasury . . . The next morning I snatched the revolver off one of the more obviously disloyal police officers and took up my position in the treasury.

Martin was forced to sit guarding his post alone until a detachment of the military arrived to save him.[130] The scenes described in this account serve to illustrate, in microcosm, the complete withdrawal of the support of the lower levels of the police and the administration from the Raj that took place in 1942, and the helplessness of the colonial state during the first few weeks of the movement. The erosion of state authority was far-reaching and revealed the ultimate weakness of the Raj, once the components of its indigenous support-structure had withdrawn their loyalty.

Two crowd actions serve to highlight the complete breakdown of authority and the absence of fear at the popular level. These were the incidents at Fatwah and Pasraha, which constituted perhaps the most severe kind of attack upon the colonial state. These were attacks on unarmed white men who, in the mass mind, represented the physical presence of colonial authority.[131] At Fatwah, on 13 August, the 19 Up express train from Howrah stopped at Fatwah, near Patna, and then proceeded west, only to be stopped near a warning signal by a crowd that approached it, throwing brickbats. Two British officers, J.H. Smith and W. Short, were travelling in the first class compartment. In alarm one of the officers fired two shots from his revolver whereupon the crowd raised an outcry in the belief that a man had been injured by this shot.[132] The crowd beat on the window and within ten minutes the train was taken back to the station and an attack on the window was launched by a number of men from the surrounding countryside who were armed with sharp weapons like *khalas*, *pharsars* and *garasas*.[133] The officers were wounded while inside the compartment

[130] H.B. Martin Papers, CSAS, Cambridge.

[131] FMBP, file no. 519/43; Pasraha Report by J.G. Shearer. See also file no. 323/43.

[132] They believed that one of the men in the crowd called Bhola Dusadh, an ex-chowkidar had been shot.

[133] These refer to spears, knives and axes.

and, to save themselves from further attack, one officer emerged from the compartment hoping thereby to assuage the anger of the crowd. However he was speared to death, as was his companion. Their two revolvers were then taken away, and their bodies were dragged along the platform. After this they were placed on a *tum tum* (cart) and taken around the town before being thrown into the Poonpoon river. An interesting feature of the composition of the crowd on this occasion was that a majority of them belonged to the lower castes and were either Gope, Dusadh, or Pasi.[134] While the ensuing prosecution mentions the crowd as being instigated to act by one Bihari Mahanth it was, in fact, the Dusadhs who had committed the act and then displayed the bodies around the town in a flagrant challenge to authority.[135] Once again the crowd seems to have been imbued with a notion of the legitimacy of their actions and motivated by a consensus and conviction that overrode all fear and deference to authority. The resulting criminal case was registered only after much delay, on 26 August, and occasioned the compiling of a list of 'absconders' which gives a useful insight into the social composition of the crowd. Most of the 'absconders' were finally arrested only in June 1943. By name they included Ramaran Teli, Siwan Dusadh, Ramdeo Singh, Satbhusan Singh, Binda Pasi, Sudbhusan Singh, Bishun Gope, Mithila Tiwari, Haridwar Tiwari and Dwarka Tiwari. The following were sentenced to death: Pearey Dusadh, Jhoparia Dusadh, Binda Thathera, Mahanth Bihari Tiwari and Trijugi Pande. From these names alone it can be deduced that the majority of those accused were either Dusadh or Pasi.[136]

The second significant incident worth alluding to at this stage also relates to the treatment of vulnerable European servicemen. On 18 August 1942 a returning bomber crashed in Pasraha near Khagaria. According to the statement of Maulvi Mahmood Hussain Khan, an Assistant Station Master, two Europeans survived the crash as the plane landed in water. They were then helped ashore by some boatmen.[137] As they needed medical help the Station Master decided to

[134] The Dusadhs were a low caste spread over northern and western India. Occupationally they formed the bulk of the village chowkidars. Many were poor labourers and also reared cattle for a living. See the Census Report of 1931 for the province of Bihar, p. 210.

[135] 'Mahanth' refers to a temple functionary. He was usually a landlord and owned many acres of land in the form of grants to the temple.

[136] FMBP, file no. 519/43.

[137] Ibid.

help them on to a railway trolley and requested some members of the crowd that had gathered to do so. Later the official noted that there seemed to be some conspiracy afoot among the people collected at the crash site. Before long some five or six persons emerged from the crowd and began to beat up the pilots. The pilots then ran into the water to escape, but were followed and killed. Another witness later saw a boat with six people by the side of the crashed plane; they first stole some articles from the wreck and then fled into the villages of Deotha and Kharauna. It was noted that all the persons in the boat (who had killed the Europeans) were residents of these two villages. Thakur Mandal, Ramrup and Dayaram were individuals later discovered as being the chief instigators while the remaining accused were named as Chaman Gope, Surji Gope, Chamru Nagar, Ratan and Jagli. Once again, as in the Fatwah case, the majority of the accused came from the lower castes and belonged to the Ahir and Pasi communities. Among the assailants were also two Kurmi, brothers of an employee of the Locomotive Department, Mahabir Kurmi. A similar incident to that at Pasraha occurred during August 1942 when a plane crashed half a mile away from the village of Ruihar. Inquiries revealed that on the day after the crash a crowd of 250 men led by Janki Singh, a Congressite, pulled the plane off into deeper water, relieved the pilots of their weapons and then murdered them.

The colonial state was particularly disturbed by these incidents since, up until mid-August 1942, the activities of the protesting crowds had focused on destroying government property rather than on killing Europeans. By these actions, in contrast, the crowds had now shown themselves prepared to kill unarmed white men. These incidents were important for they were open attacks on the physical presence of colonial rule and constituted, one may argue, a final violent assault on colonial hegemony. No longer was the British 'sarkar' a force both respected and feared at the popular level. Instead, the whole apparatus of the colonial state could be challenged and humiliated. Moreover, both upper and lower castes in the villages in question appear to have participated together in the attacks on Europeans, spurred on by a consensus that overrode notions of fear. Not surprisingly, the British administration was particularly concerned to bring the culprits to book in these three cases and to reassert some measure of authority and legitimacy in the region. However, a more subtle form of resistance was then found to be at work. The Indian officers dealing with the cases postponed taking action sufficiently long for Richard Tottenham, the Home Secretary to the Government of India, to note that 'the progress of these cases was very slow'. In this fashion, the absconders were virtually allowed to go free until 1944.[138] In June 1943, the district magistrate of Patna wrote to the Chief Secretary to the Bihar government explaining the action taken to follow up the cases and the reasons for their slow progress. 'When the planes crashed', he reported, 'for a month there was no evidence as to whether the death was due to an accident or to murder. Soon after the occurrence, however, British troops went to the neighbouring villages and set some houses on fire, killing some people, including a woman and a child.' The legal course of the case, in contrast, took much longer to get off the ground.[139]

In examining the pattern of development of the Quit India movement, it seems clear that in several areas local administration had retreated completely in the face of popular protest and parallel government. For nearly two critical weeks the British were taken completely unawares and took a considerable time to recover and restore the structure of policing and repression. Meanwhile Sub-Divisional Officers helplessly reported many cases of popular governments having been set up with, in some cases, the active sympathy of lower levels of the local administration. One such case of the setting-up and functioning of a parallel government can be picked out as exemplifying some of the strains that began to develop in the new movement at the popular level. This was the Gope and Dusadh parallel government in Barh. The Sub-Divisional Officer's report on the discovery of this structure is worth repeating:

> There is only one platoon of troops at Mokamah Ghat under a lieutenant and it was considered that a large number of troops was necessary for Malpore .... We went by military truck up to Ghoswari Ghat, a distance of three miles from Mokamah. From there we proceeded by boats to Malpore, a distance of six miles from Ghoswari Ghat ... The whole area was swampy. The result was that we could cover six miles by boat in seven hours and reached Malpore at dawn. The villagers saw us approach and batches of accused persons began running out. The troops and the police ran after them and repeatedly shouted to them to stop ... I saw about 30 persons running

138 Ruhihar Case, R. Tottenham's rebuke to district officers in Bihar. FMBP, file no. 323/43.

139 FMBP, file nos. 46/42 and 323/43. The rebuke was taken with bad grace by the lower officers.

away in different directions. Six or seven rounds were fired. Major Cole shot dead one person and the lieutanant shot dead another . . . we could arrest only one person who was a chowkidar and was given the office of assistant commander in the Congress government. From my enquiry I confirmed the report of the Sub-Deputy-Collector that the Goalas and the Dusadhs of the village had combined to set up a Congress government. They had begun to loot the Mahtons and the Dhanuks of the village. They had looted four houses in the village. They levied taxes on the Mahtons and Dhanuks and assaulted persons refusing payment . . . The village had two chowkidars both of whom were given office as assistant commanders. Both have been arrested among the accused and have been dismissed from service for failing to make a report to the *thana* for a month and a half.[140]

The petition of the Mahtons noted that, following the outbreak of the violent disturbances on 10 August (involving arson and looting), some of the Gopes and Dusadhs in the village set up a raj government in Malpore and began to intimidate, molest and loot those of the inhabitants who refused to join them. On 12 August some of the accused, headed by Sarju Gope, organized and set up the 'Raj' and styled themselves by various titles, with Surju Gope as the Raja (king), Akal Gope as the Pradhan Mantri (prime minister), and Ganesh Gope, Jhauri Gope and Harihar Gope as the Mantris (ministers). Pradip Singh was captain of the sena (army) and Jagroop Gope chowkidar and Khadiram chowkidar were his assistants. An army of Gopes and Dusadhs was also formed.[141] They proclaimed, by the beat of drums, that the British raj had been abolished and Swaraj had come. Accordingly, they started to levy taxes which the Mahtons of the village refused to pay. On the 14th the accused persons called on Narayan Mahton and ordered him to pay Rs 275 in cash and 15 maunds in grain, failing which he would be done to death. Mahton fled his house and his grain was taken away. On the same day another Mahton was raided and ornaments, grain and utensils were removed. A few days later a mob of 150-200 persons raided the house of Hulas Mahton. It was alleged that Akal Gope dealt two garasa blows and

[140] FMBP, file no. 75/42; Report of SDO Barh to W.G. Archer, District Magistrate, Patna.

[141] The group consisted of the following members: Rampati Gope, Rameshwar Gope, Jagiva Gope, Belchan Gope, Dwarka Gope, Chando Gope, Sukho Gope, Phaghu Dusadh, Lakhan Dusadh, Jhansi Dusadh, Kali Dusadh, Raudi Dusadh and Faujdari Dusadh.

Ram Swaroop Gope dealt a lathi blow and that thereafter they looted Mahton's house.[142] The actions of the Dusadhs and the Ahirs are significant here for they highlight the internal strains developing in the movement.

An explanation for the activities of the Dusadhs and Ahirs *vis-à-vis* other lower castes, such as the Dhanuks and the Mahtons, can be related to the contemporary 'caste awakening' which was taking place among lower castes and particularly amongst the Ahirs. In the early 1930s it had been noted by census officials that a caste awakening could be clearly seen among the various low castes in Bihar. Many of them, such as the Ahirs, the Koeris and the Dusadhs, formed caste Sabhas to proclaim their higher status. Of these the Ahirs were the most dominant caste-group, concerned with improving their ritual status and asserting it *vis-à-vis* other lower castes. The Ahirs also asserted their ritual purity through an increasing participation in Gaurakshini (cow protection) activities at this period. For similar reasons, they were also active in sectarian riots. Among the other lower castes, the Koeris largely accepted the sacred thread and began wearing Churi (glass) bracelets instead of the bell metal bracelets of the lower castes. So, too, did some of the Dusadhs assume the sacred thread at about this time. Attempts were made to spread new caste ideas through pamphlets and caste meetings as the Dusadhs tried to stake their claim to higher caste status, not only *vis-à-vis* the upper castes, but also in relation to other lower castes. Caste confrontation between the Ahirs and the Kurmis had already been noted by district officials at an earlier date in the sub-division. The Ahirs were, in fact, the most upwardly mobile of the lower castes, both in terms of economic status and ritual status. By asserting their power over the Mahtons, who were lower down in the ritual hierarchy, they were attempting to improve their own ritual and economic status. Thus the caste confrontation released in 1942 in Malpora village was symptomatic of this group rivalry in the ritual and economic arena.[143]

Apart from the strains at the popular level, what is more important and, to my mind, more interesting was the new and extended political participation of lower castes in general, and their mobilization, almost for the first time, in the national struggle. In the Laskari canal sub-division it was reported in mid-August 1942 that most of the

[142] Ibid. Petition of the Mahtons of the village.

[143] For earlier reports, see FMBP, file no. 71/42.

members of the crowds roaming the area were drawn mainly from lower castes. For example, about three hundred persons were seen marching, armed with spears, in the subdivision on 16 August. According to the district magistrate the villagers attacked a military party, and the combined strength of the mob then increased to a thousand people. Shots were fired, nine people were killed and six were injured, according to official figures. The caste origins of those killed can be gauged from a list quoted by the district magistrate. The majority of those killed were Ahirs, who seemed to be the most politically active among the lower castes.[144] Other government reports noted the sudden emergence at this time of gangs of Dusadhs and Ahirs in Gogra Sarju diara. These were armed and committing dacoities in Saran, Ballia, Azamgarh, and Gazipur.[145] Similarly, in Sasaram it was noted that four low-caste individuals, Harnarian Dusadh, Ramkumar Teli, Mohan Ram Teli and Ghashiram Koeri were committing dacoity together specifically to raise money for terroristic activity.[146]

The low morale of the colonial administrative functionaries contributed to the complete breakdown of state power at the local level in late August 1942. It also revealed the extent to which, in the localities, the colonial state had always rested on flimsy supports. At the village level, the weakness resulting from the lack of penetrating institutions at the local level was confirmed when, in 1942, the chowkidars in many areas shifted their loyalties from the Raj. In many areas they actually joined hands with the movement and, in Barh, as we have seen, even became functionaries in a new Congress 'government'.[147] The upper echelons of the police services fared no better.

[144] 1. Basudeo Ahir - shot while escaping after being taken in custody. 2. Sabhapati Ahir - shot while running away after having 'attacked' a soldier. 3. Jagannath Ahir - killed while 'assaulting' a soldier. 4. Sital Ahir Chowkidar - shot while fleeing after 'assaulting' another chowkidar. 5. Mahadev Ahir - killed 'assaulting' a military officer. 6. Girawar Ahir - shot while fleeing. 7. Bania Hazari - shot during the firing to disperse the mob. 8. unknown 9. unknown. See BSA, file 24/42.

[145] FMBP, file no. 38/42.

[146] FMBP, file no. 68/42.

[147] At the village levels the Kanungo, the patwari and the chowkidar were the Raj's main links to rural society. The chowkidars, as we have noted, were of low social status, yet were crucial figures in rural administration. Their importance grew as government links to the local level became increasingly tenuous in the nineteenth century. Not only were the chowkidars a major source

The collector of Sitamarhi noted the inefficiency and panic of the police officers deputed to bring the Sahibganj police station under control. 'Some officers', he noted, 'have become completely demoralised and are nervous about taking action.'[148] In Chapra, on one occasion, the police force retreated in the face of a violent crowd, even though the force was formidable and included an inspector of police, two sub-inspectors, sixteen constables and ten A.R.P. men armed with muskets and shotguns. In this instance the collector noted that the retreat of such a formidable force was due to a serious erosion of morale and to the 'bankruptcy' of the institutions of the colonial state.[149] Many police stations fell into the hands of the activists. At Hatra it was reported that the police station had fallen into the hands of the Congress and its sympathisers and that 'they have elected their own agents to carry on their work and it (the area under the jurisdiction of the police station) is treated as a district. At a meeting in Gangeya it was decided that each village should contribute ten maunds of rice and Rs 10 to carry on the hooliganism'. The rebels set up their own machinery of government in all the police stations, treating each police station as a district. They also appointed their own collector, judges and their own Sub-Divisional Officer and police officers. In one magistracy it was said that 'they have their own jails and are realising fines too. The disquieting feature is that in areas where thanas are not functioning the rebels are realising money from the public to secure funds for activities'. It was also noted that marwaris and banias were now helping law-breakers'.[150]

The complete breakdown of imperial authority which took place once the lower echelons of the British administrative apparatus had withdrawn their loyalties served only to highlight the fact that the colonial state was, in the ultimate analysis, buttressed by relatively flimsy indigenous support structures. Support by a thriving landed magnate element, committed by interest to British rule, had originally ensured that these fragile controls sufficed. Except in 1857, when

---

of information, but with the police so thinly stretched out, they were in effect the real force for law and order in the countryside. For the history of local control institutions see A. Yang, 'Between the British Raj and the Saran Raiyat: The Development of Local Control Institutions in the Nineteenth and Early Twentieth Centuries', in P. Robb (ed.), *Rural India*, p. 164.

148 Report of the District Magistrate of Muzaffarpur, FMBP, file no. 49/92.

149 Ibid.

150 Ibid.

some sections of the landed elements, (Kuwar Singh in Shahabad for example), challenged British rule, the Raj had never been shaken. In the 1920s and 1930s the big zamindari estates had steered clear of the Congress. Thus, despite the fragility of its administration in local society, the Raj had survived with no serious challenges to its authority and power until 1942.[151] In 1942, however, the richer zamindar's covert support for the Congress cause started to become apparent. Though they did not actively participate in the movement most of them rendered sympathetic help to the struggle, with only the Dumraon raj left actively assisting the authorities.[152] The activities of the Zamindar of Surajpura, in Shahabad, are interesting in this regard. Though ostensibly he was a supporter of the British and was even the District leader of the National War Front, he actually harboured many activists in his house, while both his sons were actively involved in the movement.[153]

The means of propaganda and the kinds of organization that developed in the course of the movement are of some interest. Many independent underground organizations sprang up in the wake of the first disturbances. The Lal Kranti Sena and the Azad Dasta were the most important of these. In Dinapur leaflets printed in red ink were issued by the Lal Kranti Sena and these were also found in circulation in Pirbahari. Such leaflets were deliberately dropped in the pockets of people dressed in khadi, exhorting them to continue the movement for attaining independence, and not regard 'firing' as a deterrent.[154] Further, the leaflets noted that the temporary want of leaders should not 'discourage the people' as the Lal Kranti Sena would come and 'act as the leader'. The second major organization that developed in the wake of the struggle was the Azad Dasta. Very little is known about this organization and even less is written about it. It seems to have been a disparate organization, although having some links with the CSP. Many young men, mainly from the south Bihar districts,

[151] *Shahabad District Gazetteer* on Kuwar Singh and the 1857 mutiny, pp. 29-35. See also A. Yang, 'Between the British Raj and the Saran Raiyat: The Development of Local Control Institutions in the Nineteenth and early Twentieth Centuries', pp. 149-71.

[152] See FMBP, file nos. 64/42, 42/42 & 45/42 for attitudes of various sections; and for attitudes of zamindars - personal interview with Zamindar of Surajpura estate, Uday Raj Sinha, Patna, 1987.

[153] Interview with Uday Raj Sinha in Patna, February 1988.

[154] FMBP, file no. 30/42.

joined the Azad Dasta, which had a notional headquarters in Patna. It had an informal leadership and several splinter groups emerged from within it. Often they issued contradictory directions, and minor internal confrontations developed on occasion. The Dasta also developed alliances with local dacoit gangs in the period. These informal organizations were politically significant in keeping the Quit India movement alive well into 1944. They regularly organized minor raids on police stations and sometimes even on military depots for ammunition. Thus, while in north Bihar underground activity was firmly under the control of the CSP, which took the lead in the agitation in these areas, in south Bihar it stayed in the hands of several more shadowy organizations, but particularly with the Azad Dasta and its breakaway groups.[155]

An important source of propaganda consisted in pamphlets and leaflets which were printed on hidden cyclostyle machines. One circular of the CSP (called Circular No. 5) was believed by the government to have greatly added to the tempo of dislocation and disruption. The circular proclaimed that:

> India's struggle for freedom has begun. From now on we consider ourselves independent and do not recognize the sovereignty of the British government. So we should not abide by the orders or laws of the British government.

It then listed the main items in a resistance programme, including the non-payment of taxes and the uprooting of railway lines. Another leaflet asked people to stop paying rent and revenue, to withdraw money from British banks, and slow down production and machinery. It also appealed to the police not to act against their countrymen. In Muzaffarpur secret pamphlets were distributed in *muffusil* areas which contained rumours of the following nature: (1) 'The police commissioner who arrested Gandhi has been killed', (2) 'The commissioner of Patna has hidden himself', (3) 'The governor of Bihar has fled', (4) 'Arrah jail has been broken open and all Congressmen are enjoined not to fear going to jail as they will be released'. Instructions were also issued in the pamphlets to seize arms and to keep them hidden, while the police and the chowkidars were instructed not to come to the thana or use arms. In a similar fashion a copy of the *Free India News Bulletin* stated that 'the first stage of

[155] Interview with Azad Dasta activist, Havildar Tripathi, Patna, December 1986. See also chapter 5 for a more extended discussion of the Azad Dastas.

India's war of freedom has been successful' and that 'Lucknow, Cawnpore, Delhi and Arrah are under peoples' government'. A Hindi leaflet found in Dinapore asked people to remove rails, destroy bridges and telephone communications, burn courts and hoist the Congress flag on buildings.[156]

After the arrests of August 1942, and as the Congress went underground, the official Congress radio broadcasts assumed a new significance. Usha Mehta (a Congress underground worker) remembers the first broadcast announcement which said, 'This is the Congress radio calling on 42.34 m from somewhere in India.' She further noted that, 'It was on a distinct wave length and so far as our programme went we began with Hindustan *hamara* and ended with *Vande mataram*. In-between we had news relays, speeches, appeals and instructions. We were the first to broadcast news about the Jamshedpur strikes, Ballia, and the parallel government in Bihar'.[157] These radio broadcasts provided an important source of information on the movement at a time when the official media was under state control and strict censorship. It thus raised morale and kept the Quit India movement going for a time despite heavy repression.

During the height of the movement, sympathy and support for the activists were very strong. The troops received little or no support from any section, including the zamindars. In Tarapur, in Monghyr, in November 1942 the Assistant District Magistrate recorded, in the course of combing the villages in that area for absconders, that:

> my party had to pass through several villages on the way, and whenever we came out of any villages, a peculiar disyllabic cry ensued from somewhere in the village, was taken up by someone in the next village and passed on. When we reached Dholpari which is a small hamlet on a hill of that name situated in the centre of a very big chaur, we could see big flares lighted in some of the distant villages. The other party had heard loud bells ring in a village which they had to pass. Our raid was futile but we saw fresh marks of cooking, and also some salt and onion in an uninhabited house at the top of the hill and in a temple nearby. We also saw some slogans in chalk and charcoal on the walls both at the house and in the temple. These slogans were in English and Hindi, some of them being, Do or die! Britishers quit India, *Pulis hamara bhai hai, Pulis walon naukari chhor do, Jasu raj priya*

[156] Copy of *Free India News Bulletin* and other leaflets, FMBP, files nos. 42/42, 49/42, 76/42 and 65/42.

[157] CSAS, Oral transcripts, interview with Usha Mehta.

dukhari, so nrip houin narak adhikari.[158] The chowkidar of Dholpari, from his behaviour appears to be in league with the insurgents. There appeared to be a lot of active sympathy for the insurgents among the villagers, I found slogans written in all conceivable places . . . and when questioned the villagers professed complete innocence.[159]

Strikes in the Tata Iron and Steel Company works and in Dalmianagar also developed in support of the movement. In TISCO a one-day strike was declared on 10 August to protest against the arrest of the Congress leaders. The response was keen and 20 per cent of the workers immediately absented themselves. Once the struggle in the rest of the province was demonstrably under way the TISCO workers decided to continue the strike after the 20th. On that day a complete and non-violent strike was declared and 30,000 workers absented themselves. While the management did eventually persuade the workers to return, local British officials still suspected the management of conniving with the workers to close the works.[160] In Dalmianagar the strike followed similar lines. A partial hartal took place in Dehri following the arrest of the Congress leaders and by 12 August the strike had become more general. An impression grew that the strike had been engineered with the active assistance of the supervising staff, who were sympathetic to the Congress cause. A magistrate reported at the time that, 'as regards the management, though they may not have taken an active part in the affair, nothing in their attitude shows that they discouraged the strike. The absence of material damage to the factory and the reported peaceful behaviour of the strikers in Dehri indicate a passive connivance of the management in the affair'.[161]

While the TISCO strike was still in progress the situation was further complicated when the police in Jamshedpur went on strike. At the instance of Ramanand Tiwari, a *havildar* in the local constabulary, a secret organization named Inquilabi Sipahi Dal was formed among them. It started with a membership of three constables, but as the struggle proceeded this increased to seven hundred members. Tiwari began his strike action by organizing a mass fast on 9 August. The

158 'The police is our brother, leave your duties police brothers, under this Raj the people are unhappy and all who support it will descend into hell.'

159 Monghyr D.C. report, FMBP, file no. 47/42.

160 Dalmianagar strike, FMBP, file no. 27/42.

161 Ibid. See also Hugh Kemp Papers, CSAS, Cambridge.

next day the constables under his leadership took a prominent part in making the strike in Tatanagar a success. That night he held a large meeting of all the armed constables of Bihar, urging them to not to co-operate with the government. Senior police officers looked on helplessly, unable to check the growing unrest among the lower echelons of the police force. Finally the Gurkhas and other British troops were sent to Jamshedpur and shortly after this Tiwari and thirty-three policemen were arrested and taken to Hazaribagh jail.[162]

THE REPRESSION OF THE QUIT INDIA MOVEMENT IN BIHAR

The colonial state reacted to this grave assault on its authority and legitimacy with violence and repression. At first the pressures of the war effort had distracted the government from immediate action. However, when it came, the repression was heavy and crushing. The army was given a free hand and soldiers were sent out patrolling and burning villages. The Collector of Patna later wrote to the Chief Secretary of the Government of Bihar, on 2 October, that, 'It is well known that during the process of coping with the situation, many forms of irregular action were taken by the military. There have been complaints that soldiers searched the house of one Ramnath Singh and took away jewels and gold ornaments.'[163] In Ruihar, where the pilots had been murdered, the troops vented their anger on the local populace particularly harshly. On 30 August a party of twelve British troops, accompanied by a clerk, raided the villages of Bengalia and Ruihar, while the men and women in the village fled their houses in terror. The troops then entered houses, breaking doors and meagre possessions like earthen vessels. In Ruihar they ordered some men to come forward, but none came up and in anger they set fire to several houses. In Dinapur the arrival of the troops and their terror tactics quickly stamped out further disturbances. A certain amount of looted property, including military stores, was then collected from surrounding fields and dikes. Revolvers and ammunition found in the possession of the rioters were seized. The system of collective responsibility for the upkeep of railway lines was also started and strictly imposed.

[162] Tiwari's activities in Jamshedpur; interview with Havildar Tripathi. See also Havildar Tripathi (ed.), *Ramanand Tiwari Smriti Grant*, Patna, 1985, p. 125.

[163] FMBP, file no. 71/42.

However, in many cases this crackdown was not seen as enough. In Monghyr, where the DC favoured collective fines and arrests as a deterrent, it was recorded that the troops had several times wanted to raze entire villages to the ground and had to be firmly restrained. Predictably the troops spread a more generalized antipathy and fear among the populace and emerged to replace the police as the main repressive force of the state.[164] Henceforth, the colonial state would rely completely on the army to maintain law and order, the institutions of the police and magistracy having shown themselves to be disloyal and demoralized.[165]

The repression was successful in pushing the Quit India movement underground. Despite this, however, the movement had successfully revealed the weaknesses of the colonial state and to some extent destabilized it. The state had few resources left to maintain control, and almost no support among its former interest-groups. The attitudes taken by zamindars who were covertly sympathetic to the movement had already been made clear, and the state could no longer count on them. Similarly, the colonial authorities had hoped for support from the business communities on the basis that they depended on the law and order context provided by the state to carry on business activities. To some extent, this assumption was justified. Even during the mass movements of 1920 and 1930 the business communities had proved to be reluctant recruits into the national struggle and in the 1930s, as Sumit Sarkar has shown, these communities had successfully coerced Gandhi into accepting the Irwin pact so that normal business activity could continue.[166] In 1942 all such business considerations were pushed aside to support a movement which was seen to be the final assault on colonialism. Marwaris and banias now supported the movement quite openly and contributed funds to it. Even the management of the largest capitalist enterprises such as Dalmia and the Indian Steel Company management had been prepared to close their works in support of the struggle. Among other groups, the mobilization of the

[164] It was alleged that in many cases local women had been molested. For details of the crackdown and repression, see FMBP, file no. 71/42.

[165] There was even some attempt made by the socialists to infiltrate the army. But this effort had little success. See oral transcripts, Nilubhai Limaye and Madhu Limaye, CSAS, Cambridge.

[166] Sumit Sarkar, 'The Logic of Gandhian Nationalism : Civil Disobedience and the Gandhi Irwin pact (1930-1931)' in *The Indian Historical Review*, vol. iii, no. 1, July 1976.

middle classes and the urban and rural poor in the struggle was, effectively, almost complete. In the end the Raj remained buttressed only by the army.

However, one has to qualify this assessment of a complete mobilization of the population in the 1942 struggle when referring to the views of the Muslim community, where clearly the Quit India movement had less support. Indeed, the Muslim League in Bihar had, in a resolution, actually asked Muslims to keep well aloof from the movement. Among the Muslim middle classes, by and large, this dictum was obeyed. Among the lower class, however, as the Begusarai District Magistrate had noted, all social groups, including Muslims, had participated in the struggle. This held true for most of the affected subdivisions in the province. Both Hindu and Muslim shops were closed during the hartals. Despite this, the working committee of the Bihar Muslim League was determined to make political capital out of the situation. It congratulated the Muslims of Bihar on 'the sense of discipline with which they have behaved as a class and as a community in keeping completely aloof from the rebellious and terrorist movement launched by the Hindu Congress and supported by other Congressite Hindus, which was started deliberately without consulting the Muslim League . . . '[167] The British were quick to appreciate this attitude and exaggerated the extent of the lack of Muslim support for the Quit India movement by levying punitive fines which were confined to Hindus.[168]

By late 1942, state counter-measures had managed to push the movement underground, and the panic of the earlier official reports gave way to confidence as most of the rebel areas were brought under control. In Monghyr, the District Magistrate had noted in August that 'anarchy reigned in the interior . . . not a single policeman or a postal peon could enter the sub-division and there were alarming reports that if they had come in uniform they would have been torn to pieces'. By September, however, the tone had entirely changed. Now the state devoted itself to rounding up the 'culprits' and instituting punitive measures. Heavy repression by the army was intended specifically, it would seem, to institute exemplary fear and terror into the minds of the recalcitrant populace. The burning of villages, and the deaths of

[167] Resolution of Muslim League passed on 21 October 1942 in FMBP, file no. 75/42.

[168] See FMBP, file no. 49/42, for army instructions issued by Major General Wakely, Commander, Eastern Army.

women and children occurred in several villages causing their entire populations to flee on the sight of the advancing troops. Nevertheless the possibility of a return to status quo ante was hardly a real one. In many respects a profound change had occurred. The balance of power had shifted subtly, and the unquestioned authority the colonial state had exercised over various interest groups had been permanently eroded. Hitherto the colonial state had legitimized its rule by relying on effective support in local society from the zamindars and the merchant communities and by developing the service institutions of the state apparatus and the institutions of authority in the localities. Indeed a colonial hegemony had survived, through careful imperial strategy, well into the 1930s. Earlier mass movements in 1920 and 1930 had whittled it down, but it was the movement of 1942 which despatched the final blow. Ultimately the colonial state became extensively militarized once it was clear that the services, the institutions of authority in the localities, the zamindars and other interest groups had shifted their loyalties. Even in instituting punitive measures Houlton, the Chief Secretary to the Bihar government, noted that the police were slow in bringing the culprits to book and that local administrations were dragging their feet. Indian judges in courts dismissed many cases on the grounds of lack of evidence. Moreover, the mass movement had, in reality, not been crushed completely. Much of it went underground, where its embers continued to glow until 1944, causing immense difficulties to the administration. Once the Quit India struggle had so successfully revealed the weakness of the colonial state there could be no easy reversal.

CHAPTER FIVE

The Wilderness Years: The Congress and Underground Activity in Bihar, 1942-4

The months following the August outbreak of the Quit India movement were a period of great trial for the Congress leadership, most of whom were behind bars. By September 1942 the British had completely repressed the mass movement, the Congress organization had been banned, and all legitimate political activity had come to a standstill. However, elements of the mass movement still remained active, despite the best efforts of the British to stamp them out, in the form of several disparate terrorist organizations and dacoit gangs that had sprung up all over Bihar by 1943. These groups carried on a slow and bitter struggle, attacking a police post at one place, and raiding a government treasury at another. Shielded by groups of sullen and hostile villagers, the political activity of these underground organizations continued to be a major irritant to the British up until 1944.[1] Many of the groups that sprang up in the countryside in this period had links with the Congress Socialist party, but several other dacoit groups were also active that had no such political linkages but were instead symptomatic of the disturbed war conditions of the period and the acute economic deprivation. They formed alliances with socialist-led organizations such as the Azad Dastas and carried out their

[1] No historian has, to my knowledge, attempted to study the nature of the underground activity and its implications for the Congress organization in Bihar, or elsewhere, in this period. Most of the studies of the Quit India movement examine only the few days in August when the mass movement erupted with full force and then neglect the more significant following period. As an underground activist, Havildar Tripathi, told me in an interview in Patna in March 1986, 'The mass movement lasted for only 2 weeks in August, we carried it much beyond that.'

activities in the name of the Congress. These dacoits can perhaps best be categorized as social bandits, precisely in the sense in which Hobsbawm has tried to define them in his seminal work on banditry.[2] They roamed the countryside, some with political affiliations, some with none, but all with the support of the village populations. In the context of economic crisis they filled what was otherwise a political vacuum between 1942 and 1944.

This chapter attempts to examine the nature of underground activity in Bihar. It outlines, for the first time, the dynamics of the Congress underground movement as it emerged after the imprisonment of Gandhi and the established Congress leadership. This underground activity was to fundamentally affect the Congress organization. It also precipitated the reopening of negotiations between Gandhi and Wavell, since the Viceroy quickly realized the potential threat posed for the post-war period were such underground activity to continue unchecked.

The organized Congress in this period floated in the political doldrums. The August movement had temporarily forced it from the negotiating table, and effectively disoriented the whole Congress organization. As the political stalemate drew out longer the leadership started to view the existing impasse with increasing worry. Furthermore, the fortunes of the war were changing in favour of the Allies. The Congress was now facing a British government whose morale was considerably higher than it had been in the early months of 1942, when it was feared that the Japanese were about to overrun India.[3] Gandhi was particularly nervous. A controlled mass movement under his leadership was one thing, while a disparate underground move-

[2] E.J. Hobsbawm, *Bandits*, Harmondsworth, 1985, pp. 17-18.

[3] The strategic situation, in a global sense, for the allies on the eve of Gandhi's fast seemed far more favourable than before in at least three theatres of the war. In North Africa Rommel was retreating in disarray, chased by Montgomery's forces still eager after the victory at El Alamein. On 15 January 1943 Roosevelt and Churchill met at Casablanca to plan the last stages of the North African Campaign and looked forward to opening a new front in Italy. In Burma, Wavell contemplated the re-taking of Burma in Operation 'Anakim'. On 18 January the Red Army recaptured Voronezh and simultaneously began to raise the siege of Leningrad. Even though tensions between the Americans/British and the Soviets date from this time and Rommel in February managed to burst through the Allied lines at Kasserine, the tide of the war had fundamentally changed. Such strategic developments, with their concurrent effects on morale, could not but affect the situation *vis à vis* Gandhi in India.

ment employing terrorist tactics was quite another. Initially the Viceroy was clearly delighted to let Gandhi sweat it out in prison for the duration of the war, and to let the Congress remain out of the political reckoning. To this end the final months of 1942 were spent in proving, beyond doubt, the extent of Congress responsibility for the disturbances. For his part, Gandhi was now desperate to regain the political initiative. In February 1943 he embarked on a twenty-one day fast. The British were immediately placed on the defensive and wary of the implications of his action. Indeed they offered to release Gandhi for the duration of the fast, a clear sign that they were backing down.

While a cat-and-mouse game was being played between the Congress leadership and the British, at the popular level, the agitations and the protest struggles of August 1942 had not been completely crushed. Indeed the underground terrorist organizations actually grew stronger in remoter parts of the province. In October 1942 Thomas Stewart, the Governor of Bihar, wrote to Linlithgow about the dependence of the administration on troops, stating that:

> though we are now on top we are still relying on the support of the troops and must continue to do so for some time. Sporadic incidents and the spate of subversive pamphlets indicates that there is still more ferment below the surface. In Bihar the military implications of the outbreak have been, if anything more important than the civil.[4]

In Bihar the end of Gandhi's fast was met with a sigh of relief by the officials. They were now able to declare that the political weapon had led nowhere and had proved to be a chimera. However, by early 1943 there were disturbing reports to indicate that Congress rebels were contemplating a resumption of political activity. The continued sabotage of telegraph and telephone wires seemed to be a prelude to extensive attempts by the Congress to instigate unrest among villagers in the districts. In Muzaffarpur, Saran, Monghyr and Patna there were several jail escapes from the overflowing prisons, and a sudden increase in mail-looting and wire-cutting. In Ranchi a number of Congress bulletins, as well as ingredients for the manufacture of gunpowder, were found. All the districts reported an increase in crime. The dacoity position, which had shown an improvement in November

[4] TOP, vol. iii, no. 64.

1942, (the period when the Quit India struggle was crushed, according to official reports), started to deteriorate sharply.[5] Significantly, district officers reported that most of the dacoities in the rural areas had been committed for lack of food. As the economic situation deteriorated, so did law and order. In the uneasy conditions of 1942 and 1943 a link was forged between the political activities of the Congress underground organizations and the simmering rural discontent. This created the conditions for a widespread agitation that continued in Bihar and elsewhere all through 1943-4. This underground activity proved a major irritant to the British well into 1944, making it imperative for Wavell, the new Viceroy, to reinstate Gandhi so that a rational political solution could be worked out, and a constitutional dialogue be activated by the time the war ended.

THE ECONOMIC CONTEXT, FOOD SCARCITY AND CRIME IN BIHAR, 1943

As a prelude to examining the nature of the underground agitation some cognizance needs to be taken of the changing economic conditions prevailing in Bihar during 1943. The year 1943 saw a further intensification of the scarcity introduced by the war. Bihar was, even in normal times, a deficit province in both rice and wheat. During the four years between 1937 and 1941 average annual net imports amounted to approximately 200,000 tons of rice and 71,000 tons of wheat. In addition, the Bihar government estimated that 74,000 tons of rice in the form of paddy were received from Nepal. Other sources of rice supply included Burma (through Calcutta), Bengal, the Central Provinces, and Orissa. Wheat was imported mainly from the Punjab and United Provinces.[6] Scarcity hit Bihar in 1943 when these sources of supply were disrupted as a result of the war. The Burmese imports had already stopped in 1942 when the Japanese overran the country, while the imports from Nepal stopped only in May 1943 when the Nepal government introduced a ban on the export of paddy. This caused difficulties north of the Ganges and in several districts in Bihar.[7] Rice supplies transferred from Bengal to Bihar, also stopped

[5] FR (1), March 1943.

[6] For details of the food crisis in 1943, see K. Prasad, *The Economics of a Backward Region in a Backward Economy*, vol. 1, pp. 246-8.

[7] See Famine Enquiry Commission: Report on Bengal. See also FR (1), March 1943.

as famine conditions intensified in the Bengal districts in early 1943. Here, as the Woodhead Commission discovered, a marginal shortage was magnified a thousandfold by the heartless hoarding of traders. Inter-provincial trade suffered, of course, as a consequence.

Food yields in Bihar itself, however, were not alone responsible for the crisis.

TABLE 9

ESTIMATED YIELDS OF RICE, WHEAT, MAIZE AND BARLEY IN BIHAR 1942-3 IN '000 TONS'

Bihar	*Rice*	*Wheat*	*Maize*	*Barley*
1939-40	3,178	424	441	405
1940-1	2,134	405	393	446
1941-2	2,747	485	456	445
1942-3	3,256	581	558	416

SOURCE : Report of the Famine Commission, 1945.

Table 9 indicates that the scarcity of food in Bihar, as in Bengal in 1943, was not a result of an unusually bad harvest. The harvest was, in fact, not very different from previous years and in 1941-2 was indeed marginally better than in 1940-1. The shortage clearly resulted from a dislocation of imports into Bihar and the failure of the supply and distribution system in large parts of Bihar during 1943. The situation was exacerbated by the tendency of producers and merchants to take advantage of the price rise by hoarding grain. Those involved in the grain trade, the banias, *beoparis* and wholesale merchants, kept a vigilant eye on every turn of the season because periods of scarcity were what they thrived on. At the slightest indication of an approaching bad season, grain, which during normal seasons was disposed of at regular intervals, would be hoarded, with a view to obtaining an exorbitant profit on it. Grain supplies in the province therefore fell to very low levels and adversely affected prices in almost every district.

A heavy demand on even this slender supply was imposed by the defence services causing further shortages of food. The increased purchases by the army are clear from the following figures.

TABLE 10

ALL INDIA WHEAT PURCHASES BY THE ARMY IN TONS

1940-1	88,000
1941-2	238,000
1942-3	312,000

ALL INDIA RICE PURCHASES BY THE ARMY IN TONS

1940-1	22,000
1941-2	47,000
1942-3	115,000

SOURCE : Report of the Famine Commission, 1945.

As a large part of the three million strong army was based in Bihar and Bengal, their demands impinged heavily on the short supplies of foodgrains in these provinces.[8] Army demands also affected numbers of plough cattle. There was already a great shortage of these at the time, which seriously affected agriculture. An unusually high level of slaughter of cattle for food had taken place, giving rise to widespread rumours in Bihar that the army had slaughtered 30,000 head of cattle in a few months alone.[9] The strained supply position and the increasing demands made on even these slender provisions had a dramatic impact on prices.

In Bihar the prices in late 1943 were nearly 350 per cent of the pre-war level. The Famine Commission noted that the prices in Bihar were approximately at the same level as in Bengal. They came down marginally, however, in both provinces by 1944. Early in 1944, prices in Bihar were lower than in Bengal, but towards the end of the year they tended to equalize.[10] Demand and supply conditions alone do not suffice in explaining this high price increase. Speculation and hoarding had a crucial impact on the market and served to push prices up

[8] Unfortunately I have been unable to obtain the exact figure for the amount of foodgrains which Bihar supplied to the army.

[9] The issue was first raised at the national level in 1942 by Gandhi who, in a letter to Linlithgow, condemned the slaughter of 'numberless' cattle for British and American troops in Bihar. See Gandhi to Linlithgow, CWMG, vol. lxxvi, pp. 265-6.

[10] Report of the Famine Commission, Part 1, p. 476.

to abnormal levels. The government in Bihar attempted to check speculation by setting up trade barriers to regain control. The controls sufficed to maintain internal prices in parity with prices prevailing in the principal rice-markets outside the province. However, when prices rose abnormally in the United Provinces, it made it impossible for the Bihar government to maintain controls, and local prices rose inexorably again. By May 1943 foodgrain price increases were becoming more and more critical. In many parts of the province it was now impossible for the poor to buy rice. Frequently, even the government found it impossible to make purchases so that, for example, the Patna Camp Jail Purchasing Committee could not obtain a single contract for supplying grain to the camp jail. By mid-1943 the Commissioner of Patna noted that a declaration of famine was inevitable in the district.[11]

Neighbouring Bengal had already been declared a famine area in April 1943 and in Bihar conditions were only marginally better. In Bengal, by the summer of 1943, the rural poor were driven by lack of food to flee to the cities and, as the famine conditions worsened, corpses began to pile up in the alleyways and by-lanes of Calcutta.[12] A severe epidemic of malaria, smallpox and cholera dramatically increased the death toll, which continued to climb well into 1944. By then the great Bengal famine had killed over three million people in both urban and rural areas.[13]

In Bihar, the situation was marginally better and the death rate did not increase so much. Even so, this should not lead one to underestimate the extent of real economic distress in large parts of Bihar and the fact that, by 1944, a serious epidemic (as had affected Bengal) of malaria, cholera and smallpox had broken out in Bihar, increasing overall mortality. In any case, in rural areas the reserves of the poor

[11] FR (1), April 1943.

[12] The Famine Commission noted in perplexity that the shortage of the previously reaped aman crop in Bengal was comparable to that which occurred in 1928, 1936 and in 1941 and was in fact less serious than the 1941 shortage. However, a phenomenal rise in prices occurred in 1943 making the situation different from that which had prevailed during the earlier shortages. The Commission concluded that the acute scarcity in Bengal was due to breakdowns in distribution rather than insufficient supply. Amartya Sen's entitlement theory is based on a similar premise that the famine in 1943 was not due to an absolute decline in food availability but due to a difference in entitlements to it.

[13] For a graphic eyewitness account of the Bengal famine, see H. Mottram Papers, CSAS, Cambridge.

were so low and they were so malnourished, even in normal conditions, that any scarcity was bound to hit them very hard.[14]

All the districts reported a shortage of foodgrains exacerbated by the unwillingness of large cultivators and zamindars to take their surplus stock to the market and the tendency of big dealers to hoard.[15] As scarcity intensified, starvation became common. The main sections of the population to be affected were the poorer classes in urban and rural areas. The Fortnightly Reports increasingly mentioned near-famine conditions in several districts. For example, in Patna city the District Officer noted that 4 lakh maunds of foodgrains were necessary to carry the area through until the next rice crop. The small reserves of foodstuff that had been accumulated by the government were supplied to government servants drawing less than Rs 200 a month and to police staff, who drew a limited ration of 2.5 seers of rice or wheat and 3-3.5 seers of other grains. With such limited supplies being distributed among the service classes in the city the plight of the rural and urban poor can only be imagined.[16] The bazaars were empty and in Jamshedpur and Patna the spectacle of the half-starved populace was pitiful.

Not surprisingly, the district administrations became worried at this stage that the famine conditions might provide a fertile soil for the protest movements to flourish in. It was reported that the Congress Socialist party was carrying out 'mischievous propaganda' against the government, and trying to make people believe that the scarcity was due to purchases by the government. When a few cases of crop-looting and theft were reported, district officers began to fear a widespread outbreak of food riots if adequate measures were not taken by the government to make supplies available at retail shops, particularly in urban areas.[17] These official fears were soon substantiated. By mid-1943 all districts had reported an increase in crime. In Patna district

[14] A.K. Sen, 'Famine Mortality, a Study of the Bengal Famine of 1943' in E.J. Hobsbawm (ed.), *Peasants in History, Essays in Honour of Daniel Thorner*, footnote, p. 204. See also The Report of the Famine Enquiry Commission in Bengal, 1945, pp. 144-5.

[15] FR (1), March 1943.

[16] FR (2), June 1943.

[17] The headlines in a leading provincial daily, the *Searchlight*, on 14 February 1943 proclaimed the 'country wide consternation due to scarcity of food grains. Grain shops have been looted and there has been a *lathi* charge on a procession of starving people'.

dacoity was on the increase, particularly in Sasaram and Buxar. In Purulia, several daring dacoities were committed in May and June 1943.[18] In Manbhum sub-division, crime had increased dramatically in this period. Significantly, in 75 per cent of the theft cases, grain was stolen. Indeed, the administration found that the areas where the food deficit was the worst had the highest rate of crime. In Tirhut the worst affected areas were Saran and Bhagalpur. These were always deficit areas with a dense population, so that the food scarcity in 1943 seriously worsened the situation.

By 1943 the increase in dacoity was noticeable all over Bihar. The most important factors determining the incidence of crime in the province were the availability and price of foodgrains. This correlation between crime and grain prices was particularly clear to district officials. Fortnightly reports record a dramatic increase in lawlessness. In Bhagalpur, in Banka subdivision, gangs of 'criminals' were observed to be moving about quite freely. At one village in Patna district fifty armed men appeared on the morning of 23 June in the village rice market and attempted to steal the rice brought in for sale from the neighbouring villages. They were successfully driven off. In another village a large party of women were reported to have looted a train of pack bullocks which were carrying rice and passing through the area. Many of the dacoities seemed to be committed for food and it was found that women and children accompanied dacoits in carrying it away.[19] Soon there were even recorded cases of goods trains being held up for food looting.[20] Increased levels of thefts of copper wire along the East Indian and Bengal Nagpur Railway were complementary indicators of the extent of economic distress in this period. Copper wire procured a particularly high price in the market and stiff

[18] FR (1), May 1943.

[19] It has been found in recent studies that, with the onset of a famine, most men are driven to migrate to cities in search of food. As a consequence women are often left in the village to fend for children and the old. As famine conditions increase further it is the women who suffer the most. In Bengal in 1943 they were sold or exchanged for food. An overwhelming number of starvation deaths were found to be women, as they have always borne the brunt of poverty and malnutrition. See M. Kishwar and R. Vanita, *In Search of Answers*, p. 260. See also David Arnold, *Famine, Social Crisis and Historical Change*, London, 1989, p. 89.

[20] FR (1), & (2), June 1943.

collective fines were imposed by the authorities on the villages concerned.[21]

On close examination the crime figures during this period reveal certain interesting features. Most of them were committed either through joint actions by the poor and starving in the villages or by organized dacoit gangs. Figures for murder, robbery, cattle theft, riots, and dacoities all showed a dramatic increase in 1943 over their pre-war average. All were clear indices of the extent of economic distress. As prices rose after the war there was a steady increase in all these figures. The figures for 'true theft' (i.e. theft, cattle theft and house-breaking) increased from 13,000 in 1942 to 20,000 in 1943. The figures for cattle theft also showed a similar increase.[22] Figure 1 shows figures for theft including cattle theft in Bihar for the period 1942-4.

The figure for 'true dacoity' (technically defined as robbery by five or more persons) caused the greatest anxiety to the district officials. It had risen to dramatic heights in 1943 so that the new figures were 440 per cent above the pre-war level. Figure 2 shows the increase in dacoity figures in Bihar during 1942-4.

It can be noted from the graphs that both theft figures and dacoity figures peaked in 1943 and then declined somewhat in 1944.[23] While it is probable that the marginal price decreases of 1944 had an impact on the level of crime, these declines can best be explained by the tighter police controls imposed in pursuing the underground activists. However the fact that the movement was not completely checked is clearly demonstrated by the dacoity figures for 1944. These were still above those of 1942.

The 'dacoity' category was, however, an imprecise one and could encompass phenomena right through from popular famine-looting to the organized armed activities of professional dacoits. Famine-related crime swelled the dacoity totals in the years of bad harvests and rising prices. However in 1943 and 1944, when the district officers talked

[21] FR (1), May 1943.

[22] However, by and large, cattle theft continued to be under-reported.

[23] There was probably an under-reporting of all these crime figures as the police administration, especially in the countryside was quite weak. The figures refer to the years 1935-44 and it can be argued that the extent of under-reporting did not undergo a marked change in this period. Thus, even though the exact figures may not be very reliable, the underlying trend reflects a rising crime rate.

True Theft Including Cattle Theft

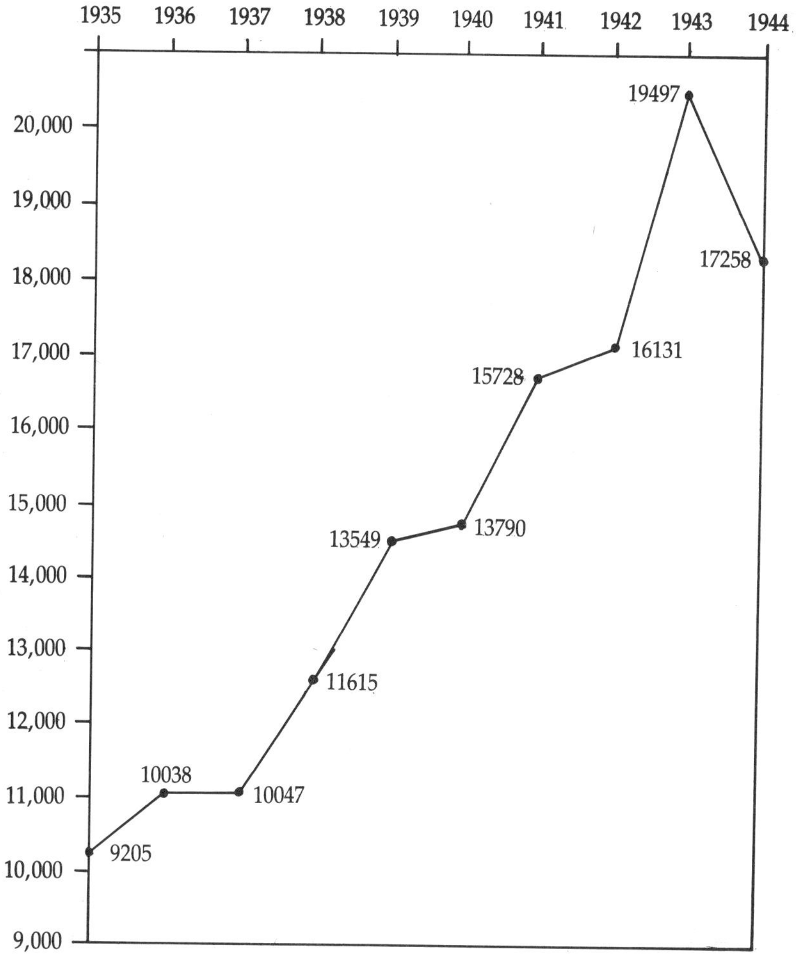

True Dacoity

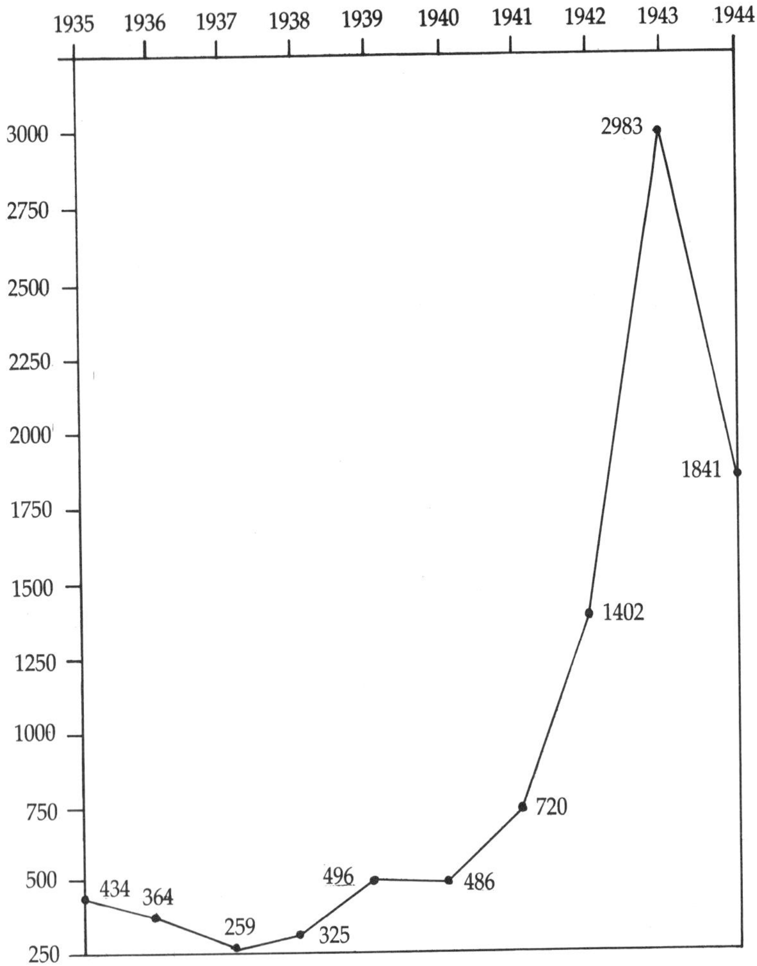

of an unprecedented increase in dacoities they tended to mean only the depredations of organized dacoit gangs. The activities of these gangs were critically different in character from food-looting by hungry villagers. Moreover, district officials were more wary of organized dacoities. As David Arnold had pointed out 'a crime so bold, violent and destructive was necessarily seen by the state as a danger not only to the security and property of rural propertied classes but also to its own authority and capacity to exercise effective control over the countryside'.[24] Organized in groups of thirty-forty members, the professional gangs drew their recruits from low-caste marginal peasants or landless labourers. In Bihar, Gope and Dusadh dacoit gangs were common. These gangs were usually armed, and made daring raids at night on villages, forcing the local landlord or bania to part with their money and grain. They were usually able to terrorize the villagers sufficiently to prevent them informing the police about their whereabouts, while informers were killed or mutilated. Sometimes the dacoit gangs obtained the covert support of villagers, as in the case of the Banka dacoits in 1943, and were able to operate with the full connivance of local people. In 1943 and 1944 the figures for dacoit 'crime' increased greatly, as they included the activities of the Azad Dasta gangs who were at the time committing 'political dacoities' designed to secure funds for running their organization.

UNDERGROUND ACTIVITY IN BIHAR 1943-4: THE AZAD DASTAS AND THE DACOIT GANGS

Apart from these growing levels of 'economic' crime, the period also saw a dramatic increase in what were known as 'political crimes'. With the Congress leadership in jail and the Congress, the CSP, and other parties under ban, an underground organization was being built up to keep the protest movement alive during this period of enforced inactivity and under conditions of state repression. This underground

[24] David Arnold, 'Crime and Crime Control in Madras, 1858-1947' in A. Yang (ed.), *Crime and Criminality in British India*, Tuscon, 1985, pp. 70, 74. Arnold tends to argue that the activities of professional dacoits were effectively suppressed by the twentieth century at least in the province of Madras. However, the situation was clearly different in northern India as my evidence for rural Bihar shows that professional banditry continued to be a problem well into the 1940s and even until today. In Banka subdivision, the activities of these dacoits was a nagging irritant to the local police administration in the countryside.

organization was dominated by the leftist groups primarily associated with the Congress Socialist party. The existence of an underground organization of the Socialists, the Forward Bloc and extremist Congress elements was mentioned in certain police reports as early as December 1942.[25] These were accompanied by growing fears in government circles that the Congress rebels were contemplating a resumption of open activity. By December 1942 there were several escapes by political prisoners from Patna, Muzaffarpur, Saran, and Monghyr jails. Soon after this a sabotage of telegraph and telephone wires occurred in several areas, followed by armed dacoit raids on ammunition depots and government treasuries.[26]

By mid-1943 these attacks had increased. Though numerous political suspects were being arrested the crimes showed no signs of abating. In Ranchi a store of gunpowder, along with a number of Congress bulletins, was found in a house. This confirmed the existence of an underground organization in the town. In Bhagalpur alone the number of dacoities suddenly climbed up to 310 in June, as against a previously recorded monthly average of 58 dacoities. In Shahabad an increase in mail-bag looting was reported. On 29 May 1943, an armed dacoity took place at a post office in Tarapur, Monghyr. The dacoits used firearms, and a chowkidar and a daffadar were injured. In other parts of north Monghyr six other dacoities were reported, including one committed to obtain arms. At Khatra police station it was reported that a daffadar had been brutally murdered, probably because of his role as an important prosecution witness to a dacoit crime. The whole village of Dhanaur was reported to be hostile and disposed to shield the offenders.[27] By August 1943 widespread rumours of a revival of civil disobedience had surfaced. One Fortnightly Report commented on the open defiance of the police by some of the Congress gangs, implicitly connecting this with the low morale of the force. Siaram Singh, one of the leaders of a socialist group, appeared in Sultanganj on 14 July before an unarmed police party consisting of a Sub-Inspector and two constables. The police made no serious effort to arrest him and allowed him and his party to leave the village unmolested and cross the Ganges. Siaram Singh then began successfully collecting subscriptions from villagers, largely on the strength

[25] See FRs (1), December 1942, (1), January 1942.
[26] FR (1), December 1942.
[27] FR (1), May 1943; (2), June 1943.

of his show of defiance.[28] The attacks of the Azad Dastas were also becoming more diversified at this time. As they began to forge alliances with the professional dacoit groups in the countryside their attacks were no longer confined only to state property. It was reported that the 'criminals' were in many cases also directly threatening to zamindars and businessmen.[29] In the Santhal Parganas the houses of rich zamindars were attacked. The dacoits set fire to those houses into which they could not force an entry, with the result that eighty head of cattle perished in one village. Interestingly, in this case, the name of Gandhi was invoked by the dacoits while carrying out the attack.[30] In Darbhanga, the zamindar's *kutcheri* was picked out for particular attention and was attacked by a mob of 500, led by CSP leader Suraj Narayan Singh. The rebels fled only on the arrival of the armed police force, which opened fire killing two men. By now all this activity had made it apparent to the provincial administration that a well-developed underground organization was behind the sporadic attacks on both state and private property.

What was the nature of this underground organization connected with the Congress, at this time? By mid-1943 some kind of organizational control had been established over the disparate underground political groups and it appears that Jayaprakash Narayan was the mastermind behind the new attempt at control. He had escaped in May 1943 from Hanumannagar prison in Nepal, where he had been incarcerated by the British since November 1942, in a daring jail breakout in which fifty underground Congress activists had broken into the prison to rescue him.[31] Once outside, he was able to direct socialist activity from his underground base in the Nepal terai region, much to the discomfiture of the British, who failed to recapture him until October 1943. In essence, Jayaprakash Narayan and other underground socialist leaders were attempting to organize the Azad Dastas as guerrilla bands to provide the main striking force of the underground movement. In the course of intelligence searches at CSP

[28] FR (1), June 1943.

[29] FR (2), August 1943.

[30] BSA (Political Special) 6, (vii)/47.

[31] FR (2), May 1943, Interview with Azad Dasta activist, Havildar Tripathi, in Patna, February 1986. His escape was a cause of much worry for the district and provincial administration at the time for he had a reputation of being an expert 'fish plate and explosive man' specializing in sabotaging troop trains. See A. Flack Papers, CSAS, Cambridge.

underground bases in Nasik and Nepal the authorities discovered two important documents containing these organizational plans. One document, called the *ABC of Dislocation*, commented on the lost opportunities of August 1942 and comprehensively revealed the future plans of the Congress Socialist party. In addition, the *ABC* document outlined the potential future role for the Azad Dastas. It argued from the starting point of one successful raid, the looting of stores at the Barauni BNW railway station. During the raid:

> the people took possession of 25,000 rifles and 30,000 rounds of ammunition. With that material a fine people's militia could have been raised and a revolutionary government organised in a few districts which would have held out against the enemy for some time. The masses will take some time to recoup their energy . . . the objective forces such as food and price situation, repression, propaganda turns in the war will also take some time to create the psychological atmosphere that would require only a spark to produce a mass explosion. We should utilise the present, firstly to remove the deficiencies that were discovered to exist at the beginning of the revolution and secondly to produce as soon as possible those factors which would again lead to a mass upheaval. One method (we should emphasise that it is only one of other methods) of doing this is to create in every district a band of shock-troopers, a sort of advance guard, who should possess the proper technical and political training and who should in the manner of guerrillas wage ceaseless war against the enemy. Let us call the bands Azad Dastas in Hindustanee. The idea of these Dastas, in the form of guerrillas is widely current. It is necessary to give concrete shape to these ideas . . . the nearest equivalents to our Azad Dastas would be European guerrillas . . . Guerrillas played an important part in the Russian Revolution and in occupied Europe and today are continuously harassing the Hitler regime. In occupied India our Azad Dastas can become the instruments of the complete paralysis and demoralisation of British rule. They can in time become the leaders of the mass insurrection and the nucleus of the revolutionary people's army of the Republic of India . . . There are 250 districts in British occupied India. In an average district (there will be) 250 Azads organised in 5 Dastas of 50 Azads, each will be further divided into 25 Dastas of 10 Azads each. Every member of the Dasta will have the sub-title Azad added to his name . . . which will be synonymous with being a member of the Azad Dasta.

This document also showed quite clearly that the Dastas were to be organized on very disciplined lines, with a strict and hierarchical authoritarian structure.[32] All Azad Dasta activists were required to take oaths of allegiance which included the following: (1) a pledge to carry on fighting the British usurper up until the establishment of a free India; (2) unquestioning loyalty to the organization and the observance of military discipline; (3) acceptance of any punishment, including death, for violation of the rules of the organization. It was also suggested that the main recruiting ground for the Dastas would be among political workers, school teachers, students and deserters from the army or police.

The organization of the Azad Dastas was gradually elaborated. Members of the lower level socialist leadership in Bihar such as Suraj Narayan Singh, Siaram Singh, Suraj Nath Chaube, and Gulabi Sonar provided the main leadership. They were particularly interested in attacking police stations and post offices, as these were considered to be state property. Soon they were able to attempt frequent gang robberies and attacks on police stations. In Shahabad in mid-1943, for example, Sonbarsa police station was attacked by a gang led by Siaram Singh himself. The gang included Nityanand Singh, ex-constable of Monghyr, who had taken part in the Hanumannagar raid and was now Siaram's chief instructor. He had deserted with his rifle from the Jamshedpur Internal Security Force. Gulabi Sonar, another CSP activist, was also reported to have taken part in the attack. An intelligence report noted that the activity of such gangs presaged an extensive programme of sabotage at a time when the military were likely to be occupied either in repelling or undertaking an invasion in the East.[33] It was also feared that many of the gangs had contacts with Bose and the INA.[34] In Bidhupur, the post office was attacked by a mob of twenty to thirty people led by an Azad Dasta activist shouting revolutionary slogans. Another post office, at Jhandaha, was also raided and burnt by twenty-five armed men. There were also several cases of cutting of telegraph wires.[35] In March 1944 Mohuiddinagar railway station in Darbhanga and Tirhut railway station were burnt

[32] Home political file on Socialist Plans, NAI, file 5/64/43.

[33] Ibid.

[34] In reality the gangs did not have any contact with Bose's INA. Many Forward Bloc members, however, also joined this underground organization and distributed pamphlets to the effect that Bose would soon arrive in India to save it. This fact was recorded at a personal interview with Forward Bloc activist Ramayan Singh in Patna, February 1986.

[35] FR (1), July 1943.

and papers destroyed. At Mohuiddinagar a mob of two hundred people attacked the station, seized the Assistant Station Master, and set fire to the station.[36]

The activities of the Azad Dasta gangs mainly took the form of raids on ammunition depots, government treasuries and other government offices. Experienced gang leaders were very important for the success of such raids, and the individual activities of some of these leaders are instructive. One of the chief activists in this period was Siaram Singh. His gang was particularly active in the Banka subdivision of Bhagalpur, where it carried out several armed robberies and attacked post offices and government armouries. As many as fifty villages in the subdivision were assisting Singh with grain, money and shelter. At one village in Budhuchak, Singh collected as much as 5,000 rupees and one hundred maunds of grain, mainly from the well-to-do in the village.[37] Referring to these activities a local district officer, B. Sivaraman reported that:

> The gang of political dacoits seems to be a group working for Siaram Singh. Most of the leading absconders are of Bhagalpur district. I am trying to keep down dacoities at the point of a revolver. Their activities usually take the form of collecting money. They have also hindered political cases by threatening witnesses and on two occasions by shooting people.[38]

Another important CSP activist, Suraj Narayan Singh of Thahar, carried out daring raids in Darbhanga, terrorizing police informers and sending threatening letters to the police. He had joined the CSP in 1940, had received Azad Dasta training at the Nepal headquarters of the CSP, and then organized Azad Dastas in Darbhanga district. Among other actions, he and his group were responsible for derailing an engine and cutting wires between Jainnagar and Khajauli in Darbhanga district. Narayan Singh then took part in the Mohuiddinagar Railway Station dacoity and arson incident and was responsible for the murder of a police officer and a chowkidar at Andama.

[36] FR (1), March 1944.

[37] BSA (political special), 95/45. The persons who paid up in Budhuchak village were the following: Tulsi Kumar-Rs 500, Buli Kumar-Rs 500, Gobind Chaudhuri-Rs 500, Ram Prasad Chaudhuri-Rs 500, Budhu Bhagat-Rs 400, Sunder Prasad Choudhuri-Rs 200. They were all either petty zamindars or rich peasants. Elsewhere in Bhagalpur, the villagers of Khartik and Telgi were also paying Siaram Singh money and sheltering him from the law.

[38] Ibid.

He was often in confrontation with the police and, in two leaflets addressed to the Superintendent of Police at Darbhanga, threatened dire consequences to officers harassing his family. See following illustration.

In pursuit of his activities Narayan Singh convened several secret meetings along with Gulabi Sonar, Ram Lochan Singh and other socialists. At one of them, in Damodarpur village on 30 November 1943, Singh was placed in charge of Madhubani subdivision. The meeting also decided to collect money and grain for party activities and to endeavour to increase party membership.[39] Individual successes on the part of some activists served to reinforce the Dasta organization. An intelligence report revealed that at one secret meeting of the inner circle of the party it was decided to organize the Azad Dastas with an active reserve branch. The aim was to organize an effective force in each thana. It was decided that these should be three times as numerous as the local police force to ensure tactical success.[40]

The Azad Dasta gangs did, however, face a constant battle for arms and money. While procuring firearms always remained a serious problem, the Dasta organization quickly mastered the art of indigenous bomb-making. In Khatra when one CSP activist, Ram Bhugawan Singh, was arrested, subsequent searches revealed Azad Dasta literature, several country-made bombs, gelignite cartridges and explosive material. Arms were procured by raiding government armouries and, in one case in Monghyr, by securing the loyalty of a Monghyr armoury employee. Guns were also smuggled across the border from Afghanistan.[41] [For the ways in which arms were distributed, see pp. 267 and 268.]

As the organization developed, funding became a constant problem. However, the sympathy of the local populace, including the gentry, to the movement (at least in the beginning) ensured some kind of financial help. As the main targets of attack by the Azad Dasta leadership were government institutions, (the post office, government treasuries, and railway stations), local zamindars and the gentry tended not to feel directly threatened. Indeed, they often surreptitiously helped the movement financially, hoping thereby to avoid becom-

[39] CID, File no. 107/44. See also BSA (political Special), 6/vii/47.

[40] FR (1), March 1944.

[41] Interview with Azad Dasta activist, Havildar Tripathi, Patna, February 1947. This fact was also corroborated by another underground Azad Dasta worker, Maheshwar Gope, interviewed in Patna in February 1987.

Threatening Letter from Azad Dasta Activist, Suraj Narayan Singh, in 1943 and Official Translation

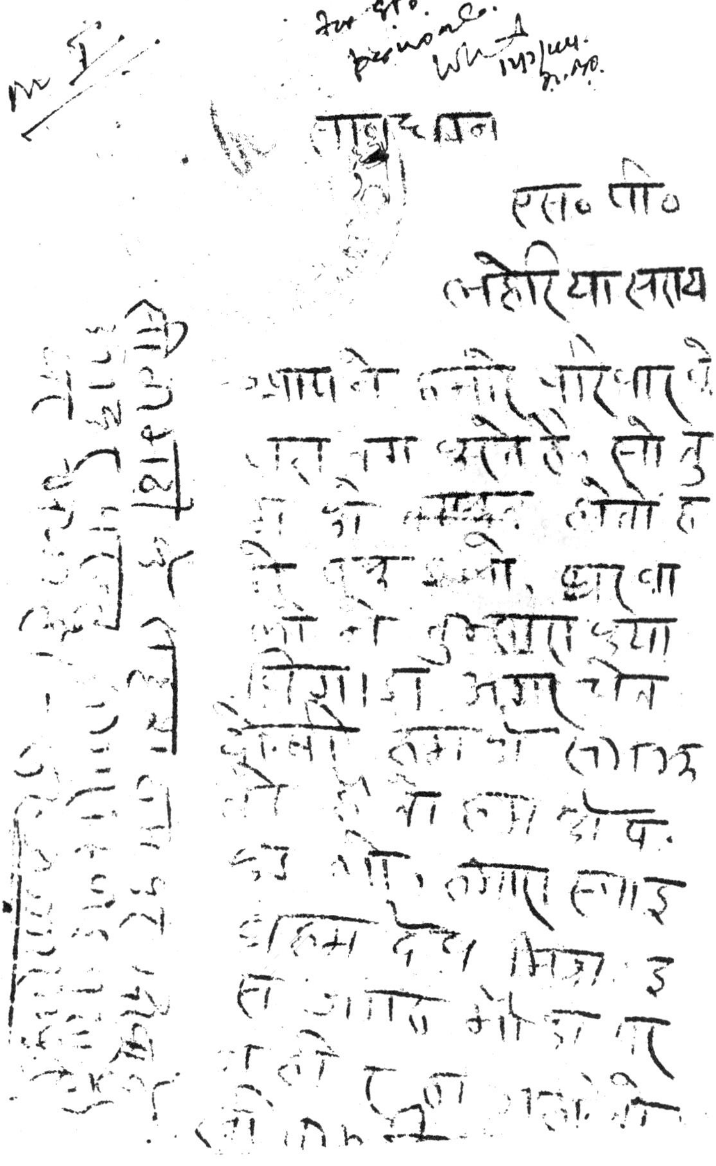

Translation of the attached notice emanating from Vindbasani Singh, Hajipur to Ramgati Singh.

Member Notice.

India Free. Be ready 3. Long Live Inquilab.

My members should know that whatever materials are ready with me being prepared from their wealth are imperatively necessary to be distributed in 4 districts to exhibit something to the white skin District Magistrates. So this is my appeal to you which member will do his work in which district. The Jap. are advancing and so I will be helped. It is essential to distribute 200 cartridges in each district. This is the time to show something to the white skins.

Vindbasani Singh,
Hajipur.

Babu Ramgati Singh
Vill. Manjhi
Mubarakpur.

[illegible] 20/7

ing targets of attack. The Muzaffarpur intelligence report revealed that Jayaprakash himself had visited Muzaffarpur with the object of raising money from the local zamindars for 'subversive campaigns'. When two leading CSP absconders, Sureshwar Pathak and Ram Swarup Sharma, were arrested in Monghyr they admitted that they had been able to get all the money they wanted for Congress work from the leading gentry of the district.[42] In May 1943 Parsuram Singh, one of the leaders of an Azad Dasta gang, was arrested whereupon another Azad Dasta leader, Siaram Singh, was able to raise money from the rich marwaris in the towns for his defence. Apart from these sources of finance, in many villages the Azad Dastas were provided with money, grain and shelter. In Shahabad, in particular, there was clear evidence that the Azad Dasta gangs were subsisting on money and grain provided by villagers.[43] The police could only very rarely guess the whereabouts of the men hidden in the villages, in the face of the hostility of local inhabitants. However, despite such practical and financial support, regular funds for underground party work continued to be a problem and some district-level directorates complained of being bankrupt.[44]

The Congress Socialist party attempted to maintain some kind of organizational control over the Dastas by arranging regular meetings and submitting regular reports. Secret meetings of the party were held constantly. On 22 July 1943, the sub-divisional officer at Muzaffarpur reported that a meeting of the underground CSP had been held in Patna under CSP activist, Suraj Narayan Singh, and that the latter had then left for Bombay to attend an important meeting with senior members of the party such as Achyut Patwardhan, Aruna Asaf Ali and others. Such liaisons continued into 1943 and into the next year. An intelligence report in July 1944 recorded a secret meeting of members of the party at Mahua. This meeting was attended by Bindhasan Singh, Bechan Rai and A.P. Verma, and made the decision to attack the police camp at Chaksaid.[45] Orders issued by the provincial directorate

[42] Havildar Tripathi in his interview revealed that he had often sought shelter in his underground days at the house of the Zamindar of Surajpura, Uday Raj Sinha.

[43] Interview with Havildar Tripathi. He records that often the villagers would keep watch over their hiding places, to warn the activists in case of a police raid. See also FR (1), September 1943 and BSA Home Political, File no. 143(340)44 and 143(350)44.

[44] Report for Bhagalpur district, 8 May to 23 May 1943, CID, File no. 22/43.

[45] CID, File no. 107/44.

Leaflet Circulated by an Azad Dasta Activist Regarding the Distribution of Arms and Ammunition in the Districts of Bihar and Official Translation

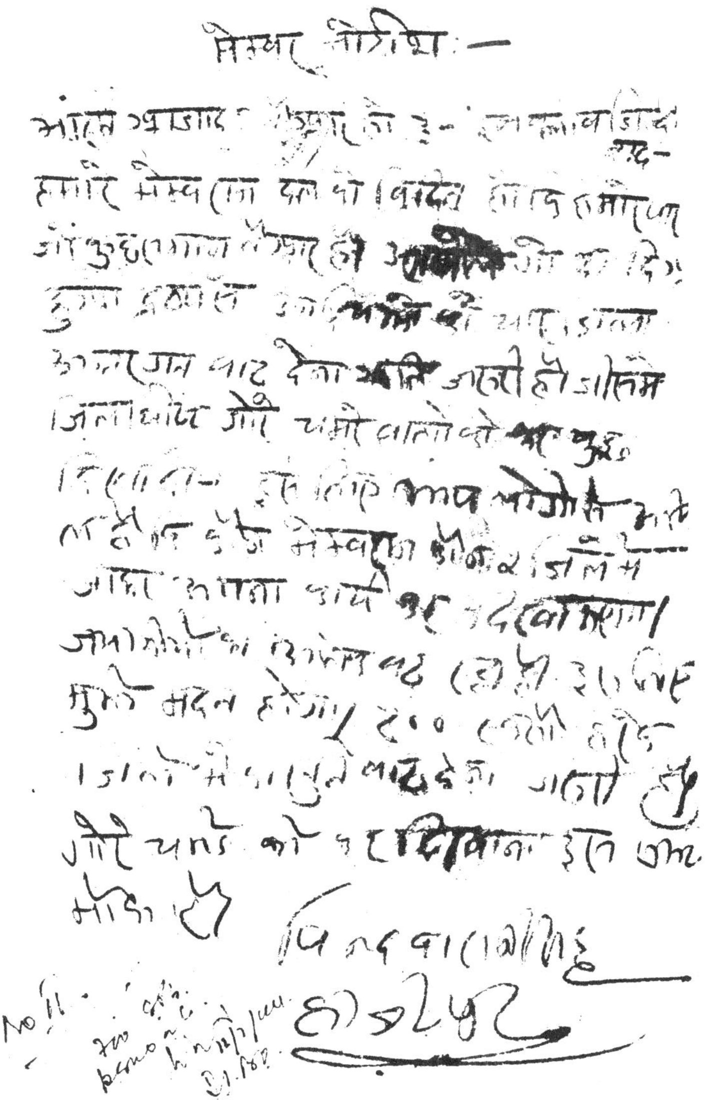

Translation of the attached letter addressed to S.P. Darbhanga.

21

Beware!

S.P. Laheria Sarai,

You are very much vexing my family members. If you have got powers, arrest me. What ill my family members have done to you. Beware. If you think me defaulter, arrest me. My spy Brahmdeo Mishra is not present here at this time otherwise I would have done like Sitamarhi. But my spy is not here. He has gone home.

Sd/- Illegible.
(Suraj Narain Singh)

K Prasad
2/7

were intended to be conveyed to the sub-divisional captain, the district captain and then to the many local Azad Dasta captains in each district. Contact was also maintained with such individuals in the national CSP leadership as Aruna Asaf Ali, who visited Bihar several times in secret. In 1943, K. Gopalan, the CSP leader from Kerala, attended a guerrilla training camp in Bihar. In March 1943 he joined the Provincial District Committee of the Azad Dasta organization. When the CSP plans for sending jathas to Poona were made known, he toured the countryside and succeeded in recruiting about a dozen volunteers for despatch to Poona. He was also in close touch with the official head of the CSP in Bihar, who at this period was Suraj Nath Chaube.[46] With Jayaprakash Narayan being constantly in and out of jail, the main orders for underground activity actually emanated from Chaube. The CSP also attempted to maintain control by issuing socialist party leaflets. These urged the people to persevere and to continue the struggle. Copies of the 'inflammatory' periodical, *Inquilab*, and other pamphlets edited by national level CSP leaders, Aruna Asaf Ali and Achyut Patwardhan, were also found in Patna. [See pp. 270 and 271][47]

However, the control of the CSP over these organizations was gradually being threatened as the Dastas began establishing links with the professional dacoit gangs active in the area. The nature of these links was highly unusual or novel. The Azad Dasta men, on the one hand, were usually members of the CSP or other extremist groups with definite political aims. Their peasant dacoit partners, in contrast, were men such as Sanjiwan Dhanuk and Mahendra Gope who had emerged straight from village society as political products of the prevailing conditions of acute material deprivation. In terms of background, then, these uniting groups had, on the face of it, little in common.

As outlined above peasant dacoities had increased dramatically during 1943 and crime was rife. It was in this context that peasant dacoits such as Sanjiwan Dhanuk and Gope carried on their depredations in, for example, Banka subdivision.

In north Bhagalpur it was noted by officials that the dramatic increase in dacoities was due to the activities of semi-political dacoit

[46] BSA (political special), File no. 331/1943.

[47] CID, File no. 107/44. These leaflets were also important vehicles of propaganda. Leaflets issued by the local Azad Dastas in 1944 gave credence to wild rumours and claimed that Bose was about to enter India at the head of a victorious Indian National Army.

Proscribed Socialist Party Leaflets in 1943

ISSUED FORTNIGHTLY

No. 9] Saturday, 25th Dec, 1943 [Four As.

TEHERAN TANGLE

There will be little disappointment at the rather brusque reticence of the Teheran communique, though a few tears may be shed over its ideational barrenness. The Allied leaders have come to believe in a sort of plainness set in Basic English. It is good that the world is not inflated with empty promises. There was no attempt at Teheran even to annotate further the Atlantic Charter already annotated into threadbare nothingness. 'As to the peace we are sure of the supreme responsibility resting upon us' say the leaders and the world has to leave it at that. It is easier to 'seek the co-operation and active participation of all nations, large and small,

महायुद्ध का स्वरूप

"एक कांग्रेस समाजवादी"

बिहार सोशलिस्ट पार्टी

पटना

मूल्य एक आना।

Map of Bhagalpur District Showing the Rugged Terrain of Banka Subdivision Where the Dacoit Gangs were Most Active in 1943

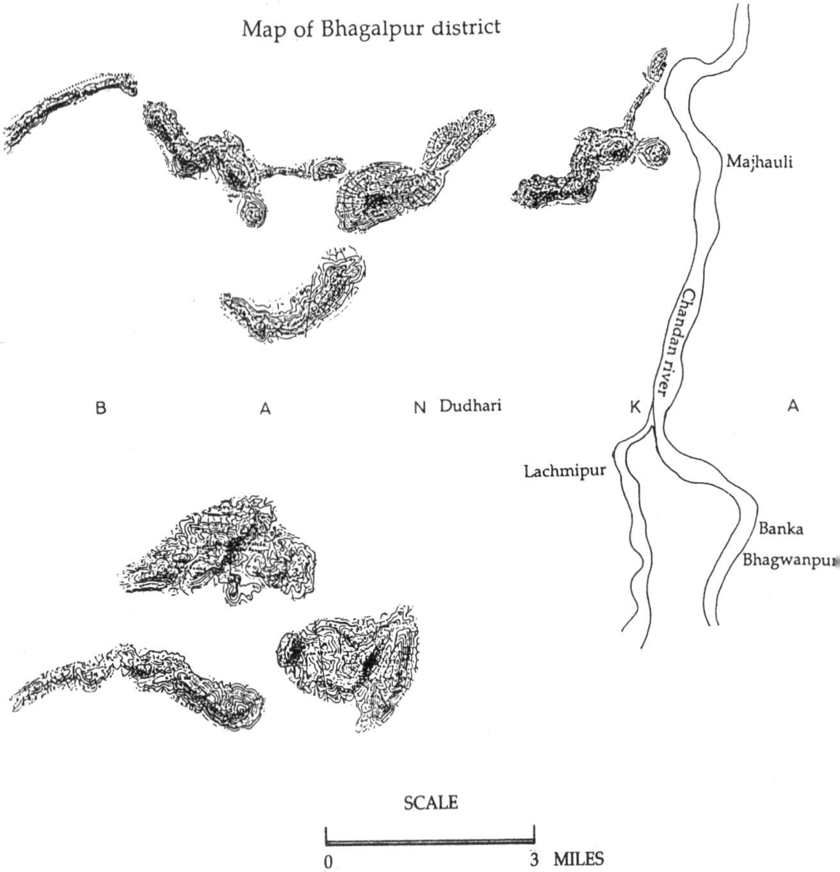

groups where 'the basin of the Kosi river provided a fertile ground for their depredations'.[48] This was, of course, a concise and accurate reference to the background of ecological marginalization which had been so decisive to popular action in the Kosi in the 1930s. Indeed the ecological inequities of the region now provided an appropriate setting for Azad Dasta activity.

As the links between the Azad Dastas and the more 'indigenous' dacoits increased it became difficult to distinguish between the two. As one district officer complained, 'It is impossible to distinguish between political and ordinary crime. Politicians become dacoits and dacoits politicians.'[49] In fact, of course, many of the dacoities were not motivated by political aims and, as one observer commented, 'many of the dacoits have no political connection and even if there had been no insurrection last August . . . there would still have been a large increase in dacoities caused by economic distress.'[50] Mahendra Gope was a case in point. He was an ordinary dacoit, according to police records, who had committed several crimes by looting the houses of the rich and well-to-do in Banka subdivision. Yet he carried out these crimes in the name of the Congress. His activities included raiding government treasuries and occasionally looting the houses of the rich. He was finally arrested while carrying out an armed dacoity, along with twenty-five other men, in the house of a rich zamindar in Patna district.[51] The duality of Gope's role was a striking one and quite unique to the period.

Some historians of crime have pointed out that 'banditry and related crimes are seldom consciously motivated by a desire to protest against social injustice and political oppression'.[52] They do not equate dacoits with the social bandits described by Hobsbawm. The case of Gope, in particular, may serve to discredit such a notion. He was, understandably, both conventional dacoit and social bandit. A low-caste peasant, he was clearly a popular figure among the rural population, who often gave him food and shelter. 'The point about social bandits', Hobsbawm has observed, 'is that they are peasant outlaws

[48] Ibid.
[49] FR (1), January 1944.
[50] Ibid.
[51] Fortnightly report (2), July 1945.
[52] See D.A. Arnold, 'Crime in Madras' and John R. Maclane, 'Bengali Bandits, Police and Landlords after the Permanent Settlement' in A. Yang (ed.), *Crime and Criminality in British India*, pp. 71 and 44.

whom the lord and state regard as criminals, but they remain within peasant society and are considered by their people as heroes, champions, avengers, fighters for justice, perhaps even leaders of liberation, and are admired, helped and supported.'[53] When the Congress movement went underground in 1942-4, it was forced to ally with Gope and other peasant dacoits like him. Gope reciprocated by committing many of his dacoities in 1943-4 in the name of the Congress.[54] It may be suggested that the fact that the Congress (that is, in its more extremist CSP wing) in this period of underground activity implemented its post-1942 policy of guerrilla-type struggle by allying with some peasant dacoits in the countryside had far-reaching effects on the nature of the Congress organization in this period.[55] The political alliance forged between the Azad Dastas and the dacoit gangs was in every sense motivated by a conscious desire to protest against social injustice and political oppression.

What was the nature of these peasant decoit groups and their links with the CSP-dominated Azad Dastas?[56] Gope's gang was dominated by low-caste landless peasants, and some of their names are known. They included Phuleshwar Gope, Birari Gope, Polish Gope, Baisakhi Rai, Gaurab Rai, Jagdishwar Baitha and others. Birari Gope and Jagdishwar Baitha, both low-caste peasants, had no income when they joined the gang and no Congress history before 1943. Since then they had carried out numerous armed dacoities and murders of police

[53] First, in his book on *Primitive Rebels* and subsequently in his analysis in *Bandits*, Hobsbawm had brought 'social bandits' to the attention of social historians.

[54] Again, as Hobsbawm notes, movements of tribal or national resistance may develop a characteristic interplay between bandit guerrillas and populist or millennial sectarianism. In 1929 Mao's red army seems to have been composed of such elements. The communist peasant guerrillas of Colombia also contained many fighters who had formerly been freebooting brigands. See Hobsbawm, *Bandits*, p. 106.

[55] FR (1), June 1945. When Rajendra Prasad was released in 1945, he was quick to distance the Congress from Gope's activities. At one stage he made the extraordinary allegation that 30-40 dacoits from Banka had been released by the SDO without justification in September 1942 deliberately to besmirch the Congress. See BSA Political Special, File no. 589/45.

[56] Yang has pointed out that, though banditry was a ubiquitous phenomenon in many parts of South Asia, little firm evidence has been mobilized to confirm the presence or absence of social banditry. The problem arises because the activities of 'criminals' and their ties to social movements are not readily discernible to the historian. A. Yang, *Crime and Criminality*, p. 22.

collaborators in the name of the Congress.[57] Baisakhi Rai (charged with kidnapping, murder and having taken part in the Amarpur armed dacoity, kidnapping a police spy) had no previous Congress attachment either. Bansi Singh, also an associate of Mahendra Gope, had similarly had no political experience before 1942. A poor peasant, with less than 2 bighas of land, he became one of the most feared Banka dacoits. Apart from conducting raids on police posts and government treasuries the activities of these semi-political gangs also involved raids on the houses of landlords and traders in Banka subdivision. Another dacoit gang which established links with the Congress during this period consisted of a group of low-caste peasant dacoits, Garbhu Dusadh, Madho Pande, Dewa Singh, and Gurcharan Musahar, all of whom belonged to the three villages of Mobarakpur, Jalalpur, Hanumangarh.[58]

Perhaps the most significant point to notice here is that most of gang members of Gope's gangs had had no links with the Congress before 1942-3. They were all low-caste landless peasants who took to crime in conditions of acute economic distress. At least part of their 'criminal activity' can be construed as political action. Such a perception of the activities of dacoit gangs in the 1940s does not imply, of course, that all forms of criminal activity can be construed as political action, nor can all offenders be equated with either nascent or mature political activists. As some Marxist historians have pointed out, there have been many ordinary criminals who have committed crime without qualification: thieves, robbers, highwaymen, forgers, arsonists and murderers. But there were also many 'good criminals' such as rioters, smugglers, poachers and primitive rebels in industry who were premature revolutionaries and forerunners of popular movements. Their acts have been categorized as 'social crime'.[59] The activities of such 'criminals' and their ties with popular movements are less readily discernible and are only now beginning to be analysed. Certainly the activities of the dacoit gangs in Bihar, with their links with the Azad Dastas, can be said to belong to this terrain of 'social crime'.

The uneasy alliance between the Congress socialists and the dacoits was soon to be beset by problems, as these gangs did not confine themselves to attacking government treasuries, but also fre-

[57] BSA Political Special, File nos. 39/45 and 42/45.
[58] BSA Political Special, File nos. 39/45 and 42/45.
[59] E.J. Hobsbawm, *Bandits*, p. 18.

quently attacked petty zamindars and small businessmen. Above all, the activities of the dacoit gangs frequently challenged property. What was more, their activities were generally characterized by clear gestures of defiance of all forms of control, including that which was sought by the CSP-led Azad Dasta leadership. Ultimately such defiance was unacceptable to the Azad Dasta leadership, which wanted to avoid alienating the propertied sections of the population from the Congress cause, and still functioned within the broad Congress parameters of class harmony. The opening-up of the Congress organization to the peasant dacoit gangs thus strained the Congress polity and gradually forced the CSP to distance itself from such dacoit activity.[60]

As the dacoities increased the rural police became more and demoralized. Thus one finds the Commissioner of Bhagalpur writing in some distress to the Chief Secretary about the situation in July 1943:

> The situation in Banka subdivision is still far from normal . . . The absconders Mahendra Gope, Sridhar Singh and Bubneshwar Missir are still at large. Now Siaram Singh has reappeared in Sultanganj. The conduct of the officers who arrested him and then allowed him to escape is terrible. The number of dacoities is staggering.[61]

In Bhagalpur alone 632 special reports and 537 dacoities were recorded. In what constituted a peculiarly symbolic step back into the early colonial past, the state resorted to the Criminal Tribes Act. Originally this had been an administrative device used to render certain castes criminal. Now it was used in a slightly different way, to control political dacoity. In 1944, 4,133 persons were registered under the act and two gangs, that of Bhuwan Ajodhya Dusadh of Muzaffarpur and Darbhanga and Marai Nonia's gang of Saran, were declared criminal tribes.[62] Clearly, a blanket criminalization of particular caste groups offered great security advantages. After April 1943 the police had quite simply been unable to catch dacoits, especially in Banka district. Of course, the use of the Criminal Tribes Act further transformed the invidious position of the police in the eyes of the dacoits. As a result those chowkidars seen as police informers were repeatedly harassed. Other chowkidars were already simply in

[60] Maheswar Gope recorded in his interview with me that in south Bihar the activities of these gangs were completely out of control and many landlords in Bhagalpur were looted.

[61] BSA, File no. 95/43.

[62] Bihar Police Administration Report, 1944.

league with the dacoits. In April 1943 members of the rural police in Banka subdivision were terrorized by threats made by a 45-member dacoit gang led by Siaram Singh. In north Bhagalpur a chowkidar who was the prosecution witness in a case connected with the looting of Bihpur was murdered while, in the Santhal Parganas, the chowkidar of one village was attacked on three instances. The district officers in most of these districts naturally reported a growing demoralization in the police, especially amongst the lower ranks. This demoralization, and even uncertain loyalty, meant in turn that even fewer dacoits were arrested. Even when Siaram Singh was arrested his subsequent easy escape was almost predictable. Rutherford, the Governor of Bihar, wrote to Linlithgow in this connection that, 'I am disappointed with the progress in cleaning up the political and professional dacoits in Bhagalpur and Monghyr. Since the drive in Banka sub-division I am contemplating putting in a large force of police, mounted and military, and charging the cost of them on the worst villages in the area'.[63] Another more extreme option suggested by Rutherford to deny the dacoits shelter was to place all the relatives of the dacoits in concentration camps. This idea was soon given up as impractical. However, it reveals how completely the British were at a loss in dealing with the situation that prevailed in 1943 to 1944.

The colonial power was, in fact, exceptionally dependent on the police as an agency of surveillance and coercion. Despite this, the resources the regime was prepared to allocate to the police service remained very restricted. The village police, (the chowkidars and the daffadars), were all weak links in the chain of administration. Recruited from the lowest castes (the Dusadhs) they were frequently so badly paid as to positively encourage links with criminals. Indeed, some rural police had long been actively involved in crime. As one Police Administration Report put it: 'They are the agency through which information regarding the interior is collected and on their reports so much that is vital to the well-being of the masses rests. Yet with rare exceptions they are totally illiterate, they are drawn from the lowest castes and consequently command very little respect in their villages, and are paid so badly that they are bound to indulge in other activities or to resort to corruption to live.'[64] Over 527 chowkidars

[63] See Rutherford to Linlithgow, 5 March 1943 in Linlithgow Papers, Mss Eur F 125/4 and BSA, File no. 95/43.

[64] Bihar Police Administration Report, 1944, p. 7.

were dismissed in Bihar in 1944, amounting to nearly 1 per cent of the total force. Given this inherent bankruptcy within the rural policing system many of the criminals had, effectively, been allowed a free reign well into 1944. In contrast to the mass movement of August 1942, however, the flourishing of sporadic guerrilla-type activities did not in fact pose any overall challenge to the British. Nonetheless, the dacoits successfully harassed the district administration well into 1944 and thus highlighted the weakness of the colonial state in the disturbed conditions following 1942.

By that year severe problems had been created by the dacoit and underground groups for the nationalist movements. The influx of ordinary dacoits and other elements into the Congress organization in this period seems to have completely changed the nature of the organization. As Havildar Tripathi has stated, only the leadership were established Congressites, the rank and file being mainly new recruits who had joined the Azad Dasta movement during 1943.[65] By 1944 the CSP leadership found it difficult, in some areas, to maintain organizational control. While in north Bihar the CSP was still able to maintain some influence over events from its base in the Nepal tarai area, in south Bihar the party had very little control and gangs like those of Mahendra Gope and Sanjiwan Dhanuk held easy sway. Very often contradictory orders were issued, and fights broke out between different Azad Dastas.[66] Given such a chaotic situation the CSP leadership became increasingly embarrassed by its links with ordinary dacoits like Mahendra Gope. After a particularly gruesome robbery at the house of a wealthy zamindar by Gope's gang in mid-1944, the CSP leadership began vigorously dissociating themselves from him. Thus a CSP leaflet in June 1944 contained a strong appeal by party activist Siaram Singh to the people not to support 'gangsters like Mahendra Gope and Sanjiwan Dhanuk for committing murders and dacoities in the name of the Congress'. The leaflet even insinuated that the government had released eighty dacoits, including Gope, from jail in 1942, with the express purpose of giving the Congress a bad name.[67] The CSP found it difficult to control the dacoit gangs, or to draw the line between robberies intended simply to supply the movement with funds and those which were just unorganized freebooting.

[65] Oral interview with Havildar Tripathi.

[66] Interview with Havildar Tripathi. He records that there were many cases of gang warfare in 1944 and killings of opposed gang members.

[67] CID, File no. 107/44.

As a result, the socialists found themselves sanctioning a series of uncontrolled dacoities on the houses of local landlords and traders, conducted in the name of the Congress.[68]

As reports of this underground activity filtered through to Gandhi and the right-wing Congress leadership in prison, they viewed the course of events with increasing worry and dissatisfaction. Bihar was not an isolated case. A summary survey of the period 1942-4 in provinces such as UP, Maharashtra and Gujarat reveals that the bulk of the underground activists refused to surrender even as late as August 1944. In eastern UP the situation was very similar to that in Bihar and socialist activity continued unhindered up to 1944, while in Gujarat the underground activists succeeded in establishing a very successful parallel government in 1943.[69] In western Maharashtra, as Gail Omvedt has noted, guerrilla activity continued all through 1943-4. In Satara district a parallel government or *prati sarkar* was set up that continued to function until 1946 in spite of British repression and the indifference of the Congress leadership.[70] Such activities occurring simultaneously in different provinces could not go uncontrolled for long.

A mass movement under an organized Congress leadership was one thing, while a disparate underground movement descending into violence in alliance with peasant dacoit gangs was quite another. Gandhi was also especially critical of the way in which Aruna Asaf Ali, Jayaprakash Narayan and others in the CSP leadership were running the movement. When faced with the argument that their

[68] Hobsbawm has noted the use of the term 'expropriation' as a tactful name for robberies designed to supply the revolutionary movement with funds. In Russia, Lenin did his best to fence off expropriations from unorganized free-booting with an elaborate system of regulations. They were to be conducted only under organized party auspices and in a framework of socialist ideology, in order not to degenerate into ordinary crime. See *Bandits*, p. 110.

[69] See David Hardiman, 'The Quit India Movement in Gujerat' in Gyan Pandey (ed.), *The Indian Nation in 1942*, pp. 102-3.

[70] In Maharashtra the movement followed a different course. Omvedt describes how, in Satara, the *prati sarkar* acted against the dacoit gangs infesting the rugged mountain terrain. This situation was quite unlike that which prevailed in Bihar where the Azad Dasta forged links with the local dacoits. One reason for this difference may have been the greater degree of control exercised over the underground movement in Satara by the underground Congress leadership. See Gail Omvedt, 'The Satara *prati sarkar*', in Gyan Pandey (ed.), *The Indian Nation in 1942*, p. 224.

tactics of sabotage and terrorism were necessary to its success Gandhi commented:

It is an old argument. One used to hear it in old days in defence of terrorism. Sabotage is a form of violence . . . sabotage activity has retarded the country's freedom. I have the highest admiration for the courage, patriotism and spirit of self-sacrifice of people, say, like Jayaprakash Narayan. But Jayaprakash cannot be my ideal. If they (the people) had responded to all the steps that had to be taken for the effective organization of non-violence and carried out in their fullness the various items of the eighteenfold constructive programme, our movement would have taken us to our goal. But today our minds are confused because our constructive work is so weak.[71]

For Gandhi it had become even more important now to secure his release and save the Congress movement from the abyss into which it had fallen. His views coincided with those of the new Viceroy, Wavell, who clearly realized the dangers of an underground Congress agitation continuing without any control and creating immense problems for any state, let alone the British, once the war was over.

In May 1944, Wavell started to press for Gandhi's release, realizing the dangers inherent in continuing to keep him in prison. The Viceroy was extremely critical of the policies followed by his predecessor, Linlithgow, coupled with Churchill's policy of stonewalling on the Indian problem.[72] With the war coming to an end he felt that the British would be faced with a mass upheaval and would urgently need a sober Gandhi with whom to reopen negotiations. Wavell felt that:

India is quiet now and can be kept so until the Japanese war ends . . . At the end of the war we will be faced immediately with demobilisation, dispersal of labour from war factories and the winding up of vast wartime clerical establishments . . . There will be no improvement in the food position and other economic troubles. Our political prisoners will have to be released and

[71] CWMG, vol. lxxvii, p. 268.

[72] See Lord Wavell, *The Viceroy's Journal*, Oxford, 1973, p. 33. Linlithgow's political acumen was in distinct contrast to Wavell's, who clearly saw the writing on the wall. When Wavell met him in October 1943 on the eve of taking over as Viceroy, Linlithgow told Wavell to proceed cautiously as he was quite certain that Britain would continue to have responsibility for India for at least another thirty years. Wavell on the other hand was more sympathetic to nationalist aspirations. See also D.A. Low, *Lion Rampant, Essays in the Study of British Imperialism*, London, 1973, p. 149.

> an explosive material will be created . . . All this will cause unemployment and discontent. I am clear that the post-war period is one of great difficulty and danger and unless we can beforehand provide an outlet for the political and administrative energy of the educated Indian they will range themselves against us . . . A positive move is essential and in view of the rapidly approaching defeat of Germany it should be made as soon as possible.[73]

He further argued that the provincial administration, the governors, the civil service and most rational political opinion were all in favour of making such a move. In line with this idea he had looked for a suitable opportunity to release Gandhi, which was provided when Gandhi fell ill in May 1944. Accordingly he cabled the Secretary of State to release him on medical grounds.[74] When, soon after, Wavell reopened negotiations with Gandhi, Churchill was furious, considering that, 'He [Wavell] has no right to negotiate with Gandhi at all, considering he was responsible for passing to us the medical opinion in which we were told that Gandhi would never be able to take part in politics again . . . Wavell has no right to enter into correspondence with him which cannot fail to wear the aspect of a great parley between the Viceroy and newly-released invalid'.[75]

Gandhi did not disappoint the Viceroy. Soon after his release he devoted time to re-establishing control over the Congress movement. He now unequivocally condemned the violence in the movement and asserted that 'our success has been mathematically proportionate to the extent to which we have adhered to truth and non-violence'. Secrecy was 'a sin and a symptom of violence' and all underground activity was taboo.[76] He sent word to the underground workers to meet him at their own risk. Several of them accordingly came and saw him at Juhu and later at Panchgani. They included R.R. Diwakar, Achyut Patwardhan, Aruna Asaf Ali and others. Gandhi's advice to them was

[73] Wavell to the Secretary of State, IOL, file L/PO/6/108B. Wavell was also exceedingly critical of the way in which the Home government had dealt with his demands for food during the Bengal famine. In a firm letter to Churchill, he wrote, 'I feel that the vital problems of India are being treated by His Majesty's government with neglect even sometimes with hostility and contempt.' See Wavell, *The Viceroy's Journal*, p. 95.

[74] See CWMG, vol. lxxvii, p. 262.

[75] *Transfer of Power*, vol. iv, pp. 1138, 1191.

[76] This marked a real change in his policy since August 1942, when he had implicitly sanctioned all forms of mass action, including general strikes, stoppages of railways and granted the possibility of civil war.

emphatic; they were to come out into the open. 'To the extent to which secrecy has crept in, it has hurt our cause . . . it is only by eschewing all secrecy and working openly that you can advance . . . today you may be two, tomorrow you will be twenty, if you come out into the open, and so the struggle will go on gathering momentum from day to day.'[77] Gandhi also spelled out a new revised policy of conciliation, by declaring in an interview with Stewart Gelder that he had no intention of offering civil disobedience and that he would be satisfied with a national government in full control only of civil administration. So far as military operations were concerned, he stated that the Viceroy and the Commander-in-chief would have complete control.[78] This was a dramatic shift from the position Gandhi had held in 1942. He had, after all, launched the Quit India movement on the issue of Indian control over defence and the right of non-participation in an imperialist war. All this seems to have changed in the new context of 1944. The war was turning in favour of the allies and Gandhi realized that a non-compromising stance, of the type taken up in 1942, would yield no results. Along with the Congress leadership he would simply spend the duration of the war in prison. Now the end of the war was in sight and the Muslim League was rapidly building up its strength, having had a period of unrestricted influence at the centre while the Congress remained under a ban. There was, too, an urgent need to resuscitate the Congress and to reclaim a place at the negotiating table. Gandhi was astute enough to realize all this and his position in 1944 differed accordingly. His conciliatory statement was soon followed by the gradual release of the remaining members of the Congress working committee and the lifting of the ban on the Congress organization.

The process of reorganization of the Congress in the localities was to be a difficult one as, in many ways, the phase of underground activity had meant that many new and unwieldy elements had entered the Congress. Essentially the dismantling of the underground organization of the Azad Dastas occurred during 1944, and many of the prominent leaders of dacoit gangs were quickly arrested. Immediately after his release Rajendra Prasad issued a press statement distancing

[77] Out of deference to Gandhi, Achyut Patwardhan decided to withdraw from underground activity. But Aruna preferred to stay underground and emerged only when the warrant against her was cancelled in 1946. CWMG, vol. lxxvii, See appendix xiv, p. 468.

[78] CWMG, vol. lxxvii, p. 350.

the Congress from the activities of the dacoit gangs and their leaders. The old Congress leadership began gradually to assert itself. A new relationship with the Congress Socialist party, which had dominated the underground organizations, had to be worked out. In practice, however, this meant that the Congress Socialist party in Bihar was to find itself increasingly peripheralized by the late 1940s. The extremist members of the party in Bihar were also soon to feel threatened by the 'clean-up operation'. The return to the status quo ante was to be difficult but by no means impossible. It so emerged that the right-wing leadership in Bihar was indeed strong enough to regain the initiative *vis-à-vis* not only the British, but also in relation to its own party organization. By 1946, once the crucial negotiations for the transfer of power were underway and Congress ministries had been elected by popular vote in Bihar and elsewhere, it had become clear that the old guard was firmly back in control.

CHAPTER SIX

Communal Rioting, Agrarian Unrest and the Second Congress Ministry in Bihar

The final stages and the immediate aftermath of the Second World War saw important changes take place in India. Above all, Gandhi's release marked the first step towards the achieving of political consensus. This was followed by the defeat of the Conservative party in the British general election of May 1945. The consequent removal of Winston Churchill and Leo Amery from office was a particular relief to the Congress. This relief was further boosted by the hope that a Labour government might prove much more sympathetic than the Conservatives had been to nationalist aspirations. In contrast, the economic situation was far bleaker. The post-war period in India was characterized by rising inflation and a dramatic increase in unemployment, exacerbated by the demobilization of over two million soldiers from the Indian army.

These developments provided a fertile ground for discontent to flourish. They had the effect, too, of releasing pent-up passions and once more lifting the lid off the underground movement in Bihar which had been simmering all through 1943 and 1944. At the elite level, the political situation was also becoming increasingly unstable. As the political arena opened up and the colonial state moved closer towards relinquishing power the confrontation between the Congress and the Muslim League became more embittered. Consequently, in this penultimate phase of the Raj, the tide of popular passions ran high and threatened to drown even the new inheritors.

By early 1946 a wave of popular action threatened to engulf Bihar. A period of intense political activity at the mass level, a series of agrarian upheavals and gory communal rioting followed the Congress acceptance of office. This period of turmoil lasted right through 1946 and 1947. In several districts communal rioting and agrarian unrest broke out within a short time of each other. Thus Gaya, Darbhanga, Muzaffarpur, Monghyr and Bhagalpur districts, all of which were major centres of Kisan Sabha activity, were also the most communally disturbed areas of Bihar. The observations of a district officer in Monghyr in November 1946 make this point well. He noted that in one village an intense agrarian agitation was gathering force with both Hindu and Muslim peasants actively challenging the landlord. However, in contrast, in an adjacent village across the river, a gruesome communal riot over a trivial issue had left many dead.[1] How is one to explain this complexity?

When one examines the popular protest movements of Bihar in this period an interesting weave of different ideologies ranging right through from nationalism and religious communalism to class-based agrarian unrest can be observed. Majid Hyat Siddiqui, in his critique of existing historiography, notes that historians of modern South Asia tend to jealously guard their own apparently exclusive historiographical preserves of 'nationalism', 'religious communalism' or 'peasant unrest' with scarcely a thought that these boundaries may be false and that different ideologies and social upwellings may overlap at the popular level.[2] It is quite clear in the context of popular unrest in Bihar in 1946 that these differing ideologies coexisted and operated at several different levels of consciousness. It is only by examining the ambivalent nature of popular actions that one can hope to understand the perceptions and attitudes of the participants in protest movements.[3]

[1] Very little work has been done on the popular actions of this period. While A.N. Das has cursorily examined the agrarian agitation in Bihar in the 40s, his work covers a wide timespan and fails to examine the agitation in any real detail or to understand its links with communal outbreaks. See A.N. Das, *Agrarian Unrest and Socio-economic Change in Bihar 1900-1980*, Delhi, 1983.

[2] Majid Hayat Siddiqi, 'History and Popular Rebellion; Mewat, 1920-33', *Comparative Studies in Society and History*, vol. 28, no. 3, July 1986, p. 442.

[3] Scott has pointed out in his work on everyday forms of resistance that 'class does not exhaust the total explanatory space of social actions. Nowhere is this more so than within the peasant village, where class may compete with kinship, neighbourhood, faction, and ritual links as foci of human identity and solidarity. Beyond the village level, it may also complete with ethnicity, language group, religion, and region as the focus of loyalty. Class may be applicable to some situations but not to others; it may be reinforced or cross-cut by other ties; it may be far more important for the experience of some than of others. Those

To date several historians have attempted to write about the historical context of the communal question. Most of them have viewed it, however, as being principally a product of elite manipulation and have concentrated in their analysis on elite-level politics. As a result no real attempt has been made to examine the internal dynamics of communal riots in the 1930s and 1940s or to study the nature of communalism in detail, on the ground.

It is difficult to identify simple causes for the particular configuration of communal and agrarian rioting that developed in Bihar in 1946. While in Bengal the communal upheavals in 1946 can apparently be explained quite easily, as Partha Chatterjee and others have done,[4] in terms of the class divide between Hindus and Muslims, the situation in Bihar was probably more complex. In the latter province, by contrast, there is little evidence to suggest that Hindus and Muslims fell into two distinct class categories in those villages where communal rioting was at its worst. While in some villages communal rioters seem to have come from outside, in many others, resident lower caste Goalas, Kurmis and Dusadhs attacked their equally deprived Muslim neighbours. It is probably true that in Bengal it is possible to say that the main target of attack for poor Muslims was the property of their class enemies. They attacked the absentee Hindu *jotedars*, Marwari shops and idols worshipped in the houses of Hindu landlords. In Bihar the situation was quite different. Here sectarian conflict actually coexisted with class conflict and, in fact, it would be difficult to argue that class was the primary or only contradiction in the locality. Struggles between the communities, over issues such as cow protection and the playing of music before mosques, had created

who are tempted to dismiss all principles of human action that contend with class identity as 'false consciousness' and to wait for Althusser's determination in the last instance are likely to wait in vain. In the meantime the multiple reality of messy identities will continue to be the experience out of which social relations are conducted. Neither peasants nor proletarians deduce their identities directly or solely from the mode of production, and the sooner we attend to the concrete experience of class as it is lived the sooner we appreciate both the obstacles to and the possibilities for class formation.' See Scott, *Weapons of the Weak*, p. 43.

[4] Partha Chatterjee, *The Land Question, Bengal 1920-47*, pp. 109-10; Sugata Bose, *Agrarian Bengal: Economy, Social Structure and Politics, 1919-47*, pp. 211-12; Tajul Islam Hashmi, 'Communalisation of Class Struggle, the East Bengal peasantry 1923-29', *Indian Economic and Social History Review*, 25, 2, (1988), p. 195.

serious tensions between certain sections of Hindus and Muslims since the early part of the century. However, at that time the nature of the communal problem was still far less complicated than that which developed by the 1930s and 1940s. This change was contingent on a clearer political articulation, or crystallization of perceptions, that came about at the provincial and national levels in the later period.

Clearly, therefore, the picture is a more complex one. It is true that mass upsurges occur at certain specific conjunctures in the historical process, and simplistic arguments cannot explain exactly when and why they happen. It is only by examining certain latent structures in social relationships and the nature of the conjunctures at which popular actions occur that we can understand or attempt to explain these upsurges. A main theme in this chapter, therefore, consists in locating these latent structures as well as in examining the nature of popular actions in the period. At another level I also examine the attitude of the Congress organization to popular outbursts and the attempts made by the new nationalist institutions of government to deal with these struggles. In 1946 the colonial state once again devolved power at the provincial level to the Congress. This ostensibly meant the democratization of institutions, more representative government and fewer authoritarian constraints on the people. But what actually occurred differed greatly from popular expectations. If one studies power relations at the local level, it can be seen that the institutions of the new nationalist state continued to operate within the paradigm of authoritarian constraints, so that the state played an ever more crucial role in shoring up urban and rural elites and controlling popular movements of opposition. Once again, as in the first Congress ministry, there was in effect a development and consolidation of older systems of exploitation by new nationalist institutions, often in alliance with the various dominant classes.[5]

[5] This process was in no way inevitable. Mass struggles and movements of opposition constantly threatened and questioned the elements of the new national state. A stronger left-wing opposition with the backing of a mass movement might have changed the contours of the new national state along socialist lines. As Brenner comments on modes of transition, 'the element of indeterminacy emerges in relation to the different character and results of conflicts in different regions. It is precisely in this zone of theoretical indeterminacy that real history becomes the battlefield of active human agents, consciously striving to change the given structure of social relations. In modes of transition there can be no final instance of structuralist determination. On the contrary if in the mode of production the economic is the determinant, in the mode of transition

This chapter will focus, among other things, on the elements in the development of the new national state. Initially, however, it is necessary to construct the basic economic background to the second Congress ministry and, in particular, to describe the lasting economic impact of the war on rural Bihar.

THE FOOD CRISIS, DECLINE IN INDUSTRIAL PRODUCTION, AND THE EFFECTS OF DEMOBILIZATION

The beginning of 1946 saw the post-war economy in India in a crisis. Prices had registered their most alarming increase in 1943-4 but had tended to settle in 1944-5. They rose again steeply by 1945-6, moving up by 40 index points between October 1945 and November 1946. [See the following table.]

At an earlier stage the fortnightly reports and the newspapers had already begun to sound notes of warning. The leading Congress daily, the *Searchlight*, noted in February 1946 that the food situation in Bihar was alarming.[6] The winter rice and the rabi harvest had failed and it was estimated that Bihar would run short of 8.5 thousand tons of foodgrain in 1946. Since rice, the staple food of the Biharis, was generally difficult to obtain the prices of other foodgrains started to soar. In Barh and in Buxar there was a hoarding panic in the market and rice entirely disappeared. In Daltonganj the black market price of rice reached between Rs 18 and Rs 20 per maund and starvation commenced in some parts of the district.[7] By February 1946, the threat of famine was recognized by senior officials and serious conditions of scarcity prevailed in Saran, Hajipur, Chotanagpur and Monghyr. One correspondent, who travelled in Hazaribagh in March 1946, recorded this pitiable, and typical, story in a village called Barkagawaon:

> It was market day. We were surrounded by starving people and in the whole of the market except for sag and mahuwa we found nothing else. For three

---

it is the political practice of struggle which is the determinant.' R. Brenner, 'Agrarian Class Structure and Economic Development in pre-Industrial Europe', *Past and Present*, 70 (February 1976), p. 52.

6 *Searchlight*, 17 February 1946. The paper noted that there had been several deaths by starvation and urged the government to realize the gravity of the situation.

7 Even in the 1943 famine, the situation had been better. See also *Searchlight*, February 26 1946.

TABLE 11

ALL-INDIA PRICE INDEX IN 1946
X—INDEX NUMBERS OF WHOLESALE PRICES, COST OF LIVING AND SECURITY PRICES

	Wholesale Prices* : week ended 19 August 1938 = 100 (Average of weekly figures)							Security Prices* (1927-8=100)		
									Industrial Securities	
	Agricultural Commodities	Raw Materials	Primary Commodities	Manufactured Articles	General Index	Chief Articles of Export	Working Class Cost of Living Bombay (July 1933-June 1934 = 100)#	Government Securities	Fixed Yield	Variable Yield
	1	2	3	4	5	6	7	8	9	10
1939-40	127.5‡	118.8‡	124.2‡	131.5‡	125.6‡	130.5‡	108	114.1	135.3	111.7
1940-1	108.6	121.5	113.4	119.8	114.8	114.1	114	114.6	147.5	117.4
1941-2	124.2	146.9	132.5	154.5	137.0	137.3	128	115.9	155.9	129.0
1942-3	166.2	165.9	166.0	190.4	171.0	161.7	174	115.0	144.9	138.6
1943-4	268.7	185.0	232.5	251.7	236.5	236.4	237	117.2	164.3	182.2
1944-5	265.4	206.0	240.5	258.3	244.2	243.9	236	118.0	176.2	197.3
1945-6	272.6	210.0	246.2	240.0	244.9	248.9	239	118.6	187.0	220.8
October 1945	268.7	210.8	244.5	242.9	244.1	248.0	242	118.7§	187.3	214.9
November 1945	271.1	210.8	245.9	236.0	243.6	246.4	242	119.3§	187.7	225.3
December 1945	278.5§	211.7	250.3	235.8	247.1	249.1	242	119.0§	189.0	232.4§

		Wholesale Prices* : week ended 19 August 1938 = 100 (Average of weekly figures)							Security Prices* (1927-8=100)		
										Industrial Securities	
		Agricultural Commodities	Raw Materials	Primary Commodities	Manufactured Articles	General Index	Chief Articles of Export	Working Class Cost of Living Bombay (July 1933- June 1934 =100)#	Government Securities	Fixed Yield	Variable Yield
January	1946	285.4§	211.2	253.9§	237.3	250.2§	253.5§	242	119.3§	189.8	238.6§
February	1946	291.5§	210.3	256.7§	239.2	252.8§	257.4§	243	119.2§	190.8	236.7§
March	1946	295.7§	206.3	257.1§	240.2	253.3§	261.0§	247	119.2§	191.3	249.7§
April	1946	294.3§	206.1	256.3§	240.6	252.7§	262.8§	248	119.8§	193.8	259.8§
May	1946	299.3§	207.4	259.6§	239.2	255.0§	266.7§	249	119.4§	195.9	267.3§
June	1946	302.8§	226.4	270.5§	240.0	263.5§	279.9§	259	120.7§	200.5	280.9§
July	1946	303.8§	245.7	275.7§	250.5§	270.1§	289.5§	268	¶121.9§	¶202.8	¶303.8§
August	1946	303.5§	240.7§	277.3§	250.2§	271.3§	291.2§	267	**122.8§	**204.0	**314.9§
September	1946	302.2§	243.2	277.7§	248.6	271.1§	289.0§	270	121.2§	200.9	282.0§
October	1946	321.7§	239.8	286.9§	262.4	281.5§	303.2§	264	121.5§	198.5	265.5§
November	1946	352.3§	237.3	302.1§	275.6	296.1§	324.3§	–	–	–	–

SOURCES : * Office of the Economic Adviser to the Government of India.
‡ Seven months ended March 1940. § Estimated.
** Relates to week ended 10 August.
Reserve Bank of India Price Index for 1946.
Labour Gazette Bombay.
¶ Average for first three weeks.

months, rice had not been selling in the bazaar and the people were living on sag. There was also a cholera epidemic in 6 villages and forty had died. In towns where there was rationing, the government staff siphoned off most of the rice. The system as it worked was most inadequate . . . 200 maunds of rice were distributed every day and out of this 110 maunds were given to 5,000 government servants and their dependants, while the remaining was distributed among 26,000 people.[8]

By June 1946, rice and wheat prices had shot up further and rice was selling at Rs 20-25 a maund. This was a very serious matter as, even according to official figures, ration shops offering subsidized foodgrains were accessible to only 18 per cent of the population. With a shortage of nearly 8.5. thousand tons of foodgrain, this meant that even if all the food in Bihar were shared out equally at the then existing level of rations [10 oz per head per day], about six million individuals would receive no food.[9]

The food crisis was not the only problem the post-war economy had to contend with. Indian industry was also hit hard in the post-war period as demand slackened. Having sustained itself during the war period with huge government orders,[10] the close of the war saw a developing market crisis as war purchases began to decline. In fact, industrial production had already begun to decline in the midst of the war itself, due to the vagaries of war demand. For example, pig-iron production dropped from 2 million tons in 1941 to 1.4 million tons in 1945, a decline of nearly 20 per cent over 1939-40. Coal production, which had risen to 29 million tons in 1942, dropped to 25 million tons in 1943 and 26 million tons in 1944. Cloth production, which had risen to 4,871 million yards in 1943-4, had declined to the pre-war figure by 1946. This decline in industrial production resulted in the closure of many factories and threw thousands out of work. Where not made redundant, the workers still had to suffer severe cuts in their wages, adding to the general economic deprivation.[11]

[8] *Searchlight*, 9 March 1946.

[9] *People's Age*, 23 June 1946.

[10] These were worth Rs 28 crore in 1939-40. By 1942-3 they had risen to Rs 247 crore and in 1944-5 stood at Rs 145 crore.

[11] Industrial statistics quoted in the *Communist Party of India Documents, 1950-57*, pp. 21-2. The price structure of industrial goods was totally at variance with the purchasing power of the people and industry had made profits only because of the large government orders which offered a huge stable market.

The return of demobilized soldiers added to the post-war burden. The all-India figures showed that 2 million members of the Indian services and staff recruited for war purposes were retrenched by 1946. Soldiers were paid Rs 60 to 70 per month and sent home Rs 40 to 50, a far too meagre sum for families to make sufficient savings to fall back on. Once they had been demobilized, moreover, soldiers received no more than 56 days pay and Rs 350 as a parting gratuity. The effect of the return of 2 million soldiers on law and order can only be imagined.[12] They were unemployed and without any visible source of income. By January 1946 it was noted in Bihar that crime figures had climbed well above the average for the preceding four year period. Official reports asserted that the deteriorating crime figures were due to food shortages, the return of the soldiers and the steady reduction in war-work. The crime statistics for this period, as seen in Table 12, confirm this increase.

TABLE 12

CRIME FIGURES FOR BIHAR, DECEMBER 1945

Year	*Murder*	*Dacoity*	*Robbery*	*Burglary*	*Theft*
December 1945	45	70	37	1383	1255
December 1940-4	31	83	37	1477	1196
Pre-War Average	25	26	16	1280	840

SOURCE : Bihar police administration report, 1945

It should be noted here that all the districts, except Purnea, Darbhanga, Ranchi and Singbhum, shared in the increase.[13] The crime rate reflected the growing impoverishment of the common people which had been exacerbated by the increasing loss of peasant land in the countryside.

With the decline of war demand the market entered a crisis. The crisis was intensified by a further decline in the purchasing power of the people, a consequence of the retrenchment of the clerical staff of the government, the extra staff drafted for war in the departments of transport, railways, docks and the demobilization of soldiers.

[12] *People's Age*, 23 June 1946.

[13] Crime figures for Bihar, FR (1), January 1946 and see also *Searchlight*, 19 March 1946.

CHANGES IN THE LAND STRUCTURE

Alienation of land was not a new phenomenon in Bihar but its scale and impact greatly increased during the war.[14] In areas like Shahabad, Patna and Gaya, even in Buchanan's days, the payment of rent, (apart from other financial obligations), universally necessitated borrowing, mortgaging of peasant land and even outright sales. After the late 1920s, during the Depression, the figures for land sales continued to show an increase despite the decline in the value of land; they were mainly 'distress' sales caused by the inability of the peasantry to meet rent or debt obligations.[15] The statistics of land alienation in Bihar for the 1920s and 1930s are available from about 1923. They show that between 1923 and 1935, out of a cultivated area of 122 lakh acres under the control of occupancy raiyats in the 10 districts, (the entire Chotanagpur division and the district of Santhal parganas being excluded) annual alienations were about 1/2 lakh acres and from 1.4 to 1.6 lakh holdings. The area and the number of holdings, however, rose sharply during the opening years of the war. (See Table 13).[16] The annual average sale of occupancy holdings for the period 1938-40 seems to have been about 2.5 lakh holdings. By 1942-4 the alienation rate had increased and the annual average in the three year period went up to 3.4 lakhs.[17]

A significant aspect of such land transfers was the numerical preponderance of the sale of part-holdings. Peasants sought in all possible ways to avoid selling their entire holding. The aggregate of the sales of part-holdings during the twelve year period, 1936-47, (as shown in Table 14) was larger than that of the sales of entire holdings by about 426 per cent and the increase in the number of sales of part-holdings was also far larger than that of the sales of entire holdings. The marginal decline of the sales in the triennium 1945-7 can be explained by the fact that the agrarian agitation in 1946-7 had had an impact and the peasants were resistant to sales in this period.

[14] For the development of the land market in the nineteenth century, see Jacques Pouchepadass, 'Land, Power and Market: The Rise of the Land Market in Gangetic India', in P. Robb (ed.), *Rural India*, p. 97.

[15] Binay Bhushan Chaudhuri, 'Depeasantisation in Bengal and Bihar', p. 142.

[16] Reports on the Administration of the Registration Department for Bihar 1935-40.

[17] B.B. Chaudhuri, 'Depeasantisation in Bengal and Bihar', p. 142.

TABLE 13

THE SIZE OF CULTIVATED AREA AND NUMBER OF HOLDINGS AFFECTED BY SALES OF PEASANT HOLDINGS IN BIHAR, 1935-40

Year	*Holdings (Part and Whole)*	*Area Transferred (in Acres)*
1935	150,281	61,835
1936	171,756	182,613
1937	180,155	203,884
1938	192,765	197,402
1939	198,170	233,183
1940	317,349	314,556

SOURCE : Tabulated from reports on the Administration of the Registration Department for Bihar, 1935-40.

TABLE 14

TRIENNIAL AVERAGE OF THE SALE OF PEASANT HOLDINGS IN BIHAR, 1936-47

Triennium	*Entire*	*Part*
1936-8	27,507	154,085
1939-41	46,508	206,164
1942-4	49,080	269,275
1945-7	35,597	207,272

SOURCE : B.B. Chaudhuri, 'The Process of Depeasantisation in Bengal and Bihar, 1885-1947', *Indian Historical Review*, 1975, p. 139.

The reasons for such a sudden increase in land transfers have not been sufficiently examined by historians. On close examination it seems that the changing economic situation from 1939 was closely associated with this development. War-time inflation, food scarcity and high prices increased the market value of land. This made it attractive for the landlords to eject tenants, buy up their land and settle it at higher rates of rent. The high prices had few beneficial effects on the poorer tenants. Forced to sell off their grain at harvest time to intermediaries (the grain trader or the rich landlord) to meet rent or debt obligations, the peasant was often unable to take advantage of the price increase. In many cases he was also forced to buy his food on the market, having sold off all the grain at harvest time. It was this impoverishment of the small farmers that forced them to sell their lands. The buyers in most cases were the traditional rural elite, that is, the landlords and the rich traders who had made money through war profits and speculative business.[18] By the 1940s the rural crisis was further exacerbated by the prospect of zamindari abolition. Tenants were threatened by the prospect of landlords buying up their lands and converting them into either zirat or bakasht to avoid losing their property. Weaker occupancy tenants were also threatened as the landlords compulsorily bought up their raiyati holdings to convert large areas of land in the countryside into sites for sugar mills, a tactic frequently used to circumvent zamindari abolition.[19] All these pressures on the rural community caused sustained agrarian rioting in large parts of both north and south Bihar by the summer of 1946. Thus on the eve of the Congress takeover of office the economic situation was a source of much discontent, both in the urban and rural areas. The takeover was immediately followed by a flurry of popular actions as different interest-groups started to agitate for their particular demands. The police went on strike first, demanding an increase in pay to keep up with rising inflation. This action was followed by working-class strikes in many factories and finally widespread agrarian agitation all over rural Bihar. With the Congress in office, it was widely hoped that popular demands would be met, and it was in such a context that popular aspirations and the hope of a better future climaxed in rural and urban rioting.

THE POLITICAL CONTEXT IN INDIA ON THE EVE OF THE SECOND CONGRESS MINISTRY

At the national level the beginning of 1946 saw the Indian National Congress contending with several new pressures. The post-war political situation was in flux, with the colonial state willing to transfer power, but with the two main contenders, the Congress and the League, unwilling to come to any agreement over conditions for the

[18] In neighbouring Bengal the greatest land-ownership disaster for the peasantry was the famine of 1943 when, according to an estimate of the Indian Statistical Institute, 2.6 lakh families completely lost their land and 9.2 lakh families had to sell part or the whole of their holdings. B.B. Chaudhuri, 'Depeasantisation in Bengal and Bihar', p. 139.

[19] Ram Manohar Lohia's enquiry into the bakasht agitation revealed that nearly 50,000 acres of land in Bihar had been handed over to sugar mills.

final handover of power. The Congress bid for supremacy had been successfully challenged by the Muslim League by 1946. The preceding years had seen a dramatic build-up in the Muslim League so that, by 1946, Jinnah loomed large in the political arena as the 'sole spokesman' of the Muslim political factions.[20] Communal propaganda and passions were running high, threatening widespread rioting. The problem was compounded by the economic crisis of the post-war years and the escalation of mass discontent. This provided the backdrop to the political jockeying for power at the elite level. The following section, then, specifically examines the Congress bid for power and the prelude to the provincial elections in Bihar.

Gandhi's release in the summer of 1944 constituted the first move by the British to end the political deadlock of the early war years. The changing situation by the end of the war had necessitated such a move and even Churchill, a die-hard conservative on Indian affairs, was prepared to concede this. With the rest of the Congress working committee still behind bars it was left to Gandhi to make the right response. Soon after his release in July 1944 he brought out a statement indicating that he was prepared to accept and to advise the Congress to participate in a wartime national government in full control of the civilian administration, leaving the Viceroy and the Commander-in-Chief in control of the British and Indian armies. He expected, however, that such a government would be accompanied by a guarantee of Indian independence after the war.[21] This was indeed an effective reversal of his position in 1942 when he had stated firmly that there would be no participation in the war-effort without adequate representation.

Gandhi's actions continued to be viewed with the greatest suspicion by the British and it was feared that he was about to launch another mass movement. His Constructive Work Programme was regarded with unease, as were the attempts to revive the Congress organization in preparation for this struggle.[22] The Viceroy began sending anxious reports to London, and the administration was asked to check 'Constructive Work' activities. A number of Congressmen in U.P. and Bihar were arrested on flimsy pretexts and Gandhi was

[20] Ayesha Jalal, *The Sole Spokesman*, Cambridge, 1985, p. 171.

[21] See CWMG, vol. lxxvii, p. 475, where he said, 'Today India would be satisfied with a national government in full control of civilian administration.'

[22] TOP, vol. v, no. 86.

forced to protest. Reiterating that there was no plan under present conditions to offer any form of mass civil disobedience, Gandhi warned the government against being too high-handed:

> There is no question of Congressmen making exciting speeches in the prosecution of the constructive programme or about non-violent resistance, or much less about resorting to violence or even approving of it. That there is no plan under the present conditions to offer any form of mass civil disobedience has been clearly stated by me and I know Congressmen all over the country are acting on the advice I have offered.[23]

These were clearly conciliatory moves by the Congress and a necessary prelude to the constitutional negotiations to follow. While the Congress was making these moves the Muslim League under Jinnah remained adamant that no constitutional negotiations could proceed without it. Between 1942 and 1945 the League had had a relatively clear run at the centre and had managed to secure enough support for its programme and for Jinnah in the Muslim provinces. By 1942 the Muslim factions in the provinces had, in effect, accepted the League as their spokesman, while continuing to press for autonomy in their local affairs. By successfully mobilizing Muslim political groups around a vague and unspecific demand for Pakistan, Jinnah had managed to steer clear of dissensions within the Muslim camp, and assert himself as the leader of the Muslim parties. On the face of it the Pakistan demand had the support of the Muslim parties in both minority and majority Muslim provinces. However, there were problems. Where Muslims were in a minority, the Pakistan demand had to be cast in uncompromisingly communal terms. Yet a patently communal line did challenge the leaders of the Muslim provinces, particularly in Punjab and Bengal, where Muslim supremacy over undivided Punjab and Bengal depended on keeping ties with other communities in good repair. Thus Jinnah continued, relatively astutely, to keep the Pakistan demand unspecific for political reasons and managed to mobilize enough support to have a vantage seat at the negotiating table.[24]

On emerging from the political vacuum of the war years the Congress was forced to concede the importance of the League in the new constitutional negotiations. Early Congress overtures to the

[23] CWMG, vol. lxxix, p. 130.
[24] Ayesha Jalal, *The Sole Spokesman*, p. 186.

League came from the fringes of the Congress when Rajagopalachari attempted to produce a formula in August 1944 for a settlement between Hindus and Muslims which conceded the principle of Pakistan. Such a Pakistan entailed the partition of Punjab and Bengal. With Jinnah holding out for an undivided Punjab and Bengal such overtures were bound to fail. The same problems bogged the Gandhi-Jinnah talks. Keeping roads open into an uncertain future, Jinnah had agreed to talk to Gandhi, though he had rejected the Rajagopalachari formula. Their talks were fruitless. Jinnah made much of his two-nation theory and demanded sovereign Muslim states based on existing provincial boundaries, with Muslims alone deciding the issue of separation. Predictably, Gandhi rejected the two-nation theory and although he was looking for a way of conceding some limited form of self-determination to Muslims in the majority provinces inside an undivided India, he could not yet commit himself to a firm guarantee about the size of the League's share of power at the centre. Thus the talks simply became bogged down over the old issues of whether independence should be settled first or whether the communal problem should be resolved as a necessary preliminary.[25]

Once the talks between Gandhi and Jinnah came to naught, Wavell and his advisors desperately sought a way out of the deadlock. Wavell was deeply concerned that, were the situation to be left as it was, popular passions would run high, creating intractable law and order problems. He therefore urged Whitehall to take decisive action and intervene in the deadlock, noting on 14 October 1944:

> I have found HMG's attitude to India negligent, hostile and contemptuous to a degree I had not anticipated . . . there are dark days ahead for India unless more wisdom and goodwill are shown . . . and I think they will have to begin from the top from Whitehall.[26]

Elsewhere in a letter to Churchill he pressed home this point:

> To my mind our strategic security, our name in the world for statesmanship and fair dealing and much of our economic well-being will depend on the settlement we make in India . . . If we can secure India as a predominant partner in the British Commonwealth, our predominant influence in these countries will be assured.

[25] CWMG, vol. lxxviii, Gandhi-Jinnah talks, pp. 136-40.
[26] Wavell, *The Viceroy's Journal*, p. 93.

He then proceeded to give instances of the indifference of Whitehall noting that, despite the lesson of the Bengal famine, he had been hard pushed to secure food imports, creating near-famine conditions and an unhealthy economic situation. Further, the sudden increase in soldiers' pay had added some Rs 50,000,000 to the inflationary position, a considerable part of the burden being paid by the Indian taxpayer, thus intensifying the general economic distress. The situation was further complicated by the return of demobilized soldiers and the closing down of war factories, both of which added to the numbers of the unemployed. These people, Wavell noted,

> will find a fertile field of agitation, unless we have previously diverted their energies into more profitable channels, i.e. into dealing with the administrative problems of India and into trying to solve the constitutional problem . . . We cannot move without taking serious risks, but the most serious risk of all is that India after the war will become a running sore which will sap the strength of the British empire. I think it is possible to keep India within the Commonwealth, though I do not think it will be easy to do so. If we fail to make any effort now we may hold India down uneasily for some years, but in the end she will pass into chaos and probably into other hands.[27]

The solution chosen by Wavell was to devise an all-India conference, and thus temporarily shelve the communal problem, with a view to forming a new executive council more representative of organized political opinion. When the war with Germany ended in May 1945, Wavell was still trying to persuade London to take an initiative. On 23 May, Churchill's coalition was replaced by a caretaker government and Churchill at last agreed to make a declaration about India, before parliament was dissolved on 1 June. On 30 May the Cabinet discussed the terms of the declaration. By now there were already signs that the Cabinet was splitting along party lines over India. Thus, with a general election in the offing, Churchill was forced to give way. Wavell then returned to India.[28]

Meanwhile efforts continued to be made to arrive at a consensus between the Congress and the League. Earlier in the year, in January 1945, Liaquat Ali Khan and Bhulabai Desai had already discussed a scheme for an interim government. It had centred around the creation of an interim administration with an equal number of persons

[27] Wavell's memorandum, TOP, vol. v, no. 19.
[28] TOP, vol. v, no. 476

nominated by the Congress and the League in the central executive.[29] At this stage Desai had sought Gandhi's support to negotiate with Liaquat Ali Khan. For his part, Gandhi had been sceptical, while remaining anxious to break the political deadlock and gain League support for the release of working committee members. He was also more in favour of Congress/League parity as opposed to Hindu/Muslim parity (which formed the crux of the Viceroy's proposals in Simla later) and noted that 'Bhulabhai's proposals interested me as one most interested in solving the communal tangle'. There was much confusion about communal parity. While Gandhi was against it completely, the Congress working committee was prepared to accept it as far as it applied to the centre, though not willing to accept the proposal that all the Muslim members of the national government would be nominated by the Muslim League. In June 1945 the Viceroy called a conference with the Congress, the Muslim League and other minority parties, at Simla. He invited the party leaders to submit to him a list of their nominees for the Viceroy's council, out of which he would make his own selection, leaving the list of names to be reconsidered by the conferees. The Congress gave him a list of names, (as did some other delegation leaders) but Jinnah refused to do so. Instead, he asked for assurances that: (a) all the Muslims appointed to the Viceroy's council would be nominees of the League and that, (b) any measures to which the Muslims objected should require a two-thirds majority. The Viceroy refused to concede either condition and the conference was declared a failure.[30]

By the end of 1945 it had become clear that both the Congress and the League had hardened their attitudes. In the Congress, the working committee of Patel, Pant and Nehru remained firm that no concessions or compromises could be made in the final bid for power. Gandhi's offer of a limited Pakistan to Jinnah during the talks in September 1944, his veto on communal parity and his support for Bhulabhai Desai were all viewed with disfavour by the committee. This was the first occasion where one becomes aware that Gandhi was no longer

[29] Liaquat Ali-Desai talks, CWMG, vol. lxxx, appendix v.

[30] Jinnah dismissed Wavell's plans for the interim government as a device to shelve the Pakistan issue, put it in cold storage indefinitely and establish a unitary government of India which the Congress would dominate. In a rare moment of candour, Jinnah admitted that a Punjabi Unionist Muslim and two Congress Muslims on the council would have questioned the very notion of the Muslim League. See Jalal, *Sole Spokeman*, pp. 129-32.

in the forefront of Congress policy decision-making. His protégé Desai was himself dropped by the working committee, for he was thought to have harmed the Congress cause by the compromise talks. Furthermore, his name was not included in the list of names put forward by the Congress for the Viceroy's Council at Simla, even though he had been leader of the Congress party in the Central Legislative Assembly since 1934. Soon he was not even considered as a candidate for the Central Assembly elections which were in the offing.[31]

The failure of the Simla Conference was followed by the Labour party's landslide victory in the British general elections in 1945. This resulted in renewed hopes and further efforts to end the stalemate between the British and the Indian political parties. In Congress circles it was felt that the Labour victory would provide a spur to quick action and remove any excuse for the delay in implementation of pledges by the British government. In principle, such expectations were not out of accord with the more liberal attitudes communicated by Pethick-Lawrence, the new Secretary of State for India, to the Viceroy with regard to both the Indian problem in its broad-based constitutional contours and to more immediate issues such as the release of detainees and the holding of elections. 'The new government (in Britain)', wrote Lord Monteath, the Under-Secretary of State for India, (to G. Abell, the Viceroy's private secretary) on 20 August 1945, 'are apparently determined to plunge into the Indian constitutional problem and settle it without further ado.'[32] The pace of release of political prisoners quickened and events seemed to be moving towards provincial elections.[33]

TOWARDS ELECTIONS

At the Governors' conference on 1 August 1945, arranged before the results of the British elections were known, the delegates had decided

[31] CWMG, vol. lxxxi, pp. viii-ix.
[32] TOP, vol. vi, no. 45.
[33] TOP, vol. vi. The question of the treatment of political prisoners preoccupied the Secretary of State in his early months of office. On his own initiative, but also in view of the political interest aroused, he showed concern about the possible carrying-out of the death sentence in the much publicized Chimur and Ashti cases and urged the quick release of detainees generally, not excluding members of the Forward Bloc or CSP (see nos. 8,9,12,13).

unanimously to hold provincial elections. They felt that any radical reconstruction of the Viceroy's Executive Council was unwise, at least for the time being, although they recognized the need for the formation of provincial governments. They also favoured, with limited exceptions, the continuation of the ban on Congress organizations and continuance of the policy on the gradual release of detainees.[34] On 29 August the Viceroy returned to Britain and, in accordance with the Cabinet decision of 20 August, reviewed the situation with the India and Burma Committee. Wavell feared that the setting-up of a Constituent Assembly, as contemplated in the Cripps offer, would be boycotted by one or the other of the two communities, by the Muslims if Pakistan was not conceded or by the Hindus if it was or by both.[35] A debate on the matter ensued and finally Attlee, the new Prime Minister, concluded in favour of an announcement reiterating the pledges already given in respect of self-government, the holding of elections and an invitation to Indian leaders, after the results of the elections were known, to state their preference for any alternative method under which a new Indian Constitution might be devised. Failing this, HMG would implement their undertaking and arrange for a Constituent Assembly on the terms of the Cripps offer.[36]

When it came the official announcement affirmed the government's intention to convene a constitution-making body as soon as possible, and authorized the Viceroy, immediately after elections had been held, to undertake discussions on this matter with the representatives of the newly-elected Provincial Legislative Assemblies. The Viceroy was also authorized, as soon as the election results were known, to bring into being an executive council having the support of the main parties. The Labour government seemed at this point only too willing to delay making any decisions about the transfer of power until after the Indian elections. The government could have forced the pace by arriving at some decision soon after their election victory. However this opportunity was missed and the problem was evaded. Pethick-Lawrence, the Secretary of State, claimed that he was anxious to avoid anything which might be considered by one side or the other as calculated to influence elections.[37]

[34] TOP, vol. vi, nos. 2,5,6.
[35] TOP, vol. vi, no. 78.
[36] Ibid., TOP, vol. vi.
[37] TOP, vol. vi, no. 90.

The Congress response to these developments was mixed. Congress had had great expectations of the new Labour government, but was ultimately disappointed to discover that, beyond holding elections and setting up a constitution-making body, no concrete proposals had emerged. The early reactions of the Congress leadership therefore took the form of angry speeches, berating the government. However, as the tempo of mass activity built up, with rioting taking place in Calcutta in November 1945, the INA strikes and finally the RIN mutiny, the working committee of Patel, Nehru and Gandhi grew increasingly worried about the implications of such a deteriorating public order situation for their own future strategy.[38] They soon realized that, having decided to contest elections and take responsibility for provincial administration, a policy of deliberate incitement would militate against their own interests. It is possible to trace this gradual change in Congress attitudes quite distinctly, once the new mass movement and agitation had emerged.

Most of the Congress election speeches in September and October 1945 had virulently attacked the British policy-makers and the civil services. In November 1945 Thomas Rutherford, the Governor of Bihar, wrote to Wavell that he found the Congress campaign worrying. A policy of attacking the police and the services was likely to boomerang once in office, he believed.[39] From the Central Provinces another Governor wrote that he could not recollect any period in which there had been such venomous and unbridled attacks against government and government officers.[40] In November, the Viceroy noted that Nehru's attitude was one of disappointment at the slow pace of change and that he was unrepentant about the incitement to violence in Congress speeches. Indeed, at this stage, Nehru informed the Viceroy that he did not see how violence could be avoided if legitimate aims could not otherwise be obtained.[41] Three days later, on 6 November, the Viceroy sent a memorandum to the Secretary of State, to the effect that they were now faced in India with a situation of great difficulty, in which a Congress ultimatum after the elections was likely to be served on the British government. In default of its

[38] The INA strikes were launched in support of the release of members of Bose's Indian National Army. The RIN mutiny was, essentially, a mutiny of ratings in the Royal Indian Navy.

[39] TOP, vol. vi, no. 159.

[40] TOP, vol. vi, no. 203.

[41] Wavell's note on interview with Jawaharlal Nehru, TOP, vol. vi, no. 188.

acceptance a mass movement on a much larger scale than in 1942 was likely to be organized with the result that the Viceroy felt he must, with the 'utmost gravity', warn HMG to be prepared for a serious attempt by the Congress, probably by the following spring, to subvert the British administration in India.[42] The Viceroy's memorandum of 6 November prompted the chiefs of staff in London to telegraph the Commander-in-Chief for an appreciation of its military implications. In reply, General Auchinleck despatched an assessment, dated 24 November, in which he stated that it was reasonable to assume that widespread trouble, either communal or anti-government in nature, could be expected in late spring.[43]

The public order situation in India was indeed deteriorating. From 21 to 23 November 1945 there were serious riots in Calcutta over the INA trials. The public trials of Bose's INA men caused particularly serious protests. These men were lionized by the public; and a sentence passed against one of the officers particularly inflamed them. On 21 November 1945 a procession of students, mainly from the Forward Bloc (also joined by the Students Federation and Islamia college students), was prevented by the police from passing through Dalhousie Square. When the students refused to disperse, they were lathi-charged by the police. The students then retaliated by hurling stones and brickbats. This in turn invited firing, in which two students were killed and fifty-two injured. The public then protested and, on 23-24 November 1945, a mammoth meeting was held in Wellington Square in Calcutta. Barricades were set up, Europeans assaulted and hartals organized.[44] These upheavals had repercussions all over the country, culminating in the RIN mutiny in Bombay in February 1946 which spread in turn to Karachi, Madras, Visagapatanam and other ports. In Bihar, in November 1945, INA protest meetings erupted in urban areas and were carried forward by students. At this stage, the economic situation had also caused a serious increase in criminal activity, while the retrenchment of employees in factories and services was causing considerable unrest and strikes. A police strike was also in progress, provoked by the discharge of a considerable number

[42] TOP, vol. vi, no. 194.

[43] TOP, vol. vi, no. 256.

[44] Report from Wavell to Pethick-Lawrence, 27 November 1945. See TOP, vol. vi, no. 246. See also Sucheta Mahajan, 'British Policy, Nationalist Strategy and Popular National Upsurge' in A.K. Gupta (ed.), *Myth and Reality: The Struggle for Freedom in India, 1945-47*, Delhi, 1987, p. 76.

of temporary constables who had been in the force for 2-3 years, engaged in war time duties.[45]

Reports from provincial governors referred to the generally unhappy effects of this deteriorating situation on the services. The Fortnightly Reports on Bihar for October 1945 show that attacks by Congress leaders were directed mainly against civil service officers, seriously affecting morale. Moreover, the lower ranks were also anxious, fearing their vulnerability once the Congress took office. Rutherford, the Bihar governor, feared that 'unbridled speeches will incite the people before long . . . We might have another outbreak of violence . . . I've told Rajendra Prasad that the time will come when the Congress would themselves need a reliable police force and magistracy and the present policy was not in their own ultimate interests'. However, Rutherford noted that Prasad had made it clear that the Congress had no intention of stirring up trouble, and that any anti-government attitude was due to the disgust at the way the Simla Conference had been staged.[46]

The British were very worried by this time. In November 1945 the Viceroy noted that 'half measures will be of no use in dealing with a movement of this kind, and the choice will lie between capitulating to the Congress and accepting their demands . . . or using all our resources to suppress the movement which would involve the use of a force of British troops'. He concluded by asking HMG for a public statement of firm intent, arguing emphatically 'that the objective of the Congress was to destroy British influence in India'. The Congress would, he said, go to any lengths to achieve this and would use violence if necessary.[47] The India Office reacted by reasserting the government's position that the setting-up of a constitution-making body was a matter of the greatest urgency, as it would give the opportunity to Indians to decide their own future. More immediately, the India Office promised to formulate a plan for an interim government, aiming at averting immediate violence by devising a transitional form of government which would reduce British powers of control to a minimum and devolve the maximum measure of self-government to the provinces. By early 1946 the Cabinet had decided to send a

[45] FR (1), November 1946.

[46] Rutherford to Wavell, 22 October 1945, FR (2), October 1945; see also TOP, vol. vi, nos. 259, 260, 265, 272.

[47] Wavell to Pethick-Lawrence, TOP, vol. vi, no. 194.

mission to India in late March, when the results of the provincial elections would be known.[48]

While the British remained engaged in debates over the deteriorating law and order situation in India and how best to control what they perceived as Congress threats of mass action, the Congress leadership itself was becoming increasingly worried about the emergence of mass activity. In fact, the evidence clearly shows that, contrary to British fears, the Congress leadership was not envisaging a mass movement at all in 1946.[49] Undoubtedly, in the course of the election campaign the Congress had challenged the government and built up popular passions, but it grew increasingly wary as these passions threatened to spill over. As the mass activity escalated in the course of the INA uprisings in Calcutta in November 1945, the Congress leadership sounded its first note of caution. The outbursts in Calcutta had had repercussions elsewhere in Bihar, Bombay, Madras, Karachi and other places. On 10 December 1945 Gandhi requested a conversation with the Viceroy. In the course of this he made it clear that he agreed with the government on the danger of violence and indicated that he was trying to reduce the temperature.[50] Furthermore, the AICC session of 7-11 December strongly affirmed its faith in non-violence, in significant contrast to the September AICC session when many members had glorified every aspect of the 1942 struggle. By this time it was clear that the right-wing Congress leadership was worried by the mass activity of November and was keen on conducting a safe election into office. The attitude of the Congress to the RIN mutiny in February 1946 further revealed their worries. While the CPI was calling for a nation-wide general strike to support the RIN mutiny, Patel told the ratings to surrender, and condemned the frittering-away of energies in what he called 'trifling quarrels'.[51] Gandhi also recorded his relief at the end of the strike in a press statement in March 1946, stating that:

> It is a matter of great relief that the ratings have listened to Sardar Patel's advice to surrender. They have not surrendered their honour. So far as I can

48 TOP, vol. vi, no 372. It was agreed that there would be considerable advantage in such a mission and that for a variety of reasons, the members of the mission should be of cabinet rank.

49 CWMG, vol. lxxxii, pp. 200-1.

50 Ibid. See also Wavell's report of the meeting, TOP, vol. vi, no. 280.

51 Quoted in Gandhi's press statement, CWMG, vol. lxxxiii, p. 184.

see, in resorting to mutiny they were badly advised. If it were for grievances fancied or real, they should have waited for the guidance and intervention of political leaders of their choice. If they mutinied for the freedom of India they were doubly wrong. They could not do so without a call from a prepared revolutionary party.[52]

By this stage, then, the Congress leadership was clearly rejecting all mass confrontation in favour of a negotiated settlement and a smooth transfer of power. While it was realized that mass aspirations could be built up in the course of the election campaign, the Congress leadership believed that this was not the time for a mass movement. Instead mass activity had to be controlled and not allowed to threaten Congress strategy.

THE CONGRESS ORGANIZATION IN BIHAR IN THE 1940s

The gradual release of political prisoners in Bihar by 1945 encouraged concerted measures by the Congress leaders in Bihar to revitalize the Congress organization. The period 1942-5 had pushed the movement underground. Gandhi insisted that a necessary step towards rebuilding local Congress organizations would be to actively give effect to his 15-point programme. A Congress election meeting in July 1945 in Bihar noted the need for and the importance of constructive work (particularly Khadi hand-spinning and weaving) and the determination of the Congress to push the constructive programme as far as possible.[53] But attempts to get the constructive programme underway were slow, as Rajendra Prasad noted in August 1945. The main activity of the Congress was centred on the preparations for elections and in this context the Congress role in 1942 and the sacrifices of its leaders were constantly reiterated to sustain the Congress image as a people's organization. In Fatwah, Jagat Narayan Lal, another Congress leader, made a spirited speech in which he congratulated the public on the bravery they had displayed and the suffering undergone by them since 1942. He exhorted them to show to the world how India could achieve swaraj by non-violence, under the leadership of Gandhi; and he called upon them to contribute generously to the Political Sufferers Relief Fund.[54] On the 9 August 1945, independence celebrations were held all over Bihar. These consisted in

[52] CWMG, vol. lxxxiii, p. 184.
[53] FR (2), July 1945.
[54] *Searchlight*, 15 October 1945.

meetings of Congress workers, peaceful strikes in schools and colleges and the holding of *prabhat pheries*, flag hoisting and salutation and spinning demonstrations. Nehru visited Patna in October 1945 and addressed a meeting attended by 80,000 people.[55] On this occasion he recalled the sacrifices made in the August revolution and charged government officials with ruthless barbarity. The effect of all this was to give new life to mass activity. Indeed, the Bihar Congress membership increased to 420,980 members at this time.[56]

The evidence outlined in the preceding chapters clearly shows that, despite its policies under the first ministry, the Congress in Bihar had retained generalized popular support among the people. Each mass movement in 1920, 1930, and 1942 had succeeded in extending the Congress base in Bihar dramatically. In a distinctly different way, underground Congress activity between 1942 and 1945 had successfully kept the movement alive despite the official ban on the organization. Moreover, the dissemination of a new nationalist ideology had been remarkably strengthened in the countryside in 1942 as peasants in large numbers attacked police thanas, railways and other institutions of the colonial government. Many 'middle' and 'poor' peasants were mobilized in this movement to an extent that they had never been before. By such means, the Congress had emerged from 1942 with a stronger base in rural areas.[57]

In the countryside new proverbs and songs became current. Gandhi's swaraj promised a hope that filtered down to the poorest peasant in Bihar. One proverb that reflected this hope and is current even today in Bhojpuri was: 'This is like Gandhiji's swaraj come true.'[58] The charkha had also become important as a symbol of rural self-sufficiency and there were many songs that recorded its importance in village life. A Bhojpuri folk song on the importance of the charkha recorded in the proscribed literature of the 1940s ran as follows:

[55] *Searchlight*, 15 October 1945.

[56] FR (1), November 1945.

[57] Individual attitudes were revealed in a personal interview with Munirka Singh, a poor peasant, at Surajpura village in February 1988, who said that he became aware of the national movement for the first time in 1942 and had participated in the Quit India struggle during which he pulled down the British flag from the zamindari estate. Middle and poor peasants participated in the struggle quite widely as is clear from the evidence cited in my chapter on the struggle in 1942.

[58] As narrated by Munirka Singh in Patna, February 1988.

Come my sisters, ply your charkha and turn the world upside down. It makes a melody as you ply it and it calms your heart. Come my sisters ply your charkha and no longer will your country mourn.[59]

Such folk songs and proverbs became part of the new popular idiom. Jayaprakash Narayan reported in 1946 that when he was returning from an election meeting he suddenly heard the cry, 'Swatantra Bharat ki Jai' (Victory to free India) followed by the age-old cry, 'Raja Ramchandra ki Jai', (Victory to Ram). Apparently a group of workers was listening to a recitation of the Ramayan which was periodically punctuated by the cry of 'Raja Ramchandra'. Narayan notes that:

The thought that a national cry had gained such respectability and popularity as to be coupled with this religious and deeply devout cry gripped my mind. It signified nothing short of a great mental revolution among the people.[60]

The popular perception in 1946, as in 1939, was that swaraj was near and that, with the formation of the new Congress government, all the problems of the poor would be solved. This universalist image was, of course, of enormous political advantage. An interview with a 'middle peasant' in Shahabad district reveals the extent to which Congress ideology had permeated the countryside. Kuldip Singh Kula Rai is a Rajput middle peasant of about 75 years of age from Burnah village, in Bikramaganj thana, and in 1989 owned five acres of land. He was first attracted by the national movement during the civil disobedience days in 1930 and remembers going to a Congress meeting then at Bikramganj, where Gandhi was present. In 1932 he had joined the Congress as a member and attended the annual Congress session at Ramgarh in Bihar in 1941. In 1942 he was in the forefront of the Quit India struggle in Bikramganj and was one of the crowd that set fire to the Bikramganj post office and then attacked the police thana. He further remembers that he and others participated wholeheartedly in the movement, for there was a general feeling that once India was free, 'rivers of milk would flow'.[61]

Some idea of the extent of the Congress mass base in the 1940s can be gathered from the preceding pages. The activities of the left-wing groups in the Congress, principally of the socialists, had helped to

[59] Proscribed Pamphlets, IOL, PP Hin B112. For other Bhojpuri songs with similar themes, see *Hindi sahitya ka Itihas*, p. 147.

[60] *Selected Writings of Jayaprakash Narayan*, p. 34.

[61] Interview with Kuldip Singh Kula Rai in Patna, 10 February 1988.

broaden this base. The Left in the Congress had been very active in 1942 and in the underground movement from 1942-5. While the communists had been discredited by their 'People's War' line, the socialists had shown themselves to be in the forefront of Congress activity. They now hoped to reap the benefits in organizational terms. This time, in contrast to 1937, the socialists had chosen to accept the official Congress policy of participation in elections and office acceptance. The right-wing leadership reacted by patting the socialists on the back and then seeking to absorb them, to dilute any chances of future radical opposition. S.K. Sinha, the right-wing leader of the Congress, described the socialists in glowing terms as the 'right hand of the Congress'. A Fortnightly Report noted in this connection that a main tactic of the Congress now consisted in trying to absorb the Kisan Sabha, the Congress Socialist party and every other party in the election campaign.[62] Ultimately the Congress Socialist party did decide to help the Congress election effort, although some members of the party retained reservations about the strategy and were aware that they would only be made tools of the Congress.[63]

In the case of the Kisan Sabha, however, there were more diverse attitudes to the question of office acceptance. In fact, by 1945 the Kisan Sabha in Bihar had been firmly split. The Congress Socialists and the Forward Bloc had left the organization on the grounds that it had become dominated by the communists. Sahajanand Saraswati followed suit soon after, resigning from the presidentship of the Kisan Sabha on the grounds of 'the *de facto* communist domination over it'. He formed a rival organization called the All-India United Kisan Sabha which came to be closely allied with the Congress and supported Congress strategy, including office acceptance.[64]

Meanwhile the communists were being increasingly opposed by the Congress and were being threatened with expulsion. This was an important move as regards the implications for the Congress organization itself. Hitherto the Congress in Bihar and elsewhere had constituted a broad-based national front, including both the socialists and the communists. In the first Congress ministry these two groupings had shown themselves to be strong critics of government policy and

[62] FR (1), October, 1945.

[63] FR (1), November 1945. As Jayaprakash Narayan noted, in this final phase of the struggle for independence, the CSP was with the Congress.

[64] Attitudes of the Kisan Sabha; Rakesh Gupta, *Bihar Peasantry and the Kisan Sabha 1936-1947*, p. 177.

had challenged the right-wing Congress leadership in Bihar both inside and outside the legislatures. During that period the Congress had already attempted to clamp down on them. These attitudes now hardened as the Congress became even less tolerant of dissent.

The attack on the communists was designed to counter the most vehement expression of dissent,[65] and thereby specifically to bolster the right wing in the Congress. That the attack was a political move to curb dissent can be gauged from the fact that during the war, at the grass roots level, the communists had conspicuously failed to toe the 'People's War' line of the national party leadership, while the Bihar communists had been particularly active in fomenting the 1942 uprising. They had also been reported as being in favour of the Congress policy of contesting elections and had actually issued pamphlets expressing the view that it was unnecessary to wait for the release of political leaders before accepting office.

Despite this, on the eve of contesting elections, the Congress had, both at the national and at the provincial level, decided to expel the communists. An added impetus was given to this decision by the CPI policy of supporting the Pakistan demand of the Muslim League on the grounds of the Soviet notion of 'self determination of nationalities'. In September 1945 the AICC asked the communists to show cause in a strongly-worded statement and declared:

> It is not without significance that complaints and accusations against Communist activity in India have been received from a very large number of Congressmen all over India. There is no doubt that the attitude of the Communist party towards the communal problem and their unqualified support for the Muslim League claim for Pakistan added to their prevailing estrangement. It is manifest that there is at present a widespread and deep

[65] SWJN, 14, pp. 525-8. See Nehru to R.P. Dutt, 12 August 1945 where he wrote, 'You must realise that it pains me to see the gulf that has arisen between the Congress and the communists in India. The gulf at present is wide and deep and has all the passions of three years behind it . . . . The gulf has arisen because of internal policy in India and the fact that communists ran down popular leaders in India at a time when there was bitter conflict between nationalism and the imperial structure, they appeared before the people as acting on the side of the latter . . . Politically the fact that has gone most against them is their attitude on the communal question. They have become full blooded supporters of Jinnah's demands (unspecified and vague as they are) and in the name of Congress-League unity they demand a complete surrender by the Congress to Jinnah.'

sentiment against the members of the Communist party in India and the Congress rank and file are powerfully influenced by it.[66]

On 9 December the Communist party response was received and turned down. In a letter signed by Nehru, Pant and Patel it was noted:

> their explanation amounted to a tirade against the Congress . . . They have altogether forfeited the confidence of the Congress and are unworthy to occupy any responsible or elective place in the Congress. They seem to have already realised their position. Conscious of their guilt, their members have already resigned from all subordinate Congress committees, and also from the primary membership of the Congress. We recommend that Shri S.G. Sardesai, V.G. Bhagat, V.D. Chitale, K.M. Ashraf, S. Sajjad Zaheer, Sohan Singh Josh, Karyanand Sharma and R.D. Bharadwaj should be expelled from the AICC and necessary directions should be sent to all provincial Congress committees so that they may take appropriate steps, wherever necessary, to remove members of the Communist party from all elective offices.[67]

In Bihar the first consequence of this directive resulted in the Monghyr DCC deciding to expel three of its members belonging to the CPI including Karyanand Sharma and Brahmadeo Sinha. By October 1945, 300 Bihar communists had been forced to resign from the Congress.[68] The communists were not, however, the only groups to be expelled from the Congress. The Royists, who had formed the Radical Democratic party, were also barred from participating in elections from the Congress platform. Nehru, in justifying the ban, recorded that 'The Royists . . . have no place in the Congress at all as they have themselves left it and attacked it'.[69] As a result of the expulsions the candidates nominated for the seats in Bihar originated mainly from the right wing within the Congress. There were few socialist nominations as there was still a state ban on the CSP and its workers were under arrest. It was emphasized by the Bihar leadership that Congress nominations would come from among those who had participated in the 1942 movement. Inevitably there were complaints that many opportunists had obtained tickets on false pretexts. For example, Sadiq Ali, in a letter to Ram Dulari Sinha, one of the

66 Ibid.
67 Ibid., pp. 546-7.
68 *Searchlight*, 23 October 1945.
69 SWJN, 14, p. 526.

Congress candidates, noted that good bona fide Congress workers were being kept out, and that some Congress candidates had allegedly indulged in black-marketeering during the war.[70] The quality of the Congress candidates was clearly deteriorating and, by the 1946 elections, they were a far cry from the type of Congressman idealistically visualized by Gandhi. This, then, was the state of the Congress organization on the eve of the 1946 elections.

THE MUSLIM LEAGUE, HINDU COMMUNAL ORGANIZATIONS AND THE ELECTIONS

The other main parties in the electoral fray were the Muslim League and the Hindu Mahasabha. The end of the first Congress ministry in 1939 had seen a rapid development of the Muslim League in Muslim minority provinces and growing communal tension at the popular level. During the war years the League had succeeded in further broadening its base by mobilizing the Muslim majority provinces around the communal banner and over a demand for 'Pakistan' that was conveniently unspecific. In Bihar, by 1944, 'Pakistan' conferences were being held all over the province. The Fortnightly Reports note that in Monghyr these meetings were attended by several thousand people. The League was aided in its efforts at mass mobilization by the activities of the Muslim volunteer organizations, the Khaksars and the Muslim League National Guards. The Khaksars had been started as a social service organization for Muslims in 1931, but had gradually become more militant by the 1940s. Their founder, Allama Mashriqui, had declared it to be a non-communal organization dedicated to the service of mankind and not affiliated to any political party. Nevertheless, there were close links between the Khaksars and the League especially at the grass root level where the Khaksars were effective in organizing the Muslims in neighbourhoods or *mohallas*.[71]

[70] Rajendra Prasad Papers, File no. 11/1947.

[71] Y.B. Mathur, *Muslim Politics in India*, Delhi, 1979, pp. 218-20. This was because the Khaksars were mainly organized around the mohallas and they felt that the work of regeneration and remoulding the people was to be carried on in the mohallas. The Khaksars gradually came to be organized along military lines, wearing Khaki uniform and saluting in a military fashion. The organization provided for a leader and commanders of the various grades called Sar Salars, Salar Idara, Sippa Salars. Their street squads were the cream of the organization and were required to sign in their own blood a pledge to undertake to guide all Muslims according to the injunction of God and the prophet.

Though the organization was active mainly in Punjab, by the 1940s, its activities had increased in U.P. and Bihar and were reinforced by a substantial enrolment of students. In Bihar the Khaksars were most active in the Purnea district, which had a large population of Muslims. The activities of this organization, especially in urban areas, had developed support for the League through unashamedly communal propaganda. The Muslim League National Guards, the official volunteer body of the Muslim League, had also tightened up its organization by extending the training of its members along militaristic lines. By 1946, bolstered by the support of these volunteer organizations, the League was well prepared to win the Muslim seats in the elections.

The main Hindu communal organization contesting the elections was the Hindu Mahasabha. In 1946 the Sabha was engaged in contesting all the general seats in Bihar and in disseminating a Hindu communal ideology. It was aided in these efforts by the activities of a Hindu volunteer organization, the Rastriya Swayam Sewak Sangh or RSS. The membership of this organization had increased dramatically all over India by the 1940s. Like the Hindu Mahasabha, the Sangh believed that India would not be able to overcome her national problems or to achieve a prominent international position until the Hindu community had organized effectively. However, unlike the Hindu Mahasabha (which was a political party), the Sangh was organized along militaristic lines and by the 1940s had branches all over India.[72] Organized mainly in urban areas, its volunteer members, the Swayam Sewaks, built up their physical prowess through rigorous physical training and intensive discipline. Nearly 60 per cent of their recruitment came from among college and university students while the rest were from the lower middle class, shopkeepers, clerks and tradesmen. In essence, they felt infused with the spirit of a Hindu rastra. By 1946, there were 40,000 Swayam Sewaks in Bihar who were actively engaged in building up Hindu communal feelings.[73] The activities of all these organizations in Bihar had resulted in a considerable deterioration of communal relations at the popular level by 1946. The increasing frequency of inter-religious incidents had al-

[72] J.A. Curran, *Militant Hinduism in Indian Politics*, New York, 1951, p. 61

[73] The attitude of the Congress towards the Hindu Mahasabha was interesting for, though the official leadership of the Congress declared firmly that they would have nothing to do with Hindu communal parties, many lower-level Congress functionaries were active in the Hindu Mahasabha and sympathetic to its ideology..

ready become apparent in 1945 when, in several urban areas in Patna and Monghyr, communal riots erupted over characteristically petty matters. It is important to point out here that, during the war years, religious propaganda was kept low so that few communal tensions had been reported between 1940 and 1945 in Bihar. All this was changing now and the Fortnightly Reports show that communal rioting was now on the increase as people seemed inclined to take matters into their own hands.

The Muslim League viewed the rising communal tension in Bihar with some satisfaction as it provided the best context in which to win elections. The League claimed that it was confident that it would sweep the Muslim seats and win nearly 32 out of 40 of them. There was also an increasing incidence, at this time, in the number of desertions of Congress Muslims to the League. By 1946 the Congress had realized that it had little hope of winning many Muslim votes at all. It had therefore decided to contest only 9 out of 42 Muslim seats and rested its hopes on the two Congress Muslim stalwarts, Syed Mahmud and Abdul Bari.[74]

THE 1946 BIHAR ELECTIONS

The elections were contested on almost precisely the same electoral rolls that had been prepared in 1936 and then hastily revised in 1942. The franchise to the Central Assembly extended to only 3 per cent of the people while that of provincial Assemblies was 13 per cent.[75] The Congress was confident that, so far as the general constituencies were concerned, they would sweep the polls. The election manifesto of the Congress for the Central Assembly envisaged a free democratic state with the fundamental rights and civil liberties of all its citizens guaranteed under the constitution. This constitution was to be a federal one with a great deal of autonomy for its constituent units, and with its legislative organs elected under universal adult franchise. Eradication of poverty was the prime goal and to this end the manifesto noted that:

> It will be necessary to plan and coordinate social advance in all its many fields to prevent the concentration of wealth and power in the hands of individuals and groups, to prevent vested interests inimical to the society

[74] Rajendra Prasad Papers, File no. 9-R/1945-6.
[75] SWJN, 14, p. 83.

from growing and to have social control of the mineral resources, means of transport and principal methods of production and distribution in land, industry and other departments of national activity, so that free India may develop into a cooperative commonwealth.[76]

On the basis of this manifesto the Congress pursued a vigorous election campaign, and the leadership toured the countryside making many promises. S.K. Sinha declared grandly that 'when the Congress comes to power the Zamindari system will be abolished'. Rajendra Prasad, too, toured Saran and Chotanagpur propagating the same kind of message.[77] In tribal areas Prasad told the people that Bihar was to have a Congress ministry and that the Adivasis would only harm their interests by choosing to put their own representative on the opposition benches.[78] In Muzaffarpur an election meeting of many thousands of villagers took place. At it, A.N. Sinha appealed to those present to strengthen the Congress by enrolling as many members as possible and voting the party.[79] Most election speeches virulently attacked the colonial state and its institutions, particularly the police. This was a considerable irony in view of the use the Congress had willingly made of the police in Bihar in 1937-9 to suppress unrest. The Fortnightly Reports commented that the new Congress attacks on Indian police officers seriously affected the morale of the services.

The election meetings were rarely held in a calm atmosphere. Often there was violence between the supporters of the Congress and the communist candidates. Karyanand Sharma, the CPI Kisan Sabha leader who was opposing Krishna Sinha, alleged that in several places in Monghyr his election meetings had been disturbed, abusive language used and party workers assaulted.[80] In Begusarai constituency, where the opposing candidates were Sarju Singh of the Congress and Brahmadeo Singh of the Communist party, the communist election meetings had been disrupted. Brahmadeo Singh alleged that a 3,000 strong kisan rally organized by the communists was disrupted by Sarju

[76] Ibid., pp. 96-7, Draft Election Manifesto.

[77] *Searchlight*, also FR (2), November 1945.

[78] *Searchlight*, 15 October 1945 and 24 October 1945 where he is quoted as saying 'people who had bared their chests to bullets and their homes and possessions to the ravages of unheard-of force in the name of the Congress, despite the absence of a definite programme, could be safely depended upon for the insignificant something of a mere vote'.

[79] See Rutherford to Wavell quoting the speeches, FR (2), October 1945.

[80] *Searchlight*, 24 October 1945 and 20 January 1946.

Singh together with the lathials of the local landlord.[81] In Monghyr the first election meeting to be addressed by the communist, P.C. Joshi, in Lakhisarai, was broken up by an organized attack led by the thana Congress president, Rajeshwar Singh and other Congress workers. The communist candidate was so seriously injured that he had to be removed to hospital. In many villages the zamindars' private armies were used extensively to break up communist meetings and *Jathas*.[82]

Since the main competition for the general seats came from the communist candidates, the Congress was, by and large, assured of success in the elections. Nehru, who had long perceived that elections and office acceptance could introduce stresses and strains in the Congress organization, commented that he was

> not doubtful of success in the polls; but sometimes when I look at the vast pulsating life in the country, I get frightened. This is because I fear that our energies might get frittered away in petty provincial and other conflicts. The Congress should utilise its strength in the right direction, and I am sure that the country will respond to the Congress whole-heartedly.[83]

The results, when they came, confirmed the predictions of the Congress leadership, at least for the general seats. There they achieved a resounding victory, capturing 98 seats. The Congress had been routed in the Muslim constituencies and even Abdul Bari had lost his seat while only Syed Mahmud won on the Congress ticket. Thirty-two seats had gone, however, to the Muslim League, 6 to the Momins and 2 to Independent Muslims.[84] The communal divide thus seemed more

[81] *People's Age*, 24 February 1946.

[82] *People's Age*, 17 February 1946. These attacks on the communists were carried out first on the communist headquarters in Bombay, and Nehru was to comment, 'I cannot believe that any Congressman can be responsible for these attacks, but whoever they are, they have done an ill service to our cause. No cause, least of all the cause we stand for, ultimately prospers by hooliganism. I should like to add that instances have been brought to my notice when hooliganism was indulged in by the Communists and their sympathisers.' See SWJN, 14, p. 545.

[83] SWJN, 14, p. 90.

[84] FR (2), March 1946. For debate on the failure of the Congress to win Muslim seats in Bihar, see Rajendra Prasad Papers, File no. 9-R/1945-6. Yunus wrote to Prasad after the defeat in the Muslim seats on 10 December 1945: 'I hope the eyes of the Congress have been opened by the result of the Central Assembly elections from Muslim Constituencies. Nobody is to be blamed

unbridgeable than ever and the stage set for the bitter communal killings of November 1946.

THE CONGRESS RETURN TO POWER, STRIKES, AGRARIAN UNREST AND COMMUNAL RIOTING

The new ministry was sworn in soon after the elections on 16 April 1946. The ministers, including the prime minister, S.K. Sinha, came mainly from the Congress right wing, and included men such as Ram Chritar Singh, Krishna Ballabh Sahay, Binodanand Jha and Baidyanath Verma.[85] Though Mahamaya Prasad was elected leader of the Provincial Congress Committee as a concession to the socialists, it was only a matter of time before they felt they were being sidelined within the Congress organization. In the elections to the thana, Congress committees in September 1946, the CSP and the left wing suffered serious setbacks, and though officially the socialists left the Congress only in 1948, they were already on the decline soon after the Congress took office.[86]

The Congress in Bihar assumed office during a political and economic crisis. Prices continued to escalate in the early months of the Congress takeover and a scarcity of foodgrains quickly developed. The food situation was considered especially precarious in north Bihar and Chotanagpur.[87] One of the first acts of the Congress in office was to attempt to rectify the rice scarcity by enforcing the Paddy Levy Order. These attempts were, however, plagued by inefficiency. Before the Congress took over there had been virtually no supplies in the hands of the government Food Department. The entire rice harvest had been seized by traders and landlords while government food officials stood looking on. Congress attempts to remedy the situation by implementing the Paddy Levy Order failed to have any impact. According to the order, those who owned 100 acres and above of land

---

except the Congress for it talks and writes about Muslim mass contact and does nothing.'

85 For list of ministers, see FR (2), April 1946. The list of AICC members from Bihar, however, included more representatives from the left wing, seven members of the CSP including Jayaprakash Narayan, Sheel Bhadra Yajee and two others from the Forward Bloc, and Sahajanand Saraswati from the Kisan Sabha. See *Searchlight*, 7 June 1946.

86 Jayaprakash Narayan, *Towards Total Revolution*, p. 82.

87 FR (2), July 1946.

or had a grain surplus of 500 maunds were to give up half the amount to the food administration. Those who had 1,000 maunds were to part with three-quarters. Few landowners, of course, actually parted with the surplus. Indeed, apart from the Mahanth of Bodh Gaya who had handed over 25,000 maunds of paddy, most paddy growers evaded the levy altogether. Even those Congress MLAs who owned middle-sized zamindaris failed to deliver any stocks to the government. The Hindi daily, *Rashtravani*, owned by the associates of S.K. Sinha, noted that less than one lakh maunds of paddy had been delivered in Bihar as a whole till 24 May 1946. The District Magistrate of Shahabad, a big surplus district, had expected 10 lakh maunds of paddy, had served notices for 2 1/2 lakh maunds and then received only 4,000 maunds. In Monghyr, where 3 lakh maunds were expected, the delivery was only 2,500 maunds. In Patna, where 16 lakhs were expected, only 5,000 maunds had been delivered.[88]

The failure of the Congress in office either to tackle the food crisis or bring prices down resulted in widespread agrarian and urban unrest. Labour strikes erupted all over the province in several departments. Notable among them was the police strike, caused by the acute unrest among police constables over the increase in the prices of food and cloth and a cut in police rations. The police strike, which broke out a day after the new ministry was sworn in, continued to plague the Congress for its first few months in office.[89] Trouble first started in Bhagalpur and spread to Monghyr, Ranchi, Saran, Patna, Darbhanga, Champaran, Purnea and Saharsa. In eleven out of seventeen districts armed police also stopped work and refused duty. A police pamphlet issued just before the Congress took office demanded (a) minimum pay of Rs 100, (b) free residential quarters for each of the Bihar constables, (c) foodstuffs at controlled rates for police constables, (d) a free ration card for all constables. The police also claimed that the ration of cloth of 8 yds per year and the foodgrain ration of 3 chhataks of rice and 1 chhatak of other grains per person was insufficient. The strike was organized by an ex-constable, Ramanand Tiwari, a Congressman who had been dismissed from service for organizing an earlier strike in 1942.[90]

[88] *People's Age*, 23 June 1946.

[89] FR (2), April 1946. Rutherford wrote to Wavell that the new ministry was sworn in on 16 April and soon after called a cabinet meeting to consider the question of dealing with a police strike.

[90] FR (2), April 1946. See also BSA, File no. 149/46.

The Congress government attempted to end the police strike through a carrot and stick policy. At first it decided to take a firm line and refused to consider any representation unless the men returned to duty. It then attempted to use its influence with Ramanand Tiwari to ask the men to return to work. At first Tiwari attempted to compromise the government by asking the striking constables to return to work and show more flexibility by, for example, accepting a Rs 40 pay increase. In a speech to the constables, he asserted that:

> the strike was not launched after due deliberation. The organisation of the constable brothers had not been strengthened by this time. Still it appeared as if it was launched against the Congress government though it was not the fact. I therefore saw the Prime Minister of the Congress government, Babu Kesari S.K. Sinha, and others . . . and they gave consolation to our constable brothers that if they called off the strike their demands would be immediately fulfilled.[91]

Tiwari's attempts at conciliating the Congress government backfired, as he soon lost credibility in the eyes of the striking constables, who then turned to socialist leaders such as Basawan Singh to help continue the strike. The CSP had made considerable inroads into the police unions so that Tiwari was eventually forced to retract his compromise solution and shift allegiance to the CSP in favour of direct action. This time the Congress ministry reacted with active disfavour, and thirty-nine police constables were dismissed from Hazaribagh and Ranchi. This, of course, made the strikers more adamant. In a speech in September 1946, Tiwari voiced the general mood of the constables, when he said, 'The Congress government is our own government, it is a government for the poor but they are letting us down repeatedly. Why? Because, as you know, even a mother will not give milk to her child unless the child cries or struggles. It is therefore necessary that we should let the government know of our determination.' Subsequently, strike actions by the police continued with short breaks well into 1947, without reaching any settlement. In March 1947 a section of the armed police in Patna, Monghyr and Gaya again went on strike. The military had to be called in and shots were fired.[92]

The activities of the socialists in leading and organizing such

[91] Ibid.

[92] FR (1), March 1947.

strikes further alienated them from the official Congress leadership in Bihar. Naturally, the Congress ministry was extremely critical of the activities of the socialists in the police strike. Furthermore, the authorities realized that Monghyr was a hotbed of socialist activity and that the release of Jayaprakash Narayan and Basawan Singh had given great encouragement to the socialists. It was asserted that Tiwari's latest speeches had been intended to malign the Congress government and that they had subverted the confidence and morale of the Monghyr constabulary. It was on this basis it was said that the new Bihar government had attempted to repress the strike. Congress leaders also sent a deputation to Jayaprakash Narayan, the socialist leader, asking him to present his case before Rajendra Prasad.[93] Once again, as in 1937-9, the battle-lines were being drawn between the socialists and the official Congress right wing. The Congress ministry was clearly worried that the actions of the left wing were deliberately intended to confront the Congress government. In this situation the new government leaders could not resist the temptation of resorting to the machinery of the colonial state to clamp down on their critics as a wave of unrest spread both to urban and rural areas.

The ministry also faced a postal strike in May 1946. It began when, at a number of places in Patna, Saran and Monghyr, striking postmen started to obstruct the movement of mail. Their demands were framed in terms of a wage increase and an increase in dearness allowance to meet the rise in commodity prices. Almost immediately the ministry was forced to order the arrests of those postmen obstructing the movement of mail. Several strikes also erupted in other industries.[94] In Tatanagar foundry, a dispute over wages resulted in the workers calling a strike which then dragged on and showed no signs of a settlement. In the coalfields in Jharia there were a series of disputes that were referred to the arbitration committees.[95]

These labour strikes, in both government departments and in industries, were not the only symptoms of the economic malaise affecting Bihar. General crime also increased dramatically, as indicated by the statistics for Bihar quoted above [page 292]. Many incidents of foodgrain looting were reported, a reliable indicator of the growth of poverty and malnutrition levels among the poor majority. Even goods

[93] Police file, BSA, File no. 149/46.
[94] FR (1), July 1946.
[95] FR (1), June 1946.

trains on the Bengal-Nagpur railway were being constantly stopped and there were seventy-two cases recorded of the theft of food between Purulia and Chandel stations in October 1946. A news report on one such incident described how a rice-laden goods train had been stopped on the BNR railway by 35-40 men who proceeded systematically to loot it.[96] The number of strikes and the general increase in crime were symptomatic of the high prices and the food scarcity prevailing all over Bihar, following the end of the war. With regard to labour unrest, more specifically, the strikes were a result of a general feeling among the working classes that the Congress in power would be more amenable to labour demands and the repressive machinery of the state would not be used as freely as under the British.

In this respect the Congress was clearly in difficulties. Early populist gestures had included a complete release of the political prisoners of 1942 and a refund of the collective fines imposed between 1942 and 1945. These were clearly attempts to establish credibility as a 'government of the people'.[97] Subsequently, with the outburst of labour agitation in urban areas and the gradual renewal of the bakasht struggle in the rural areas, the Congress ministry was forced to resort increasingly to the tactics of its colonial predecessor, in clamping down on agitation and introducing highly conservative legislation to ameliorate the situation. The application of this new reactionary approach made its greatest impact, however, in the context of agrarian struggles. It is to this impact that one may now turn.

THE LAND QUESTION AND THE BAKASHT AGITATION IN 1946

In June 1946, a long drawn-out struggle over bakasht land broke out once again in several districts in Bihar, the Congress legislation of 1938-9 having shown itself ineffective in stemming the onset of agrarian unrest. As outlined above, social pressures had been building up steadily in the countryside during the war years. A rapid increase

[96] BSA, File no. 336/46.

[97] Rutherford to Wavell, 9 April 1946, FR (2), 1946, where he noted that he had had the most difficult arguments with the Prime Minister over the release of the 1942-3 cases. He further noted that the Prime Minister 'next proposed the refund of all the collective fines of 23 lakhs and as there was no special responsibility I had to accept this open justification of what happened in 1942-43'.

in peasant disappropriation had taken place leading to the proletarianization of large numbers of hitherto poor peasants due to alienation of land. The main benefits of war-time price increases had gone to the economic intermediaries—the landlord, the moneylender or the petty trader. At the same time, peasants were forced to buy their own food on the market at the prevailing high prices and take advances from moneylenders at high interest to sow seeds for the following year's crop. The post-war crisis was then exacerbated by landlords who used the agrarian price increase as a pretext for increasing rents. When the peasants failed to meet these new demands, they were evicted and their lands settled with other tenants. The landlords also sought to convert tenant lands into bakasht or zirat, as in 1937-9, to take advantage of the price increases.

In the mid 1940s a new development encouraged the rising incidence of landlord attacks on tenant land. This development consisted in the acceleration of plans for the legislative abolition of zamindari under the second Congress ministry. The zamindars reacted with extreme disfavour to these new proposals and hoped to avoid the consequences of the Congress legislation on the issue by (a) claiming to be owner-cultivators, (b) using the threatened lands as sites for erecting sugar mills[98] or (c) resettling the lands at higher rents with tenants so that if compensation was to be paid on the basis of the rent roll they would stand to gain. This attack on tenancy rights reached its culmination in the attack on occupancy holdings and the conversion of peasant lands into either bakasht or zirat.[99] By claiming occupancy tenant holdings as their bakasht or zirat the landlords hoped to circumvent zamindari abolition. The result of these attacks on occupancy holdings and the increase in land transfers resulted in the dispossession of the tenantry and depeasantization on a large scale. Even where old peasants remained, the conditions under which they

98 Ram Manohar Lohia, in his survey of the bakasht problem, noted that the sugar mills occupied a substantial area by June 1947 (a total of 50,000 acres). In fact, a local proverb noted that *Nil Zulum* (indigo exploitation) had been replaced by Mill *Zulum* (exploitation by sugar mills). The Lohia commission was set up by the CSP in 1946 to examine the whole question of peasant rights. Its report was never published. BSA, File no. 6 (ii) 1947.

99 As noted in chapter 2, bakasht lands were those on which the zamindars exercised more direct control over cultivation and, though occupancy rights could accrue here in certain circumstances, the zamindars normally cultivated the land through short-term tenants or sharecroppers. Zirat lands were the homestead lands of the zamindar on which no tenant rights could accrue.

worked were substantially altered as a consequence of the land transfers. They remained as sharecroppers (bataidars), a position inferior to that of tenants. Many of these sharecroppers, by the late 1940s, had rights only to their homestead lands, averaging about 0.04 acres, which they obtained from the landlord rent-free in return for working on his land. In effect, therefore, they were being pushed into the ranks of landless agricultural labourers.[100] By the 1950s the number of landless families had increased from 26 per cent to 45 per cent in some villages, an extraordinary rise by any standards.[101] In such circumstances attacks on peasant land and the consequent erosion of tenant rights could not remain unchallenged for long. As early as June 1946 a serious agrarian movement had erupted in several districts in Bihar. The timing of this unrest was directly related to the most recent attacks by landlords on tenancy rights, following fears of imminent zamindari abolition in 1945-6.

The struggle to reclaim bakasht lands affected hundreds of villages in Bihar in 1946. Most villages had some bakasht land and in 1946 there were nearly 20 lakh acres of bakasht land in the whole province. In Bhabua subdivision, for example, there were 1,789 villages with some bakasht land. There was a minimum of 2 lakh bighas in the sub-division, distributed unevenly among the villages. About 2 per cent of the villages had over 1,000 bighas each, 5 per cent had over 500 bighas, 10 per cent had over 100 bighas, 25 per cent had 50 bighas, and 58 per cent had 25 bighas or less of bakasht.[102] The peasants cultivating these bakasht land had few rights. The lands they tilled were given to them in plots in villages to cultivate either on high rates of cash-rent (Rs 8 or 10 per bigha) or on produce-rent, the latter system being more prevalent. The zamindars frequently deprived the actual tillers of nearly 80 per cent of their produce through claims of rent and false loans. Peasants were also kept under constant threat of ejection and rarely managed to obtain any proper receipts from the landlord. Most of them were often compelled to work on the zirat lands of the landlord and forced to do begari without wages, while they had clear rights only to bari or homestead lands. In order to claim

[100] Report of the Congress Agrarian Enquiry Committee which was set up in 1949. See Rajendra Prasad Papers, File no. S 89/49.

[101] Mitra and Vijayendra, 'Agricultural Labourers and Peasant Politics, Rural Proletarianisation in Purnea, Bihar' in A. N. Das (ed.), *Agrarian Movements in India*, p. 103.

[102] *Bakasht* land disputes in Bhabua subdivision, BSA, File no. 336/46.

peasant holdings as their bakasht or zirat landlords had to evict long-term tenants. In Mokama it was reported that the landlords were anxious either to settle their lands at higher rents or to claim possession of the land, for if abolition of zamindari was undertaken then compensation would be paid only in accordance with the rent roll as shown by proper settlement. Landlords stood to lose both ways if they neither settled the lands nor retained *de facto* possession of it.[103] In many villages in June 1946, therefore, the landlords started preventing peasants from tilling their lands, claiming them as their bakasht land. The peasants reacted with anger and serious rioting broke out all over Bihar.

In Bhabua subdivision in Shahabad struggles erupted in many villages. In Mahuar village the tenants protested against the landlord evictions and began ploughing the disputed land of the malik, Indrachand Kezarawal. The landlord attempted to retaliate by using his lathials. The police Inspector-General reported that disputes involving land were so frequent that police pickets and posts had to be set up in Buxar and Bhabua to deal with the struggles. The peasants also protested by stealing the crops of the zamindars. Rural crimes were recorded as on the increase in several thanas at Nokh, Karaghia, Rajpur and several other places. Burning of sacks of straw, the stealthy removal of crops at night from fields and cattle rustling were common.[104] The paddy straw of Mahalal, a Congressman and petty zamindar, was burnt by a number of persons. Some political parties took advantage of the situation and renewed activity in the villages. In particular the Kisan Sabha, the CSP and the Radical Democratic party had all joined hands to work among the kisans. The Radical Democratic party toured the villages in Bhabua asking the peasants to resist exploitation and exhorting them to 'till the bakasht lands and pay actual and justifiable rent. Do not pay in kind and do not pay without receipt. Form people's committees in every village and obey this committee as authority to settle dispute over bakasht'.[105]

By 4 July 1946, in Bhabua subdivision, the agitation had spread to 75 villages, involving 20,000 acres of land in six police areas. As the bakasht disputes continued the landlords became increasingly worried and formed an association to resist trespass by the tenants. On 7 July

[103] Ibid.
[104] Ibid.
[105] Ibid.

1946, fifty landlords and petty zamindars met at Mokama and resolved to defend their bakasht rights. In connivance with the local police, the Sub-Inspector and the Sub-Divisional Officer they began issuing notices under section 144 of the Criminal Procedure Code in two dozen villages to prevent the tenants from tilling the land.[106] In the entire subdivision the extent of land under dispute and under section 144 amounted to one lakh bighas. The District Magistrate warned that if the lands were not sown within the following fortnight a loss of 10 lakh maunds of paddy would ensue. In Bhabua half a dozen zamindars, headed by Ramsurath, marched from village to village terrorizing the kisans. Their activities in Tamarhi village reveal the kind of terror tactics they used. Tamarhi was a low-caste Pasi village with 25 houses and a population of 250. On arrival in the village on 10 July 1946, the zamindars seated themselves under the banyan tree, and their lathials rounded up all the men, women and children and brought them before the zamindars. The three zamindars, Ramsurath Chaudhury, Babbhan Singh, and Tribeni Singh all had guns and belts full of cartridges. They proceeded to warn the peasants of dire consequences were they to plough the bakasht land. Of course there was an outcry from the peasants, upon which Ramsurath fired two blank shots shouting 'Gandhi ji ki jai' and waved the tricolour.[107] The villagers began fleeing in terror, whereupon the zamindar shot dead one peasant, Sheonath Pasi. It is significant, although not unusual, that the zamindars seem to have used Gandhi's name while attacking the tenants. In fact, Gandhi was frequently appropriated in different ways by opposing classes and the zamindars were quite happy to use his name to justify their brutal actions. However, the full contradictions of this rarely came to light in such a blatant fashion as they did at Tamarhi. The report of this incident by one of the peasants, Deonandan Pasi, deserves to be quoted to complete the picture. He stated:

All the people in my village used to cultivate the *bakasht* lands of malik Ramayan Singh and Shree Singh Missir. On Tuesday we ploughed the field and sowed seedlings. On Wednesday Ramayan Singh and the *Tahsildar*

[106] Section 144 of the Criminal Procedure Code implied that there would be restricted movement on the disputed land, not more than 2-3 people would be allowed to congregate on it and the land would lie fallow till the dispute was settled.

[107] BSA, File no. 336/46.

forbade us from cultivating the field. Then we replied that we had ploughed the field and sowed seedlings and where would we go now in the month of *asarh* and our children will die of starvation. Then their *lathials* threatened to assault us if we went near the field . . . We replied we would. Chaudhury Ramsurath Singh and Tribeni Singh then opened fire.

Despite such hard evidence of malpractice, the case was handled with extreme negligence by the local police and the courts. At first the Sub-Inspector refused to record the case and only after a visit by the District Magistrate was a warrant of arrest issued against four persons, two zamindars and two rent collectors. Even then, the Bhabua police did not attempt to arrest them. Finally, when the case came up in the courts the accused produced alibis that they had been elsewhere during the occurrence and the case was closed.[108]

Such incidents of peasant protest and zamindari retaliation also occurred in other districts in the province. In Gaya district the Commissioner wrote to the Chief Secretary, Bowstead, that reports of the forcible cutting of crops on *bhauli* (produce-rent) lands by tenants had come from several villages. The Sub-Divisional Officer reported that 291 cases had been filed in the subdivision and that lawlessness was increasing.[109] In Narma village, the landlord, Hiralal Singh, had regained possession of 100 bighas of bakasht land, while the tenants were trying to reclaim the same land by forcibly cutting paddy and sugar-cane. In Sisamba Mahal, which consisted of 5 large villages, the landlord, Ratneshwar Singh, was notorious for not granting rent receipts, while the peasants had also agitated in the past over their rights to irrigation. They now decided to defend their rights over the bakasht lands and also attempted to loot the zamindar's kutcheri (record office). On one occasion in Paleya village, under the Tekari raj, (a big zamindari estate), one hundred peasants were engaged in cutting the paddy of the bhauli lands as they had done in the past. On 25 November 1946 the raj filed a petition for distraint of crops. However, the peasants managed to cut the crops on an area of 30

[108] Ibid. The three main accused, Bhabban Singh, Ramsurath Chaudhury and Tribeni Singh all had alibis stating that they were in Dhanbad and Arrah during the occurrence and were later allowed to go scot-free.

[109] Agrarian incidents in Gaya, File no. 6(1)47, In this subdivision, the Commissioner reported to the Chief Secretary that cases of bakasht disputes had come in from the villages of Nama, Paleya, Imadpur, Sarla, Masurul, Habalipur, Tali, Mohar and others.

bighas. The police, clearly in connivance with raj officials, were then called in to prevent the kisans from harvesting the paddy. In fact, the police had actually taken up residence in the raj kutcheri and were seen enjoying the full hospitality of the raj. In Sarla, another village under the Tekari raj, there was serious tension over the division of crops on 700 bighas of land, and orders had to be issued under section 144.[110] In Jehanabad a serious clash developed between the landlord's party and the tenants, as the latter succeeded in cutting away the paddy crop on the disputed land. In this part of the province local officers noted that the troubles in Gaya in 1946 were compounded by widespread communal disturbance, which had already caused a deterioration in the public order situation.

The troubles in Gaya district, as elsewhere, continued all through 1947. Attempts by magistrates to settle the disputes often came to naught as many of them openly favoured the landlords, quickly diminishing the law in the eyes of the kisans. In August 1947 the kisans in Kurihar decided to launch a satyagraha to restrain another group of tenants with whom the attached lands had been settled. The kisans claimed that they had sowed crops before the magistrate's order settling land with the new tenants was passed. As the month dragged on without any apparent hope of settlement the agitation deepened and there were reports of violence. In many villages in the Laksamba area of Makdimpur more than a thousand people were charged with breaking bandhs (irrigation embankments). Armed with lathis, bhalas, and *garasas*, they approached the bandh shouting 'Mahatma Gandhi ki jai' and '*Inquilab zindabad*'.[111] The situation was allegedly uncontrollable and a police party fired eleven rounds. Bandhs were also cut elsewhere in the villages of Jamuwan and Mandil, where the villagers, who belonged to the lower castes of Goala and Kurmi, challenged the upper-caste Bhumihar Brahmins. There were reports by many zamindars that the lower castes were restive and that in several villages dacoities and thefts by low-caste Dusadhs were on the increase.[112] In some other south Bihar districts, serious bakasht disturbances were also in progress, particularly in Monghyr in the Barahiya Tal area. In Baruna and Mahiral villages the

110 Ibid.

111 Ibid.; Report from the Sub-Divisional Officer, Nawadah, 20 October 1947.

112 Ibid. See the petition of Baleshwar Prasad Singh and others of Konchi village complaining about the Dusadhs.

kisans won back their rightful claims to 150 bighas of land after a long agitation. However, in this case the Prime Minister himself actually intervened in a brazenly partisan way and ordered that all land which was being tilled by the zamindars themselves should be returned to them; by this means the zamindars recovered as much as 65 bighas.[113] In another village in south Monghyr, both Hindu and Muslim peasants joined hands in defending their lands against landlord incursions, thus succeeding in reclaiming 25 bighas of bakasht land.[114] Further north, in Muzaffarpur, agrarian trouble was also on the increase. In Mahiha Sadr police district a petition was filed by one Ramprasad Singh, who alleged that his tenants had been creating trouble over 30 bighas of bakasht land.[115] At another village in Muzaffarpur, the kisans looted the maize crop from the disputed lands of Shyamnandan Sahay, a zamindar and a Congressman. There was evidence here to show that the tenants had receipts in their possession on behalf of Sahay for the last twenty years. There had rarely been such excellent proofs of ownership by the tenantry and yet, even then, twenty-one kisans were arrested, allowing the landlord's lathials to moved about freely on the disputed land.

In north Bihar the most extensive actions against the landlords were in Darbhanga.[116] Rioting broke out in many villages, the Superintendent of Police of Darbhanga recording that the activities of the socialists had increased greatly in the villages and asserting that most of the struggles had been incited by them. In Jagdishpur (in Bahea police district) the kisans forcibly harvested the paddy crop in spite of the fact that Section 144 notices were posted against them. Here, however, it was reported that the zamindars were reluctant to punish tenants. In fact this was unusual, as the zamindars normally retaliated

[113] BSA, File no. 6(1)47. In other villages, as at Azizpur, 1,000 bighas of disputed land were attached by the Sub-Divisional Officer till further settlement.

[114] *Hindustan Times*, 3 November 1946.

[115] BSA, File no. 6(1)47. In some villages the landlords complained that the peasants were claiming the whole village as their batai for 'it was being hammered into their minds that Swaraj has come and the whole land is to pass to them'.

[116] BSA, File no. 6(vii)47. Twenty-one kisan agitations were recorded in Darbhanga and, at Samstipur, the Sub-Divisional Officer was of the opinion that the tenants were in actual possession of the lands under dispute and that the landlords were trying to dispossess them by trying to settle their lands with new tenants.

with violence.[117] In Kokal village, in Darbhanga, peasants looted 44 maunds of paddy and other foodgrains in protest against their lands being seized from them. In Bahera, the Sub-Divisional Officer recorded that he personally saw one hundred persons cutting paddy. 'What the raiders do', he reported, is that 'they select some land without previous intimation and start cutting the paddy. When an officer approaches, the peasant activists escape and shelter in an adjoining village.' In Andama the peasants organized themselves into a kind of guerrilla force which they called Tiddi Dal. They burst onto the fields of the zamindar with sickles and then carried away the paddy, resisting all those who stood in their way. They even regrouped to the call of a bugle so that, whenever a police party was seen, they disappeared on hearing the bugle and concealed themselves. Such guerilla tactics were said to be particularly demoralizing to the zamindar's men.[118]

The part played by the socialists in organizing the resistance was constantly referred to by local officers. In Pandual the socialists had held a large camp, where the kisans were congratulated on their fight over bakasht lands. Here the political message was drummed home that the zamindars were directly responsible for the unrest through their attempts to subvert zamindari abolition by settling lands with other tenants, thus causing extreme and rightful disaffection among the peasantry.[119] As the socialists increased their political activity, the looting of crops was widely reported in many villages in Darbhanga all through 1946 and 1947. In Chapra it was reported that 300 to 400 men were looting paddy under socialist guidance, shouting 'Inquilab zindabad, loot the wealth of the malik, zirat lands belong to the peasant . . . from now on we shall not starve, we shall have blood for blood.' In Mehur village, one hundred persons, significantly including both men and women, forcibly cut crops from the fields of the Mahanth of Sonhi.[120]

The Sub-Divisional Officer in Darbhanga noted that, as the number

[117] Ibid. In Andhri (P.S. Benipathi), the Mahanth of Andhri was the landlord. The disputed area was about 594 bighas and the mahanth succeeded in setting the kisans' houses on fire and killing some peasants.

[118] Ibid. Armed gangs went around looting crops in many villages in Darbhanga and the zamindars reported that unless protection was given not only would standing crops be looted but harvested crops would also be carried away.

[119] Ibid.

[120] Ibid.

of cases of forcible cutting of crops increased in Bahera, Biraul and Sadr thanas, the landlords started to retaliate with violence. Most of them had armed themselves to the hilt and were attempting to spread terror among the kisans, often in collusion with the police. There were many cases of even small landlords forcibly ousting tenants in order to acquire possession of their lands.[121] In some cases the police intervened in the most overt way possible. In Samastipur, for example, the Sub-Inspector of police and eighteen armed constables themselves actually ploughed 25 bighas of land belonging to the tenants. The entry of socialists into the village was banned as it was reported that the kisans had been incited by the CSP cadres. Most often the local officer's reports found against peasant claims as, for example, in the villages of Kokat, Palore, Nehra, Jagdishpur and Andhri, where trouble flared up in 1946. These villages were dominated by harshly exploitative petty zamindars who wanted to evict the tenants cultivating their bakasht or zirat lands on sharecropping terms. In these villages the zamindars claimed that the lands belonged to them, while the tenants asserted that they had been cultivating the same ground for thirty years. The Sub-Divisional Officer's report, however, found against the tenants in the following terms: 'from my enquiry I am convinced that the land in village Kokat had been under the possession and cultivation of the landlords and the claim of the kisans over them is baseless and totally false.'[122]

Such connivance by local officers in negating peasant claims was most clearly revealed in the Sathi estate disputes in Champaran in 1946-7. Sathi estate was an area of 200 bighas that belonged to Bettiah raj, a large zamindari estate.[123] The lands on this estate were part of the holdings of the occupancy raiyats. In 1943, the manager of Bettiah raj, which was under the Court of Wards, decided to evict some of the tenants from a major portion of the 200 bighas, leaving only 25 bighas in the possession of the tenants. When the latter protested, there were initially hopes of a settlement in their favour. However, the matter dragged on and the peasants began an agitation to prevent the lands from being tilled. In October 1946 the disputed lands were settled in favour of R.P. Sahi, who was the new Commissioner of Excise, and

121 Ibid. The police, in most of the cases, were on the side of the landlords and were quick to impose either section 144 or to arrest the kisans.

122 Ibid. The kisans in Kokat expressed their anger by looting 44 maunds of paddy from the zamindar's carts.

123 BSA, File no. 6(2) 2/1947.

his brother Ramrekha Prasad Sahi. This settlement was clearly engineered by the Government of Bihar, which had control over the affairs of the estate and was made in favour of Sahi, simply because he was a personal friend of S.K. Sinha, the Prime Minister. The tenants on the estate then organized a strong agitation against the sale on the grounds that it contravened their rights on the land and transferred the lands which had been in their cultivating possession to the zirat holdings of the landlord, thereby totally dispossessing them. When R.P. Sahi, as the new settler, began cultivating the land as part of his zirat, the tenants held several meetings. On 24 January 1947 about forty men, women and children of the Danger *toli* (the Dangar caste houses), headed by CSP leaders, obstructed the zirat cultivation of Sahi.[124] They then stood in front of the tractors and prevented Sahi's hired labour from ploughing the fields. Eighteen tenants were immediately arrested for 'rioting'.[125] The case then developed in a remarkable way. The Bihar government, and more importantly S.K. Sinha himself, were embarrassed lest it should become known that the government had been instrumental in negotiating the illicit transfer. Rather than let the matter be revealed in the course of the proceedings of the case they intervened to adjourn it. Subsequent High Court investigations revealed that the government had intervened to adjourn the case as many as fourteen times and had actually sent telegraphic orders signed 'Bihar' to the magistrate concerned, asking him to postpone the case.[126] The eventual High Court judgment on the

[124] Ibid. The list of the accused included the following socialists, Harihar Singh Patel, Gulabi Sonar, Kailash Missir, and 400-450 persons from the Danger toli.

[125] Ibid. A brief description of their offence recorded: 'Rioting armed with deadly weapons, criminal trespass, house trespass, having made preparation for causing hurt with brickbats and stones, mischief causing damage over Rs 50/-, and grievous injury by sharp pointed weapon.'

[126] H.R. Meridith, the high court judge, commented on the amendments: 'On 23rd July the D.M. received a telegram from the Bihar government to the effect that the government desire that the case be adjourned. The magistrate sent a copy to the sub-divisional officer, Sadr, for communication to the trying magistrate. A second telegram was received to the effect that the case should be kept pending until further orders. Since then there have been no less than 12 adjournments. It is of course rudimentary that a magistrate trying a case cannot receive instructions from administrative authorities. There has been quite unjustifiable interference with the magistrate's judicial action and if the magistrate had not the moral outrage to insist on proceeding with the case . . . considerable

scandal condemned the action of the Bihar government in attempting to postpone the case, being appalled that 'the government of Bihar as an amorphous individual or a body of individuals hiding behind the name "Bihar" on a telegram should have the temerity to interfere with the rights of the citizens of this country to have the charges against them tried with all possible expedition'. An enquiry into this incident was also conducted by none less than Sardar Patel, who concluded decisively in favour of the tenants' rights to their land. Despite all this the case was still quashed, and a settlement out of court arrived at, so that the culprits in this case were never brought to trial. Neither did the peasants recover their land.

In examining the nature of the agitations some conclusions can be firmly drawn. With regard to participation, the struggles affected all sections of the peasantry, right through from the middle peasant cultivator to the small low-caste sharecropper. Low-caste poor peasants participated in the movement in a particularly significant way. Being weaker tenants, they were the first targets of the landlords. Many of these tenants were agitating for the restoration of their lands and for the re-establishment of their rights to the land. Other more radical claims were also asserted in the course of the agitation. These included the right of the peasants to refuse to pay more than the usual share of the produce to the landlord and the right to refuse to pay *nazrana* (or cash gifts) to the landlord. Landless labour also played an unprecedented part in the agitation and in some areas these groups used the opportunity to boldly demand more land from the landlords in return for begari. In Rampur village, in Buxar, the normally submissive Chamars (landless labourers eking out a miserable living on 1/2 bigha of the land) were demanding three bighas. The zamindars tried to use lathials to force them to work. In Nawadah, in one village, some landless labourers offered satyagraha to force the landlord to give them some land.[127] Thus, in many places, the zamindars were now facing organized protests against begari and illegal extortion. Considerable community cohesion appeared in the course of the protest. Muslim peasants, for example, often joined hands with their Hindu neighbours in attacking their landlord. There is plenty of evidence in Monghyr, for example, to show this. The participation of

excuse can be found for him that the magistrates are at present under direct government control and looking to the government for protection.'

[127] For activities of landless labour, see BSA, File no. 336/46.

women was also common, as in the peasant agitations in 1937-9, and there were many instances of women participating in peasant satyagrahas by lying in front of the plough. The involvement of women further bridged caste divisions. Not only did lower-caste women agitate but upper-caste Bhumihar women came out of purdah to protest against their lands being taken away.[128]

Apart from the widespread incidence of the forcible cutting of crops on disputed lands, the pillaging of grain hoarded by landlords and cattle rustling both increased dramatically in the period.[129] In Konchi the landlords complained of the goondaism of the local Dusadhs and reported that Dusadh dacoities were on the increase. However, the landlords making the complaints were frequently hoarders themselves, whilst the 'goondas' were often starving landless labourers.[130] One case recorded the fact that the property of Jaldhar Issar, a rich peasant secretary of the Bakhari thana Congress committee, had been looted by the kisans. In Noakholi village, when five leading kisan workers were arrested for looting crops, other kisans collected the crops and then refused to allow them to be taken away. The next day the kisans again seized a cartload of maize and distributed it. In some parts of north Bihar the peasants seemed to have organized themselves into a guerrilla force, as reported in Andama.[131] They would fall in at the call of the bugle, loot the landlord's land, and disappear into the neighbouring village. Breaching irrigation bandhs and cutting wood from the zamindar's forest were other methods of protest. Where the peasants chose to participate non-violently, it was common to engage in peasant satyagrahas. In these cases, where the landlord was ploughing the disputed land with hired labour, the peasants protested by preventing the ploughing. Here again the women would lie in front of the plough.

Slogans that were common in the period related to the rights of the kisans and the fact that they would fight to reclaim them. These included: '*Malik ka dhan loot*'. '*Zirat zamin raiyat ka hai*' and '*ab ham log bhuke nahi marenge*'. (Loot the lands of the landlord; zirat land belongs to the peasant; never again will the peasant starve.) It is interesting that in many of the struggles the peasants also shouted

[128] BSA, File no. 6(vii)47.
[129] BSA, File nos. 336/46, 6(1)47, 6(vii)47, 6(ix)47.
[130] BSA, File no. 6(vii)47.
[131] Ibid.

'*Inquilab zindabad*' and '*Gandhi ji ki jai*'.[132] The cries of the national movement had thus become an integral part of the struggles against their landlords. All the specific demands converged around the abolition of the zamindari system. The socialists who were active in the agitation demanded that the abolition should be proceeded with as soon as possible and without compensation. In furtherance of this they issued pamphlets urging kisans to continue the strike and loot the crops on disputed lands. This demand was taken up seriously by the agitating kisan, as is clear from the slogan 'the zirat land belongs to the peasant'. This was a political demand in so far as it implied a political vision and an analysis of social and political evils that stemmed from the zamindari system.

In analysing the leadership and organization of the 1946-7 struggles, it appears that the socialists and the Radical Democratic party were very active in several areas and less so in others. This involvement had antecedents in the Quit India agitations whose structures had never really died away. In Darbhanga area the collector's reports frequently mention that Ramlochan Singh, the socialist leader who had been active in the Azad Dasta movement, was actively inciting the tenants. Another socialist who specialized in terrorizing the petty zamindars in Warisnagar thana was Suraj Narayan Singh. Both these men were said to have a large number of followers and to have organized the local peasants to loot the paddy from the bakasht and zirat lands of the zamindars. In Chakdah village, the communist leader, Bhogendra Jha, was also active in mobilizing the kisans in the village against Amawan Raj.[133] In recognizing the leadership of the Radical Democratic Party and the socialists in some of the struggles, I am not suggesting that the mobilization of the peasantry was contingent on the intervention of these alone. On the contrary, the peasant in Bihar, as elsewhere, acted on his own volition and was the main maker of his own rebellions.

As in 1938 the peasant actions of 1946 seem to have been caused by the violation of that moral code which E.P. Thompson has called

[132] Ibid.

[133] In Chakdah the raj wanted to settle a piece of land with new tenants. This land had been used by custom as an area where the villagers would retreat to when the lands around Madhubani were submerged under water due to the Kamla floods. The villagers opposed this move by attempting to construct some huts on the disputed land. These were later pulled down by the Sub-Divisional Officer and the raj's men.

the 'moral economy of the poor'.[134] James. C. Scott has tried to apply Thompson's concept to the peasantry of South East Asia in his *Moral Economy of the Peasant.* Scott argues that there are two moral principles of particular importance to peasants, which he calls the norm of reciprocity and the right to subsistence. The first is a moral obligation incumbent on both the peasants and dominant classes in which the former provides the agricultural produce while the latter reciprocates with services. The exchange cannot be quantified in any obvious manner being based as it is on customary practice. The second moral principle consists in the belief that everyone in a society has a right to a basic subsistence. Thus exploitation which leaves a peasant hungry will be considered to violate this moral code. Peasants therefore tend to revolt when their traditional rights are challenged. Deprived of the use of many of their basic resources, in particular their customary access to cultivable land and the use of forests, they tend to rise in protest. Such risings, it has been argued, are defensive movements caused by indignation at the breaking of the moral code rather than by starvation. Scott, following Thompson, thus goes beyond the crude economic determinism of many scholars, both Marxists and non-Marxists, who have sought to explain peasant movements in terms of starvation and hunger.[135] Hunger by itself was not the only cause of revolts. More important was a sense of moral outrage. The timing of the revolt, the mobilization of the peasantry and the targets of attack were determined by such a consciousness.

In the case of the bakasht land struggles, it was clear that the action was fuelled by the systematic attacks of landlords on tenant lands. Peasants were evicted from those lands which had always been under peasant cultivation. Attacks by landlords aimed at converting peasant lands into bakasht lands and zirat lands were not a new phenomenon. In fact, landlords had been extending their bakasht and zirat lands since the rise in prices in the 1930s to take advantage of the price increase. In 1937-9 the peasantry had sought to protect themselves against the landlords' onslaughts in an intense agrarian agitation under the Kisan Sabha. This was a period when the politics of nationalism and socialism had begun to penetrate the countryside on

134 E.P. Thompson, 'The Moral Economy of the English Crowd in the Eighteenth Century', *Past and Present*, no. 50, 1971, p. 79.

135 James C. Scott, *The Moral Economy of the Peasant, Rebellion and Resistance in South East Asia*, Yale, 1979.

a significant scale and the experience of agitation and class struggle had built up the consciousness of the peasantry. Thus by 1946, when the landlords renewed their attacks on peasant holdings for fear of zamindari abolition, the peasants were more aware of their rights and more ready and anxious to defend them.

Some historians would see the struggle as mainly economic. However, the peasant actions were not simply incoherent attempts to redress economic grievances. They were in every sense political. In undertaking to challenge his relationship with the landlord the peasant engaged himself in what was essentially a political task and a task in which the existing power nexus had to be turned upside down as a necessary condition for the full redress of any particular grievances. One has, for example, only to examine the unusually courageous peasant agitation in the Pasi village in Tamarhi, where poor low caste peasants challenged the zamindar malik, Ramayan Singh, and ploughed the disputed land despite the threat of dire, or even murderous, consequences. In Nawada, the peasants ploughed a staggering 500 bighas of bakasht lands of their mahanth landlord, in an attempt to reclaim their rights. Engrained attitudes of fear and respect for the repressive authority of the landlord seem thus to have been overcome in these struggles to such an extent that, in many cases, the landlords actually feared to restore their authority in case of further peasant reprisals.

As regards the leadership provided by the Radical Democratic party and other groupings, the political articulation they gave to the grievances of the peasantry was also important. The new political vocabulary of the socialist leadership had originated in the expression of mass grievances. However, as I noted earlier, by recognizing the leadership of these parties in organizing the struggle in some areas I am not falling victim to the elitist approach, nor am I suggesting that the mobilization of the peasantry was contingent on the intervention of these leaders. I am only recognizing the fact that by the 1930s and 40s these political parties had begun to penetrate the countryside on a large scale and were actively involved with the peasant struggles.

COMMUNAL RIOTS AND KILLINGS IN RURAL BIHAR

The above account allows one to construct a comprehensive picture of peasant protest. Widespread agrarian agitation over the restoration of bakasht land and against the exploitative incursions of the zamindars was occurring in several districts in Bihar from the summer of 1946 onward. These peasant actions included both Hindu and Muslim peasants, who frequently made common cause against the zamindars.[136] On many occasions Hindu and Muslim peasants collaborated (in Monghyr, as well as elsewhere), in the forcible cutting of crops on disputed land.

In November 1946, however, a strange new dimension developed in the popular actions in rural Bihar, that of communal killing. Between October and December 1946 Bihar witnessed one of the worst communal outrages in its history. 400,000 Muslims were affected in the districts of Patna, Chapra, Monghyr, Bhagalpur and Gaya in rural and urban areas. By December 1946, 50,000 Muslims from Bihar had migrated to Bengal in the wake of gory communal strife. A correspondent who met four Muslim families on a train noted the sad plight of a migrating Muslim peasant.

> When the train passed through Fatwah one old man with tears in his eyes pointed out the road leading to his village, where his forefathers had lived and where he had been living in peace and amity with his neighbours for fifty-five years. He had his three acres of land and two bullocks and now he was leaving it all behind.[137]

All through November and December 1946 there was an unusual configuration of both agrarian and communal rioting. Newspapers noted that, while parts of south Monghyr and Patna were smouldering with communal strife, just across the Ganges in north Monghyr, Hindu

[136] According to the census of 1931, 83 per cent of the population of Bihar and Orissa were Hindus. The Muslims numbered about 4,284,306 and formed 10.1 per cent of the population. Their maximum strength was in the north of the province and grew steadily less as one travelled from the north to the south. In the district of Purnea, the proportion was rather high. Here the population of Muslims was 886,368, comprising 40.5 per cent of the inhabitants of Purnea. In Kisanganj subdivision they outnumbered all other religious communities by 2:1. In other north Bihar districts the proportion of Muslims varied from 16.5 per cent in Champaran to 11.2 per cent in Bhagalpur. Outside north Bihar the five districts where the Muslim population was as high as 10 per cent were Patna, Gaya, Monghyr, Hazaribagh, and Santal Parganas. The Muslims had a greater percentage of their population in towns, about 8.8 per cent as against 3.6 per cent for Hindus. As a result, the relative proportion of Hindus and Muslims in urban areas was 3.5:1 while in rural areas, excluding Purnea, there were eleven Hindus to every Muslim.

[137] *People's Age*, 26 December 1946.

and Muslim tenants participated together in a widespread seizure of the landlord's crops. In another village in north Monghyr, Hindu and Muslim tenants captured 25 bighas of land.[138] In Chapra, after the fury of the communal riots, a new phase in the situation developed with the cutting of crops and the looting of properties by gangs of tenants, both Hindu and Muslim.[139] However, despite these instances of joint activity *vis-à-vis* the landlords, the rural areas in south Bihar continued to be rocked by communal carnage of the greatest magnitude.

Neighbouring Bengal had been smouldering since the great Calcutta killings and the disturbances in Eastern Bengal in August 1946. Published reports inspired by the Mahasabha in the newspapers had, however, made out the disturbances to be far worse than they had actually been. Moreover, the Congress ministry in Bihar made no efforts to scale down the inflated rumours of the numbers who had died.[140] Instead, it attempted to make political capital out of the matter and on 28 October the provincial ministry authorized the celebration of Noakhali day, to mourn the death of Hindus in East Bengal.[141] This ill-advised action further provoked the feelings of the communities against each other. Stabbings started in Patna town itself and by 31 October Bhagalpur, further to the east, succumbed to rioting and forty Muslims were massacred at Teragna station. By 2 November disturbances were widespread throughout north and south Bihar and a threat of related labour riots in the industrial areas of Jamshedpur and Dhanbad started to emerge. Tuker noted the difficulty the troops had in controlling 4,000 sq. miles of violently troubled territory. The number of Muslims killed in this short and savage killing was 7,000-8,000 men, women and children.[142]

139 *Hindustan Times*, 3 November 1946.

140 The Congress working committee commented thus on the Bengal riots: 'The committee find it hard to express adequately their feelings of horror and pain at the present happenings in East Bengal. The reports published in the press and statements of public workers depict a scene of brutality and medieval barbarity that must fill every decent human being with shame, disgust and anger. The committee are aware that facts have been exaggerated but communiques and statements of the local officials themselves paint such a picture of ghastliness that no exaggeration is necessary to add to the effect. The committee hold that this outbreak of brutality was the direct result of the policies of hate and civil strife that the Muslim League has practised for years past and of the threats'. See *Hindustan Times*, 23 October 1946.

141 F. Tuker, *While Memory Serves*, London, 1950, p. 184, and *People's Age*, 28 October 1946.

142 Tuker, *While Memory Serves*, p. 182.

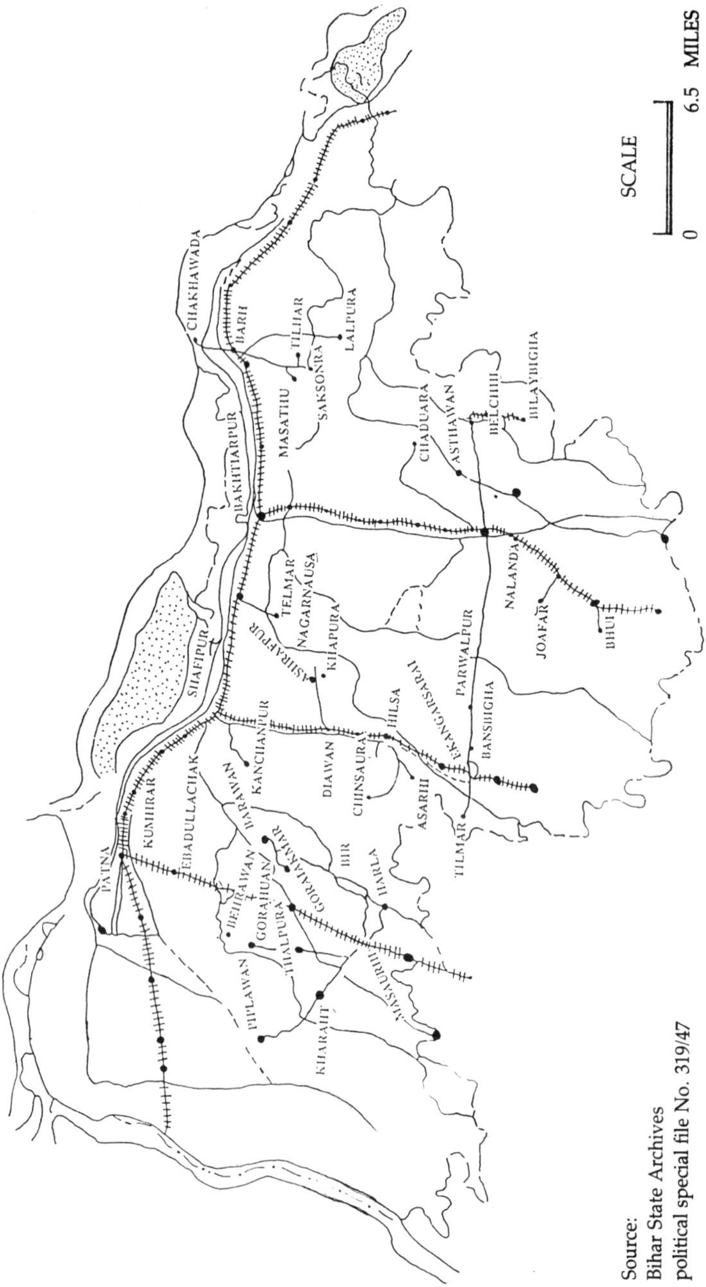

Source:
Bihar State Archives
political special file No. 319/47

Map Showing the Places Affected by Communal Riots in Patna District in 1946

The atrocities in Bihar were only a foretaste of what was to come at Garhmukhteswar in the United Provinces a few weeks later.[143] The military reports of the Bihar killings give useful details of the riots:

> It was 1st November at 9.30 in the morning that Area Headquarters rang up Colonel Venning to push two companies forward to Bihar Sharif . . . Before dawn the following morning he pushed on to Bihar Sharif and sent troops on to Fatwah, Islampur and Hilsa. His forward patrols found that murder, loot and arson had started in remote villages . . . Venning went on to Fatwah itself and found an attack going on at Juthli . . . Returning he heard that Bikalpur was being attacked so with another officer and a small force he sped off in that direction, the mob melted away, leaving several women and children in badly mutilated condition. While he was dealing with this, another report came in and he was off to another village, too late to prevent the burning of the village where about twenty women and children and a few old men lay dead on the ground, stripped naked and burnt. Other patrols went out over a wide area through the night, one of them reached Kanchipur just after midnight to find the attack over and a terrible slaughter. Two hundred Muslims lay dead. The whole village was in ashes . . . The heaviest attack being one at a village called Tehara. Nearly 400 were counted, mostly killed with shot guns or knives. Many had previously been burnt alive while some wells were full to the top with dead. Of the hundreds evacuated about a quarter were badly wounded.[144]

Bihar Sharif subdivision consisted of an area of 800 square miles with a population of 750,000, of whom 90,000 were Muslim.[145] The greatest number of casualties were in the Hulsar circle in the western part of the subdivision, where the casualties amounted to 2,500 killed. Here it was reported that at least two large villages and about 100 small villages in the subdivision had been wiped out as a result of the mob frenzy that swept the province. The butchery of Muslims in Nagarnausa area in the subdivision was one of the worst. Colonel Venning, who had moved into the area with some of his men, again noted the extent of the destruction. He could hear the cries of the mob

[143] At Garhmukhteshwar town in UP, at a customary fair, hundreds of Muslim petty shopkeepers were butchered. A few days before the riots a big Rashtriya Swayam Sewak Sangh, (hereafter RSS) rally was held in Meerut presided over by Modi, the owner of several mills, and many provocative speeches were made. See *People's Age*, 2 March 1946.

[144] Tuker, *While Memory Serves*, p. 186.

[145] BSA, File no. 319/47.

attacking the village and as he hastened forward, saw smoke rising from several burning houses. On reaching the village his party saw that a mob of between five and ten thousand were surrounding the village, burning and killing every Muslim they could find. They noticed that many of the corpses were in various stages of putrefaction so that it was obvious this attack had been in progress for several days. The temper of the mob was extremely aggressive and they attempted to attack again and again. H.B. Martin, the district officer, also recorded his impressions:

> I arrived in Bihar Sharif on the afternoon of November 4th. Almost immediately I went out to a village 6 miles from Bihar Sharif where it was reported that the local Hindus were obstructing the evacuation of Muslims to Bihar Sharif by the military. We arrested some ringleaders and shortly afterwards the Hindus of all the neighbouring villages gathered at a short distance beating drums and shouting slogans. These mobs appeared to be possessed of an almost fanatical desire to exterminate the Muslim population wherever possible. This induced them to continue attacking parties of Muslims who were being evacuated to places of safety under police and military escort even after they had been fired upon and casualties had been inflicted. At Nagarnausa, the mob returned to attack the Muslim evacuees twice, even after having been dispersed and suffering casualties of up to 50 people on two occasions.[146]

The damage to property was heavy. Not only was damage done during the riots, but a considerable amount of looting of the houses of those Muslims who had left the villages also occurred. It is perhaps a salutary comment on these events and on similar horrors of the period that no official enquiry was ever conducted into the rioting.[147]

Later, Gandhi visited the affected areas and demanded that information be collected on the riots in the areas he was to visit. Hence

[146] H.B. Martin Papers, (Centre for South Asian Studies, Cambridge).

[147] Nehru and Patel rushed to Patna as soon as rioting erupted. Nehru noted in surprise that most of the rioters were ordinary peasants. He wrote, 'In the affected areas, that is in Patna district, in a part of Monghyr district and in Gaya, there has been a definite attempt on the part of Hindu mobs to exterminate Muslims. They have killed indiscriminately men, women and children en masse . . . I have addressed some large crowds in the rural areas and I have no doubt that many of them had participated in this bad business. They were ordinary peasant folk of Bihar, very simple, unsophisticated and rather likeable. They shouted "Mahatma Gandhi ki jai".' See SWJN (second series) vol. 1, p. 63.

some picture of the nature and scale of the rioting can be pieced together. Gandhi was to note after he visited the riot areas that the extent of the killing in the riots in Bihar shocked him far more than the scale of the killing in Noakhali in Bengal. Some of these case studies of particular villages are especially useful in that they record the exact extent and the nature of the mob attacks on the Muslims.[148]

The District Magistrate recorded the following facts in respect of Andhari village (Fatwah police district) to be visited by Mahatma Gandhi. The population of Hindus before the riot was 462, while that of the Muslims was 168. After the riot, the population of Muslims was recorded, straightforwardly, as 'nil'. Some had fled and many had died. Muslims from the neighbouring villages of Jaitya, Qazibigha and Nandechak had also sought shelter in Andhari village and 200-480 Muslims were thought to have been killed. The report also recorded that the majority of the people who took part in the attack were lower-caste Kurmis, Goalas and Dusadhs. The riot had been triggered off by a rumour that one Hindu of Bendauli village had been killed by the Muslims and this angered the Hindus. On 2 November 1946, Hindus in Andhari village were joined by other Hindus from the surrounding villages and began to attack the Muslim houses. Some of the Muslims then attempted to fire on the advancing mobs, while others resorted to the extraordinary measure of actually killing their women with their own hands to prevent the attackers from dishonouring them.[149]

In Kanchanpur village (Fatwah police district) detailed statistics of the death toll were recorded. The population of Hindus before the riot was 1,079 while that of the Muslims was 309. After the riot the population of the Muslims was, again, 'nil'. The number killed was between 100 and 156 Muslims. Originally the village had been populated by 33 families of upper-caste Hindus, 3 Harijan families and 15 other caste Hindus—all cultivators. The Muslim families had also mainly been cultivators and had numbered 33 families. On the communal history of the village the report recorded that 20 years previously there had been some trouble between Hindus and Muslims over cow sacrifice. The military had had to open fire and four Hindus had been killed. In more recent times the Hindu Mahasabha and the Muslim League had been active in the area, especially before the riots.

148 Mahatma Gandhi's tour programme in Bihar, December 1946, in BSA, File no. 319/47.

149 Ibid.

On 31 October 1946, a mob crying *'Jai Mahabir'* had approached the village.[150] On that occasion, however, both the Hindus and Muslims of the village went to stop them and averted any riot, but the news that some Hindus at Bhergaon had been killed incited the Hindus of Kanchanpur. The next day, on 1 November 1946, a mob came from the west and set fire to the houses of the Muslims in Kundli which was a tola (neighbourhood) of the latter village. The Muslims of the tola came out and all assembled on the roof tops in Kanchanpur village. Even so, the mob succeeded in attacking and killing many of them. The Sub-Divisional Officer of the area noted, 'When I visited the village on 17th November I found a smell coming from the house of Pahar Mian. The well in this house was full of dead bodies, and we filled it up with earth . . . It is said that some of the rioters raped the women and killed them. Seeing this some of the other women threw themselves in the well of the house and committed suicide.'[151]

In another village in Chahnawada, in Barh subdivision, the implicated Hindus were mainly upper-caste middle peasants while the Muslims were service holders or petty businessmen. Feelings had been strained in the village some months before the riots when the Muslims in the village claimed some land in possession of the Hindus as a Muslim graveyard. There was naturally some trouble over this and section 144 of the Criminal procedure code was imposed. This had happened a few months before the riot and the feeling between the two communities in the village was already strained. Fifty-one Muslims died in the riots that followed.[152]

Such horror stories were repeated in several villages. In Ghoshu village the population of Hindus and Muslims before the riots amounted to 200 Muslims and 80 Hindus. After the riots the Muslim population was nil. Again, some had fled, while 49 had died. On this occasion the Hindu population had consisted mainly of untouchables and other low-caste people and the village had had no history of communal ill feeling in the past at all.[153] In Harla village the popula-

150 The cry 'Jai Mahabir' had become a rallying cry for Hindu militants who saw Hanuman, the monkey God, as representing the power and vigour of Hinduism.

151 The rape of women (in this case Muslim) during communal riots symbolized the attempt by one community to defile and humiliate the other. It was rarely an act motivated by sexual desire, rather more by vengeance.

152 Ibid.

153 A large number of untouchables seem to have participated in the riots.

tion of Hindus and Muslims before the riot was 100 Hindus and 300 Muslims. After the riots the entire Muslim population was wiped out. The population in the village were mainly Hindu low-caste Goalas and Harijans. The Sub-Divisional Officer who visited the village noted that many of the wells in the village were full of dead bodies which could not be taken out, and that twenty-one wells had to be filled up. This figure alone gives us an idea of the scale of the killings. In Talpura village, the Hindus were predominantly low-caste poor peasants, while the Muslims were also poor peasants. After the riots the village was completely deserted by the Muslims, who had formerly numbered 200.[154] In the case mentioned above the involvement of the low-caste Hindus was conspicuous. They were not always the instigators, however.

In some villages there was evidence of complicity by the local elites in fomenting the riots. In Safipur and Khusrupur areas on 1 November 1946, a mob of 500 Hindus (mainly Goalas) of the villages of Khusrupur, Shafipur and Harampur raided Shafipur village. They looted Muslim houses, set fire to 36 houses and killed 4 women. Sixty Hindus, mostly Goalas, were accused. However, one of the rioters was armed with a gun belonging to a zamindar called Deobi Babu and it would seem that the zamindar had encouraged and even supplied some weapons. Nearby, in Masaurhi village, it was reported that the looting of Muslim property was widespread. A distinctive aspect here was the direct involvement of the rice-mill owners in the village, for the crowd gathered on hearing the whistle in the mill.[155]

In examining the evidence on the dynamics of the riots a problem arises. Why did the rural areas of Bihar, which had been in the midst of agrarian upheaval since the summer of 1946, suddenly erupt into sectarian strife in November 1946? Such a massive outbreak of communal rioting had never occurred in the rural areas in Bihar, even during the cow protection movements in the period 1880-1917 which had given rise to the first substantial sectarian tension in the countryside. More significantly, the historian of 1946 in Bihar has to examine the specific configuration of agrarian and rural rioting that set rural Bihar alight. Can such a sudden and widespread outburst of mass action be explained away as the outcome of the manipulative activity of some elite leaders reaching down with a communal ideology and

[154] Ibid.
[155] Ibid.

successfully building up sectarian strife in the countryside? Such an interpretation would be too simplistic and would ignore the social roots of populist actions. The sustained and intense mass actions in 1946 cannot be explained away by the machinations of leaders alone. Instead it seems essential to understand the history of sectarian relations prior to 1946 before examining the social context of that year and the strains and tensions of that specific historical juncture.

The beginning of sectarian tensions in the rural areas in Bihar can be located in the late nineteenth and early twentieth century, when there had been a massive outbreak of sectarian strife in the Bhojpuri region over the protection of cows. This movement, generated by the activities of the Arya Samaj in the region, acquired substantial support among both upper- and lower-caste Hindus in the villages. Eventually it had resulted in the institution of 'Gaurakshini Sabhas' in several places, specifically to protect cows. A leading part in the Gaurakshini agitation was taken by lower-status agricultural castes, Ahirs and Koeris, who joined hands with the zamindars to rally around the cause of the cow. Gyanendra Pandey has argued that the extended participation of these castes in the Gaurakshini sabhas represented an attempt to improve their ritual status and to use the cow protection movement as a vehicle of upward social mobility.[156] These castes were increasingly conscious of their low ritual status and had returned themselves in various census operations as higher castes. The growth of the Goala movement among the Ahirs of Shahabad and the formation of caste sabhas had served to indicate the nature and force of the caste movements in the nineteenth and early twentieth centuries.[157] Though the motive force of these caste organizations was a challenge to the upper caste landlords, the sabhas also had strong links with the cow protection movement which was seen as a means of raising their ritual status.[158]

The anti-Muslim nature of the Gaurakshini sabhas had become clear as early as the 1890s, when the sabhas in Azamgarh (United

[156] Gyan Pandey, 'Rallying around the Cow', p. 74.

[157] Ibid., p. 75. See also Kalyan Mukherjee, *Bhojpur Naxalism on the Plains of Bihar*, pp. 25-7.

[158] Hetakur Jha has shown that in the 1920s the Goala caste sabhas were organizing Goalas to challenge their upper caste landlords and to refuse to do begari in Hilsa, Paithana, Islampur and other places. See Jha, 'Lower-caste Peasants and Upper-caste Zamindars in Bihar 1921-25', *Indian Economic and Social History Review*, XIV, 4 (1977) pp. 550-4.

Provinces) began to demand bonds from Muslim villages to the effect that they would not slaughter cattle under any circumstances. Muslim weavers were threatened with boycott and no one would buy their cloth nor would any bania supply them with grain. Two years later, during the Baqri-Id of 1893, severe inter-religious tension resulted in the Bhojpuri region, and the call for revenge against those Muslims who had performed *qurbani* became widespread. The excesses of 1893 led to increased police vigilance and efforts to bring about agreement between local Hindu and Muslim leaders. This temporarily subdued the cow protection movement in the region. Nevertheless, a significant revival of the demand for cow protection in the eastern Bhojpuri districts in UP and Bihar took place in 1910, culminating in a massive outbreak of violence in Shahabad at the Baqri-Id of 1917.[159]

In Bihar, Shahabad had first felt the power of the Gaurakshini movement in its nascent form. In 1890-1 the important cattle fair at Behrampur was the scene of major demonstrations by crowds of Hindus and the confiscation of some heads of cattle from the Muslim butchers and contractors acting for the military. Later the movement had spread and, in 1911, crowds of Ahirs and Bhumihars had collected to prevent the sacrifice of cows at various places in Patna district, while, in 1917, there was an outbreak of violence in Shahabad on an unprecedented scale. The riots in Shahabad were triggered off when meetings to protect the cow were organized in Ibrahimpur in 1917 by barristers, pleaders, and other influential elites in the town. This then spread to other parts of the district and Muslim houses were threatened. In due course nearly all the Muslim houses in Piro, Ibrahimpur, Bhagalpur and other places were attacked and 124 villages were looted in Shahabad district alone. Again, Ahirs and Koeris were recorded as playing an important part in the riots. It seems clear that their motives in participating in the movement on a large scale were directed at improving their ritual status and at emphasizing the purity of their faith and the strictness of their ritual adherence to it by reference to the issue of cow slaughter.[160]

[159] Gyan Pandey, 'Rallying around the Cow', p. 79. See also the *Shahabad District Gazetteer*, pp. 44-5.

[160] Ibid. Involvement in the Shahabad riots was not confined to the Koeris and Kurmis alone. Pandey cites evidence of the participation of upper-caste zamindars in attacks on Muslim villages in Shahabad in 1917 and argues that here, the sense of Hindu community was determined by notions of caste and zamindari honour. See Gyan Pandey, *The Construction of Communalism in Colonial North India*, Delhi, 1990, p. 192.

The suspicions and fears that grew out of these sectarian agitations and clashes manifested itself in the distancing of Muslims from Hindus in the celebration of festivals in the localities by the 1920s. This was a reversal of earlier trends towards social inter-mixing. Muslims who had traditionally participated as bearers in Hindu processions in the Ramleela were replaced by Hindu troupes.[161] These troupes were organized and trained in physical culture centres and came to signify the vigour and strength of a new militant Hinduism. In essence they were designed to rescue Hinduism from an emasculated existence and to enable Hindus to become a 'living nation'. Both culture centres and troupes were given ceremonial sanctification and names associated with divine or saintly figures. The name in most common use was Mahabir Dal, which connected the physical culturists with the epic hero and deity, Ramchandra, and his monkey army led by Hanuman. The primary service of these troupes consisted in displays and shows of strength. Many of them toured cities and fairs; but it was the Hindu processions in religious festivals which provided the main opportunity for shows of force.

In the 1920s these troupes were increasingly involved in tensions over putative religious rights. Communal rioting was often sparked off by minor issues such as the playing of music before mosques and by official attempts to change the routes of processions. It is not my intention to argue here that the bitter communal riots of the 1930s and 1940s were an inevitable outcome of these complex developments. The communal question in that period was also affected by the political articulation of the Muslim League, the Mahasabha in the 1930s and the jockeying for power at the elite levels of the Muslim League and the Congress. The political articulation of the Muslim League and the Mahasabha, in particular, fed on the suspicions and fears first nurtured in these early agitations and clashes. With the Congress in power in 1937-9, the Muslim League was able to build up the horrors of a Hindu raj and a sudden increase in incidents of rural rioting took place.[162]

[161] G.R. Thursby, *Hindu Muslim Relations in British India*, Leiden, 1975, p. 91. Before the build-up of sectarian tensions, Hindus and Muslims especially in rural areas, participated in several local festivals together. One local festival in rural areas was the 'Jal Jatin' where Hindu and Muslim women danced together with pitchers on their heads.

[162] See V. Damodaran, 'Office Acceptance and Some Aspects of the Congress Ministry in Bihar 1937-39', (Unpublished M.Phil. dissertation, Jawaharlal Nehru University, 1985), pp. 263-4.

Such a perspective helps to give a better understanding of the nature of the sectarian grievances and the rioting of 1946. As in the earlier periods of sectarian tension, persons from a wide range of castes and classes were involved in the organization and spread of disorder. The initiative in these early stages of the disturbances tended to originate with the petty bourgeois elements in urban areas which generally organized the religious processions in the towns. The riots in Noakhali in East Bengal infuriated these petty bourgeois elements in the small towns and cities. Almost immediately some U.P. Hindus in the cities proposed raising a volunteer force of 50,000 men and the sending of a batch of 500 workers to Bengal to study conditions in Noakhali.[163] Communal reports of the killings further incited the public. Indeed very often the nationalist newspapers themselves gave gory reports of the killings and this encouraged reprisals. Thus, an editorial in the *Searchlight* newspaper suggested on 22 October, 'If there must be a civil war let there be one'. Certain groups in the Congress also planned 'hate days'. In this connection hartals and demonstrations took place in Arrah, Gaya and Darbhanga against the killings in Noakhali. In Patna 10,000 people participated in a procession and 40,000 gathered at a meeting in Bankipore. The newspapers reported, apparently approvingly, that this was a record gathering for a town of 225,000 unsurpassed since Nehru's visit in January. In Patna young men who brandished shoes and sticks shouted, 'Hang goonda Jinnah, we will avenge Noakhali blood for blood'. In certain towns, such as Chapra, the official office-bearers of the Muslim League and the Mahasabha assisted in deliberately fomenting the trouble. The escalation of tension in urban areas reached its climax on 25 October which the Congress government observed as 'Noakhali day'. As noted earlier, stabbings started in Patna on that very day whilst in other places, such as Chapra, Muslim-owned ration shops were looted. Soon after this the killings began in earnest.

The rioting spread rapidly in the countryside where in several villages Muslims were killed in an orgy of violence. What is remarkable is that communal violence erupted even in those areas where an agrarian movement was at its heignt, with both Hindus and Muslims arrayed against their common enemy, the zamindar. In Monghyr, for example, I have cited several pieces of evidence of this type.[164] In

163 *Hindustan Times*, 22 December 1946.

164 *People's Age*, 8 December 1946 and *Hindustan Times*, 3 November 1946.

most villages both upper and lower castes participated in violent excesses. The evidence clearly shows that the lower castes of Kurmis, Goalas and Dusadhs were active in the riots and the rioting was in no way confined to the upper Hindu castes like Bhumihars and Rajputs. In villages dominated by higher castes, lower-caste tenants joined hands with their upper-caste landlords in the attack of their Muslim neighbours, as in Kanchanpur village in Fatwah. Often, even after the rioting had subsided, the Hindu population continued to be hostile. As the zamindar of Pachrukhi noted, 'The Hindus in my village are looting the houses left behind by fleeing Muslims and are determined not to rehabilitate them.'[165] Clearly the contradictions involved in massacres taking place precisely in those places where inter-religious co-operation had developed as a response to exploitation cannot be accepted at face value.

What caused this general outburst of anger among the Hindu population in the rural areas against their Muslims neighbours in 1946? In fact, there is much evidence which indicates the complicity of zamindars in fomenting the riots in several villages. One newspaper noted that, in Gaya, the president of the Congress committee had informed the ministry that the riots were the creation of the zamindars. They quoted evidence to show that the zamindars held conferences and made preparations for the rioting. In Nawadah, the zamindar's conference began on 1 November and the riots broke out on 2 November. In Gaya the arms were supplied by local zamindars and a mahanth. Moreover, the police were actually forced to arrest one of the zamindars in Gaya, Sachoo Singh, on the charge of organizing the rioting. Another warrant of arrest was issued against the Mahanth of Loghar. The newspaper alleged, perceptively, that the zamindars were trying to defuse agrarian tension by lighting the communal torch.[166] This might well have been the case. In actively fomenting sectarian strife amongst the peasantry the zamindars hoped to turn attention away from themselves. Indeed, at a zamindar conference held at Patna under Maharaj Adhiraj of Darbhanga, some zamindars actually suggested, quite openly, that it would be 'good' to make kisans and *khet majdoor* (agricultural labourers) fight among themselves. Soon confirmation of the efficiency of this strategy was provided by the experience of the communists in Gaya. In that district the communist

[165] Agrarian disputes in Monghyr, BSA, File no. 6(1)/47.
[166] *People's Age*, 22 December 1946.

leader, Bhim Deo Singh, reported that he had had a difficult time trying to organize a band of 150 red guards into different squads and to keep the kisans' attention focused on the coming battle of the *Aghani* harvest, which the zamindars were conspiring to confiscate by deliberately setting Hindus against Muslims. In Shafipur Khusrupur area, where 60 Hindus (mostly Goalas) were accused of looting 36 Muslim houses and killing 4 women and 2 children, one of the rioters was armed with a gun belonging to the local zamindari.[167] By such methods the zamindars planned to disrupt the mounting wave of anti-landlord kisan upsurges and divert it into disruptive channels. The landlords were not the only elite group actively fomenting the agitation. Traders and merchants were also actively involved in financing some of the operations. In Chapra, it was proved that the key part in the riots was played by profiteers and black marketeers. In Masaurhi, in Patna, where thousands lost their lives, the local rice and flour mill proprietor organized the rioters. As indicated earlier the mill sirens were used to collect and disperse the rioters. Clearly, the manipulative possibilities of communal ideology were starkly realized in the attempts made by the zamindars to use it to defuse the agrarian agitation.

The spread of communal propaganda in both urban and rural areas was also aided by the organizational build-up of Hindu communal bodies like the Hindu Mahasabha and the RSS in Bihar. From 1942 onwards, with the Congress leadership in jail and the Muslim League extending its influence, the RSS could not have asked for a more suitable atmosphere in which to conduct their activities. In the prevailing atmosphere of uncertainty the membership of the RSS grew steadily and, as riots broke out in East Bengal, the party steadily strengthened its organization.[168] As a body able to offer some protection to the non-Muslim civilian population the RSS enjoyed considerable prestige. In Bihar the number of Swayam Sewaks in 1946 had increased to 4,000 active members. The RSS press, too, was a virulent and widespread source of propaganda. Banias and traders contributed heavily to its coffers.[169] At the same time, the Muslim League Nation-

[167] BSA, File no. 319/47.

[168] Home Political, 28/4/46, and 28/5/46. See also J.A. Curran, *Militant Hinduism in Indian Politics*, pp. 10-15.

[169] In the neighbouring United Provinces, Govind Sahai, the provincial RSS organizer for UP noted that since 1944 in his province, banias contributed to

al Guards were also building up an organization and soon their numbers had increased to 3,000. Along with the Khaksars the activities of such organizations contributed in a critical fashion to communal tensions.

The activities of the zamindars and the communal propaganda of communal groups do not, however, suffice entirely to explain the virulent spread of communal rioting in the countryside. At the level of peasant consciousness it is still difficult to fathom why sections of the Hindu peasantry, who had combined with Muslim peasants *vis-à-vis* the landlords, should suddenly have chosen to turn against them and use them as scapegoats for real or imagined grievances. While it can be asserted that the zamindar, the mahanth and the bania had fomented the agitation and in many cases encouraged the rioting, it would be incorrect to claim that the peasants in many villages who joined in the rioting were simply swept helplessly along in the tide. Instead it can be argued that different levels of peasant consciousness were at work and that the anti-landlord confrontation was not the only conflict perceived by the peasants. In the course of the Gaurakshini agitations suspicions and fears between the two communities had already developed. This sectarian conflict had remained alive in peasant memory. As I have already suggested, the tradition of sectarian conflict in south Bihar among the Ahirs, Koeris, and Kurmis was not new. It had originated in their active participation in rioting to improve their ritual status, especially in Shahabad in 1917. Communal propaganda at the elite level in the 1930s and 40s preyed on such fears and conflicts in local society and, at the level of popular consciousness, the contradictions in the localities remained many-sided.

The possibilities of loot provided by the extended periods of disorder also encouraged lower-caste peasants in some villages to join in rioting. Especially at a time when the peasantry was under severe economic pressure, loot constituted a major temptation. The zamindar of Pachrukhi for example, had recorded that the Hindus in his village were systematically looting Muslim houses and were determined not to rehabilitate them. In some other villages local grievances provided sufficient reasons for rioting, as Hindu low-caste peasants used the

---

the Sangh much more than they had contributed to the Congress in 30 years. Even a small town gave 4,000-5,000 Rupees. See J.A. Curran, *Militant Hinduism in Indian Politics*, p. 61.

opportunity to strike back at their Muslim neighbours. In Lalpur village in Barh, where the majority of the Muslims were zamindars and cultivators, while the Hindus were low-caste tenants, Muslims had not allowed Hindus to blow conch shells at the time of their religious festivals. Feelings were further strained when the Hindus performed a *Jagya.* In Telmar village, where Muslims were affluent and influential, a land dispute 4-5 years before had strained relations between Hindus and Muslims.[170]

Thus the scale and the nature of the communal rioting seem to have been affected by many factors. It originated in the spread of communal propaganda and the deliberate nurturing of 'hate' campaigns in the countryside. It was fuelled by the manipulative possibilities of a communal ideology, in so far as the zamindars were, in many cases, able to invoke sectarian hatred between Hindus and Muslims and foment communal rioting in order to defuse agrarian unrest. Unrest also fed on older traditions of sectarian conflict in many south Bihar districts. This was so in those areas where the cultivating castes of Ahirs, Koeris and Kurmis had attacked Muslims in the course of their Gaurakshini activities. These activities had built up an atmosphere of suspicion and fear which could be exploited subsequently when the opportunity offered itself. Furthermore, rioting provided an opportunity to indulge in the loot and plunder of Muslim houses. This was an attractive proposition for many poor low-caste Hindu families in villages pressed hard by the economic pressures of 1946.

Only an understanding of all these different causes can help to explain the simultaneous emergence of agrarian and communal conflict in rural Bihar in 1946. The picture is a complex one. By concentrating on elite manipulation alone and on the role of the zamindars in fomenting the riots little can be learnt about the participation of large sections of the lower caste peasantry in the riots and the different levels of popular consciousness. The identity of interests shared by sections of the Hindu and Muslim peasantry and their joint actions *vis-à-vis* their landlords did not prevent them from acting quite differently during a communal riot. At the level of popular consciousness the contradictions in the locality remained many-sided. It would be wrong to assume that class was the primary or only contradiction perceived at the popular level. Instead, the evidence from Bihar shows that most of our methodological assumptions about the dynamics of class and communal conflict need to be reassessed.

[170] BSA, File no. 319/47.

THE PARTY IN POWER: CONGRESS REACTIONS TO THE COMMUNAL UPHEAVALS

The reactions of the Congress ministry to the communal rioting and the killings of 1946 were markedly inefficient. The provincial ministry had, in fact, itself authorized the celebration on 25 October, (Noakhali day) specifically to mourn the death of Hindus in East Bengal, thus directly provoking the most serious communal troubles Bihar had ever witnessed. The Bihar police proved inadequate to deal with the scale of the rioting and in many cases seemed dispirited and incompetent. The Muslim League alleged that the police in many places either connived with the rioters or remained silent spectators of the tragic drama and offered no protection to the Muslims.[171] In particular, in the Beniabad massacre which had first sparked off large-scale rioting, the Sub-Divisional Officer and the Deputy Superintendent of Police had failed to take prompt action. In Chapra town, where it was quite clear that trouble was brewing on 25 October, the Sub-Inspector was accused of actually instigating the riots. Moreover, the Deputy Inspector-General visited Chapra but left without actually leaving any force behind at all. The commissioner was forced to request the 43rd Gurkha Regiment to take over the town. A company of the battalion therefore went to the aid of the police. By 31 October one of the two companies of the North Staffordshire regiment was directed to Bhagalpur. At this point S.K. Sinha, the Prime Minister, stepped in to say that he did not want British troops to be used.[172] When he finally agreed to the troops being deployed in view of the gravity of the situation, he further complicated the issue by putting forward a military plan of his own, involving splitting the military units into small bands. He had to be persuaded against this and was asked to impose a curfew and proclaim section 144 to ban assemblies of any sort. Sinha then prevaricated from taking any action of this kind until 3 November and it was only after Nehru arrived on the scene that a curfew and section 144 were imposed. The Muslim League alleged that they had met the Prime Minister on 31 October and asked him to call in the troops to check the rioters. However, the premier had declared to them that he would not permit firing on Indian rioters

[171] Allegations of the Muslim League, Rajendra Prasad Papers, File no. 6-B/1946.

[172] Ibid. See also F. Tuker, *While Memory Serves*, p. 184.

by British troops as a matter of principle and that the problem of mob violence would be solved not by firing, but by prayer. The League therefore claimed that the public and formal declaration of this anti-firing policy was directly responsible for the worsening of the situation all over Bihar after 31 October and the resultant massacre of thousands of Muslims.

The local officials and the local Congress committee also seemed incapable of dealing with the crisis. In Monghyr, where some of the worst riots occurred, the Congress committee exhibited a studied inefficiency in calming the people. Furthermore, the socialists who were conducting relief work often found themselves hampered by local officials acting in connivance with the local Congress leadership, who did not want the socialists to make political capital over the issue. In some villages in Monghyr, no one from the Congress visited the locality for 3-4 days after the riots. A Congressman, Dinanath Sahay, who visited the area later, was distressed to find dead bodies being devoured by vultures and he asked the district magistrate to have the bodies buried collectively.[173] The number of criminal cases filed after the riots was also pitifully few. In Andhari (Fatwah police district) where 400 Muslims had been killed, 29 people were accused and only one arrested. In Kanchanpur, where 160 Muslims had been killed, only 17 people were arrested. In Chiksauri village where 106 Muslims had been killed, 2 were arrested and 9 houses searched. No collective fine was imposed in any of the cases. It was in connection with this lack of action that Sultan Ahmed, a Congress Muslim of Bihar, wrote to Rajendra Prasad on 24 December 1946:

> S.K. Sinha has told us that 2,500 arrests have been made in all. This is most insufficient as there are villages from which no arrest has been made and in 9/10 of the cases those who have been arrested have been let off on bail. The house searches are also very few and stolen jewellery and grains are being transferred from place to place and being sold freely without fear of conviction.[174]

[173] File on the Bihar riots, Rajendra Prasad Papers, 6-B/46.

[174] Ibid. Hugh Dow, the governor of Bihar, noted in connection with the riots at north Bhagalpur in which many Muslims had died, that out of the 100 persons arrested, the magistrate acquitted 80. Fifteen more were acquitted by the sessions court. The high court freed four more and the ministry was likely to pardon the only one who failed to get through the net. See H. Dow collection, Mss Eur E 372.

The Congress government also imposed a strict censorship on the news of the tragedies, so that coverage of the Bihar massacres was blacked out by the press. Surprisingly, the Congress-dominated Associated Press and other news agencies fully complied with the Congress directive in this regard. All in all, the Congress ministry seems to have dealt with the riots completely inefficiently and without the seriousness that it warranted. It instituted no official enquiry, imposed no collective fines, and arrested very few. The police had also shown themselves to be hopelessly biased and had almost entirely lost the trust of the Muslims. This was the one main opportunity the Congress organization was to have to strongly attack Hindu communalism, to build itself an image as a secular party and to create a politically and communally unbiased police force. It failed on all counts, and lost what little support it had once had amongst the Muslims. There was a political corollary to this. Henceforth, until the late 1950s, Hindu communal parties like the Hindu Mahasabha and the RSS were able to ally themselves closely with the Congress and thereby increase their popular support in Bihar.[175]

THE CONGRESS INTERVENES: THE DEBATE ON ZAMINDARI ABOLITION

The main area in which the Congress had to intervene to quell popular unrest was that relating to land reform. In earlier sections of the study the extent and scale of the bakasht agitation have already been discussed. Popular hopes had risen with the forming of the Congress ministry. The Congress had promised many things in its election campaign, the most important being the commitment to the abolition of zamindari as soon as it was in office. Once in power, and with a large-scale peasant agitation on its hands, the Congress could not hope to prevaricate any longer. The socialists were also clamouring for some kind of radical change. Freshly emerged from the popular movement of 1942, and British jails thereafter, Congress leaders felt it necessary to make some attempt to redeem the pledges to the people. Since the early 1930s the Congress had made very considerable inroads into the countryside among the rich and middle peasantry. These were the groups enfranchised in 1937 and in 1946, and they

175 The disastrous consequences of this Congress policy are still keenly felt today, particularly by the Muslim population.

had voted the Congress into power. In the Kisan Sabha agitation of 1937-9 they had demanded significant rent reductions and the scaling-down of landlord demands. The Congress had managed to assuage protest among the richer occupancy raiyats who had benefited from the rent reductions decreed by the legislation of 1937-9.

The war years and the changing economic situation had intensified landlord pressures on all sections of the peasantry by 1946, and new legislation was now demanded to challenge the whole basis of zamindari. After the war Congress appeared willing to concede the popular demand to abolish zamindari. However, the party was clear from the very outset that it could legislate to check zamindari only if the effects were confined to the larger landlords. This was because the Congress had too much at stake among small zamindars to be able to challenge them. It should be remembered that petty zamindar rights at this time were very complex. Many rich peasants had a zamindari interest as well, and often leased out part of their lands to sharecroppers and bataidars. Above all, many of the Congress politicians themselves were also small zamindars. Most landholders could not, therefore, be touched and any zamindari legislation envisaged would have to implicitly steer clear of them. Once this was accepted, it became clear that the rights of thousands of sharecroppers, many of them bataidars or under-raiyats, would ultimately be ignored. Their rights, coupled with the rights of agricultural labourers, went largely by the board as the true constituency of the Congress revealed itself.

By late 1946 several provinces under Congress ministries throughout India were ostensibly examining the merits and demerits of the zamindari system with a view to abolishing it. To this end, Zamindari Abolition Committees were set up in most provinces in the country, under Congress and non-Congress governments. In Bihar, however, the government decided to take decisions on the basis of office memoranda, and on the basis of a report of the Land Revenue Commission for Bengal in 1940. This report had come to the conclusion that the Permanent Settlement and the zamindari system should be replaced by a raiyatwari system aimed at improving the economic condition of the cultivators, and that the government should be brought into a direct relationship with the actual cultivator by the acquisition of all superior interests in agricultural lands.[176] In Septem-

[176] Quoted in the Report of the United Provinces Zamindari Abolition Committee, vol. 1, p. 235.

ber 1946, therefore, the Bihar Legislative Assembly passed a resolution recommending the abolition of the zamindari system. Hugh Dow, the governor of Bihar, noted caustically, 'An ICS officer has been placed on special duty with the title Zamindari Abolition officer. Undoubtedly hopes are being raised among the tenants which the ministry encourages but has no intention of fulfilling.'[177]

Due to the widespread agrarian unrest in the province, the Congress ministry in Bihar had taken up the question of zamindari abolition soon after coming to office. Two measures were conceived with a view to establishing direct relations between the state and the tiller of the soil. These were the Bihar State Management of Estates Bill 1947 and the Bihar State Acquisition of Zamindaris Bill. On 8 April 1947, K.B. Sahay, the Minister of Revenue, wrote to the General Secretary of the AICC of his intentions to present the bills before the house by May. He also predicted that the negotiation of the legislation would be protracted and that it would be ten years before the zamindaris in the province were acquired and nearly twenty-five years before the income from the zamindaris would accrue to the state.[178]

The bills were finally presented before the Legislative Assembly in April and May 1947. In presenting them Sahay emphasized that the bills were formulated in accordance with the promises that the Congress had made:

> It is the policy of the government in the interests of the raiyats and the general material and social advancement of the province to remove intermediaries between the government and the raiyats . . . for this reason the government

[177] Dow to Wavell, 3 October 1946, Dow collection, Mss Eur E 372.

[178] He wrote, 'I am today sending you 15 copies each of the Zamindari Acquisition Bill and Private Estates Management Bill. We intend to introduce the Private Estates Management Bill by the 24th April in the legislative assembly and to move for its circulation to elicit public opinion, and to introduce the Zamindari Acquisition Bill in May. It is necessary for us to have a Private Estates Management Act because if we get power to manage the estates we shall secure the necessary facilities to administer them when we actually acquire them, under the Zamindari Acquisition Act. Besides we feel that even when the Zamindari Acquisition Bill is passed into law it will be 10 years before all the Zamindaris in the province are acquired . . . . Financially within 25 years all the income from the zamindaris which is appropriated by private landlords will come to the government treasury. True the government will not get separate road cess, agricultural income tax or land revenue from landlords but these will be covered by the gross revenue from zamindaris which will be about 12 crores.' See Rajendra Prasad Papers, Abolition of Zamindari, File no. 7-A/46-47.

has decided to do away with the existing outmoded agricultural framework of *zamindari* and replace it by ryotwari. With this object it is provided that on the publication of a notification by the provincial government the interests of property and tenure holders in respect not only of rents but also *hats*, bazars, forests, fisheries, mines and minerals shall rest with the government. The government will pay to the outgoing zamindar or tenure holder compensation, calculated on the net income of the property that is to say the income after deduction of certain specified items according to the slabs (sic) given in the bill. The compensation will be paid either in cash or in bonds as may be prescribed. Special arrangements will be made for compensation to public trusts to ensure that an annual income adequate for the purposes of the trust is provided in perpetuity. The homestead and other lands used for agricultural or horticultural purposes and in the *khas* possession of the proprietor or tenure holder at a time when the notification is issued will be deemed to be settled with a proprietor or tenure holder, on rates of rent fixed by the government. The machinery for the proposed state acquisition is provided in the Act.[179]

In accordance with these ideas the Zamindari Abolition Bill stipulated that, with qualifications, all zamindari interests were to rest with the crown. [See footnotes for details of clauses].[180]

[179] Ibid. See Legislative Assembly debates on the bill.

[180] '4(a) Subject to the provisions of sections 5 and 6, such estate or tenure including the interests of the proprietor or tenure holder in any building or part of a building comprised in such estate or tenure and used primarily as office or *kutcheri* for the collection of rent of such estate or tenure and his interests in forests, fisheries, hats, bazars, mines, and minerals shall with effect from the first day of the agricultural year, following the date of publication of such notification rest absolutely in the crown free from all encumbrances. Provided that nothing contained in this clause or in any other provision of this act shall in any way affect the rights of a holder of a lawful lease of mines and minerals in respect of any area comprised in such estate or tenure. 5(1) . . . Such proprietor or tenure holder shall be entitled to retain possession of the land comprised in such homesteads and to hold it as a tenant under the crown subject to the provisions of sub section (2) and (3).

(2) When such homesteads are the homesteads of a proprietor he shall hold them (a) if the estate of which the homesteads form part is a revenue free estate free of rent and (b) if the estates of which the homesteads form part is a revenue paying estate, subject to the payment of rent to the provincial government which shall bear the prescribed proportion to the revenue of the estate and shall be determined by the collector in the prescribed manner.

(3) Where such homesteads are the homesteads of a tenure holder he shall hold them (a) if the tenure of which the homesteads form a part is a tenure which

The bill also provided that adequate compensation would be provided to the landlords for their loss of lands. For the purpose of preparing the compensation assessment roll, the net income of the proprietor was computed by deducting his land revenue, agricultural income tax, chowkidari or municipality tax, and other costs of management of the estate. After the net income had been computed the compensation officer would determine the amount of compensation to be paid in respect of the acquisition of the interests of each proprietor or tenure holder in accordance with the following provisions. [See Table 15.]

TABLE 15

RATE OF COMPENSATION

	Amount of Net Income (Rupees)	*Rate of Compensation*
(A)	Where the net income so computed does not exceed 2,500.	Twelve times such net income.
(B)	Where the net income so computed exceeds 1,500 but does not exceed 30,000.	Seven times such net income.
(C)	Where the net income so computed exceeds 50,000.	Five times such net income.

SOURCE : Section 15, Zamindari Abolition Bill, 1947.

This scale, dealing with the cost of compensation, became an issue that was debated at some length. It was at first decided to have a fixed scale at twelve times net income but estimates showed that the cost of abolition of zamindari would then be too onerous. It was therefore adjusted to income and could attain twelve times in one case and five times in another. Such rates came as a heavy blow to the richer zamindars, who protested vociferously. In a letter to Rajendra Prasad, C.P.N. Singh, the leader of the landholders' party in the legislature,

has been recorded for the time being in force as a tenure which the tenure holder is entitled to hold free of rent or has been expressly recognized as such by a revenue authority empowered in this behalf by the provincial government, (b) if the tenure has not been so recorded or recognized subject to the payment of rent to the provincial government which shall bear the prescribed proportion to the rent of the tenure and shall be determined by the cultivator in the prescribed manner. See Government of Bihar, The Bihar State Acquisition of Zamindari's Bill, in Rajendra Prasad Papers, File no. 7-A/46-47.

pleaded for postponement of the bill, arguing that 'the local government has brought before the legislature all kinds of tenancy legislations, estate management legislation, and is bringing a bill for the abolition of zamindari on the 24th. May I therefore hope that through your kind intervention these measures would be kept in abeyance till we have represented matters before a parliamentary board'. They appealed to Mahatma Gandhi as well saying, 'before we invite your attention to the various legislative measures which have been enacted to cripple us and to many more measures . . . on the legislative anvil to bring about our complete annihilation as a class, we would venture to state that we have never lagged behind in discharging our obligation to the Congress by being members of this organisation and helping it morally and materially. We have aspired to this cherished goal in the same manner as other countrymen and there is no sacrifice which we are not prepared to make in the larger interests of the country'.[181]

The Congress high command quickly knuckled under to this implied threat from the landlords and Rajendra Prasad wrote a hasty letter to Sahay asking him to stay the bills until they had been discussed more thoroughly:

> Your whole scheme was incomplete because it was not enough to get rid of *zamindari* but what was needed was a comprehensive scheme which would include steps to be taken thereafter both for the purpose of managing the estates and what should be done to improve the position of the tenants. I have never been able to understand the justice or fairness of depriving a man of the management of his property. I can understand abolition of zamindari. The Congress has sanctioned that but I do not think there is any sanction in the Congress resolution for forcibly dispossessing people of their property before they have been compensated for it . . . To single out the *zamindars* for depriving them of the right of managing their estates would appear to me to be wholly unjustified and without precedent. This feeling is shared by Kriplani, Patel etc . . . The real objection of the members was that you are going to take such a big step which is going to affect the entire economic life of the province, and in a matter of such far reaching consequence it was only fair that you should have consulted the working committee or at least the parliamentary sub-committee.[182]

[181] Ibid. It continued, 'We have always believed that the government which will be inspired by your ideals will be a *Ram Raj* where everyone will get his just dues. We find, however, from the trend of events that justice, fairness and enquiry are not propositions to be applied to the landlord community.'

[182] In many ways this postponement was similar to the postponement of the

Sahay reacted to this reprimand cautiously arguing that in order to rehabilitate the position of the Congress it was necessary that the bill should be proceeded with.[183]

You wrote to me that I have submitted a number of bills without giving you and the AICC due notice. I feel that I am not guilty of the charge . . . since the abolition of zamindari will take some time before it is completed it is our intention to take over private estates under government management *suo moto*. This will prevent mismanagement (of the estates), kill kisan and socialist agitation and enable the government to make acquisition easy . . . I informed you even as far back as August on the lines on which I was moving. If you did not approve of my action you should have written to me accordingly . . . I have received no direction from the AICC asking the Bihar government not to introduce the bills relating to zamindari acquisition without their previous sanction. The only directive that we have received is contained in the election manifesto. I therefore thought that if the good name of the Congress is to be preserved and saved from the onslaughts of its opponents the bill relating to zamindari abolition should be placed before the legislature as soon as possible. The State Management of Estates Bill is a step towards acquisition and had the support of the entire cabinet.

To his credit and in the face of some opposition, Sahay proceeded with the bill. The bill formally became an act, known as the Bihar Abolition of Zamindari Act, in 1948. In fact, the debate and controversy following the bill only increased once it became an act. The zamindars soon began to manoeuvre in the courts to achieve an injunction restraining the state government from implementing the act, while they also lobbied in the state legislature to try to ensure the future repeal of the legislation.[184] The Congress high command was already overly sympathetic to their demands. Sahay's insistence did see the bill through the house, but the active zamindar lobby eventually ensured the repeal of the legislation as early as 1949. Other legislation on the abolition of zamindari was then also forestalled. In particular, the Bihar Land Reform Act of 1950 was challenged as contravening a provision of the new constitution of India that guaran-

---

Bihar Tenancy Amendment bill in 1937-9. File on the Bihar riots, Rajendra Prasad Papers, 6-B/46.

[183] Ibid.

[184] A.N. Das, *Agrarian Unrest and Socio-economic Change in Bihar 1900-1980*, p. 188.

teed all people 'equal protection under the law'. The Supreme Court finally upheld the constitutionality of the act only in 1952. Five years had then elapsed since legislation had first been introduced in Bihar to abolish zamindari.[185] The zamindars and other intermediaries had used this time not only to delay implementation of any legislation which might alter the state land system but also to minimize the eventual impact of such legislation by exploiting existing loopholes in the law, well in advance of any future legal restrictions.[186] Even after 1952, the land reform legislation was never effectively implemented and continued to reflect and be weakened by the many compromises that were necessary to assure its enactment. Indeed, the act was filled with escape clauses and loopholes sufficient to assure the nominal acquiescence of the intermediaries to its implementation.[187] Given these problems, the first phase of the abolition of intermediary interests in Bihar, which ended in 1952, affected only the 155 zamindars whose gross annual income from their estates had been in excess of Rs 50,000.[188]

The period following the zamindari abolition legislation saw one

[185] According to Sahajanand Saraswati, one excuse which the ministry, supported by a 'reactionary' bureaucracy, used in delaying the bill was their claim to examine the financial implications of the abolition schemes. This became the basis of long drawn-out secretarial notes, thereby gaining five valuable years for the zamindars.

[186] For further details, see Francine Frankel, 'Caste, Land and Dominance in Bihar' in F. Frankel and M.S. Rao (eds.), *Dominance and State Power in Modern India*, vol. 1, Delhi, 1989, pp. 90-9.

[187] T. Jannuzi noted, 'It is an act which supports minimum change in the agrarian structure of Bihar . . . a structure dominated both before and after the Act of 1950 by those landholding elites who have traditionally enjoyed superior rights in land. The Government of Bihar also lacked data essential to the Act's implementation. The latest records of rights in land were thirty years old and the most obsolete were over 50 years old. Revisional survey and settlement operations would take time, as would the development of administrative machinery. The problems of implementation of the Act of 1950 were made even more complex by the failure of the intermediaries to provide the government with copies of their personal estate records.' T. Jannuzi, *Agrarian Crisis in India: The case of Bihar*, see pp. 34-42.

[188] Ibid. The power of the landed classes to undercut land reforms was further demonstrated when they successfully delayed the passage of a land ceilings bill from 1955-62. They arranged fictitious transfers of nearly 500,000 acres of land; so that, in 1966, the Bihar government was able to designate less than 8,000 acres as surplus land under the terms of the act. See Frankel, 'Caste, Land and Dominance in Bihar', p. 91.

of the most repressive periods in the history of the Bihar government. An alleged 'communist plot' was revealed and any attempt by the lower orders of the tenantry, under-raiyats, sharecroppers or agricultural labourers to organize themselves was seen as a sign of 'red menace'. Central government sponsored raids on the offices of the CPI took place and a Public Safety Act was passed to help prevent all mass movements by the peasantry.[189] Quite apart from the effect of these repressive measures, the cosmetic agrarian reforms had no hope of transforming the land structure in Bihar, while the zamindari system remained virtually intact in practice, if not on paper. Even more importantly, sharecropping and forced labour continued to prevail in many villages in Bihar, making a complete mockery of any 'reform' legislation. Later, the Land Ceiling Act and other measures also failed to attack the concentration of land in the hands of the privileged rich. Though several apparently useful measures were subsequently enacted, they continued to exist only on paper, while little real change came about at the grassroots level.

This is illustrated quite incisively by a field study I carried out at Suryapur village, Bihar, in February 1988. It serves as a sad, although probably fitting, conclusive note to the present work.[190] Suryapur village in thana Bikramganj originally belonged to zamindar, Raman Sinha, as part of his large estate spread over most of Bikramganj. Raman Sinha was, by the 1930s, a cautious supporter of the Congress while his son and heir, Raj Sinha, was more actively involved with the party in the 1940s. When the Zamindari Abolition Act was passed in 1952, Raman Sinha's estate came under the new regulations. In spite of this, the zamindari managed to retain the three villages of Surajpura, Kharoj, and Barul as part of the zirat holdings.

The three villages today have a population of 5,000 adult voters. The 'ex-zamindar', Raj Sinha, continues to own most of the land in these villages, which he cultivates partly through sharecropping, but mainly by employing agricultural labourers who are not paid any wages, but are instead given meagre portions of land as their bari (homestead) lands for subsistence purposes. They are also forced to work on the landlord's land. By these standards nearly 80 per cent of the population in the village are landless, and drawn mainly from the

[189] See A.N. Das, *Agrarian Unrest in Bihar*, p. 202.

[190] The field work was undertaken over two periods in January 1987 and February 1988. In what follows, some of the names have been changed.

low Kharwar, Dusadh or Goala castes. One of my interviews with some landless labourers revealed the startling extent of their deprivation. Munirka Singh, of the Yadava or Goala caste, had 5 *khathas* of land as his bari land (approximately 1/5 of a bigha) and he worked on the landlord's land free for this privilege. He earned a frugal living with his two cows and a buffalo secured through a government programme. Madusudhan Singh, another Goala labourer, held 1.5 bighas of land as his bari land and as payment for cultivating the landlord's fields. He earned his living by hiring out his labour to other peasant cultivators, for which he was paid a meagre Rs 8 (on average) per day.

An incipient and grievous land hunger prevails among the landless labourers in the village, many of whom were deprived of their holdings and their rights to the land during the landlord's attacks on their tenant rights during the days of zamindari abolition. Their situation is not without hope, however. Only twenty kilometres from Suryapur an entire subdivision, Bhojpur, burns with a sustained agrarian agitation fuelled by the Naxalites. As a result, the petty Rajput zamindars throughout the district live in constant fear for their lives. Fortunately, the organization of landless labour in neighbouring areas, such as Sahar, has begun to result in somewhat increased wages and temporarily subdued the exploitation of the landlords.[191] It is true that this process of patchy improvement has been a painfully slow one. However, in these villages the upper-caste landlords are now facing the slowly rising anger and protests of the landless low castes. Moreover, the toll of violent injury and deaths is beginning to tell, in a vivid way, the full story and the likely eventual consequences of the failed agrarian reforms of the Congress governments in Bihar.

One needs, then, to consider the constitutive elements and motivations of the new national state in Bihar in 1946. The Congress had set out to be a popular government. However, beyond some cosmetic changes, the institutions of the national state diverged critically away from popular expectations. No dramatic reversal of the policies of the colonial state followed. The year 1947 did not mark a watershed in the history of India. The new Indian state instead continued to operate within a paradigm of authoritarian constraints and further con-

191 For an interesting study of agrarian unrest in Bhojpur over a long period, see Kalyan Singh Mukherjee and R. Yadav (eds.), *Bhojpur Naxalism on the Plains of Bihar*.

solidated systems of exploitation by its continued dependence on urban and rural elites. This was most effectively revealed during the agrarian agitation in 1946-7 and the debates over zamindari abolition. We have noted in this context how quickly the leading personalities of the Congress high command, (especially Patel, Prasad and Kriplani) acted to postpone the Zamindar Abolition Bill. The repressive forces of the state were also brought to bear to crush agrarian agitation in several districts in north and south Bihar. Many incidents of police firing on agrarian rioters were reported, while in other cases the full force of the repressive section 144 legislation was imposed with scant regard for tenant rights. Often, the police and the forces of law and order worked hand in glove with the landed elites, as the sad case of Tamarhi village in Bhabua subdivision showed.[192] Indeed the police force in rural Bihar continued to be the feared institution it had been under the colonial state, but was now trusted even less. Other local officials, characteristically Sub-Divisional Officers and district magistrates, continued, almost always, to take the side of the landlord against tenant rights on disputed lands.[193] After 1947 almost all the local institutions united to aid the control, domination and power of the landed elites, if anything even more than had been the case under colonial rule.

The Congress also failed to tackle institutional corruption. This had first been extensively revealed in the context of the 1946-7 food crisis and the failure of the notorious Paddy Levy Order. Very few of the agrarian elites, which included many Congressmen, had delivered their share of the levy and virtually no grain was collected in many districts. This proved to be an accurate guide to the treatment to be expected by the poor in subsequent years. Within the Congress party itself, malpractices simply mushroomed as the new breed of politicians developed their haste to support the landed elites. The Sathi estate scandal, in the course of which the new Prime Minister actually wrote off the peasant rights to the Sathi estate in Champaran, and his subsequent efforts to intervene in the normal procedure of the law by attempting to adjourn the Sathi case fourteen times, were prime examples of this new kind of chicanery, practised even at the very pinnacle of government. To sum up, the early years of the Congress

[192] Here the Sub-Inspector of police simply refused to record the case of a peasant shot dead by the goondas of a landlord.

[193] The report of the SDO in village Kokat in Gaya district [see above] is a case in point.

government in Bihar revealed a steady trend towards the development and consolidation of older systems of exploitation, bolstered after independence by the new national institutions in alliance with various other dominant classes.

In an earlier chapter I hinted at some of the reasons why the new right-wing leadership of the Congress in Bihar acted as it did. It is possible to argue that the Congress in the 1930s and 1940s was worried lest it alienate the small landlord elements from the Congress organization by following too extreme an agrarian policy. This may well have been so. The growing small zamindar support for the Congress may have made it a more conservative organization. The right-wing leadership consistently emphasized the need for a united anti-imperialist front and Rajendra Prasad, the guiding force of the Bihar Congress, strongly believed that, rather than quarrelling with capitalists and landowners, the Congress should try to offer them compromise solutions. Ultimately, however, the way in which the Congress leadership implemented even the moderate legislation it enacted showed a strong bias in favour of the landed elites, and a rank dishonesty with regard to the political objectives actually being pursued.

This attitude, which had manifested itself quite clearly during the first Congress ministry, was strengthened and found more supporters in the second Congress ministry. It is this extreme bias in favour of the landlord elements that one needs to understand. The explanation probably lay not only in the family and class origins, but also in the ideological predilections of the majority of its leadership. Many of the Congress leaders came, like Rajendra Prasad, from small landlord families with a major stake in the prevailing social order. However, unlike the socialist leadership, many of whom had somewhat similarly privileged class origins, (and Jayaprakash Narayan is a case in point), the right-wing leadership was, in reality, ideologically committed to preserving the status quo. This agenda lay behind Congress efforts to crush the relatively moderate threat posed by the Kisan Sabha in the 1930s and to purge the Congress organization of politically troublesome left-wing elements later in the 1940s.

The negative effects on the Congress organization of the change of role from a movement of opposition to a party in power had become apparent as early as the time of the 1946 elections. The Congress then moved quickly to check all potential opposition and curb dissent by striking out at the communists in the party and expelling them. Soon after this the socialists were also being systematically attacked at the level of the Congress committees, and they had left the Congress by 1948. This tightening of the Congress organization completely transformed the party. In the 1920s and 1930s, the party had prided itself on being a movement with several contending ideologies ranging right through from Left to Right. In Bihar especially, it now became a rump party dominated solely by the orthodox right wing in conjunction with the landed elites. This development was not without consequences. Within four months of the Congress coming to office, Rajendra Prasad noted the deterioration that had set in within the organization and the inefficiency that had developed within the government. He became particularly aware of the inefficiency and corruption within the party itself in the context of the 1946-7 food crisis which the Congress had signally failed to tackle. He wrote:

> People had high hopes that with the advent of the Congress ministry effective steps would be taken which would ease the situation . . . on the other hand . . . it appears that there has been considerable deterioration . . . there has been an appreciable and marked rise in the prices of food grains. In many districts and particularly in the villages rice is selling at any thing between Rs 30 and Rs 40. This is higher than what prevailed during the worst period of the Bengal famine in 1943 . . . In the muffussil areas the people are undergoing very much hardship if not actual starvation . . . Black marketing and corruption are rampant . . . the ministry should recognize that the policy it has inherited from the late government . . . has not proved successful and it should now think out a new line of action.

The disillusionment present in this commentary speaks for itself. Prasad even went on, without apparent awareness of the irony involved, to advise the ministry to solve these problems 'along Gandhian lines', to encourage the purchasing power of the people through village industries and a sustained government commitment to small village projects. The government in Bihar, however, failed to take Prasad's advice and made no attempts in this field or in any other to devise new lines of action or really to dismantle any of the old colonial institutions at all.

In the communal arena, too, the government failed to combat the threats posed by Hindu communal ideology to the institutions of the state. Above all, the institutions of the police emerged from the riots of 1946 having failed completely to regain the trust of the Muslims. Many of the force at the lower levels were rightfully charged with having sided with the attacking Hindu mobs. In avoiding an official enquiry and in its failure to attack Hindu communal elements, the new Congress government ultimately lost its one major opportunity to develop and strengthen notions of secularism. As a result, Hindu communal organizations continued to enjoy popular support in Bihar, in close alliance with the Congress.

Such government mistakes, or policies, in the communal field were to be as disastrous in the long term as the failure of agrarian reform for the future well-being of the state of Bihar.

Conclusion

This study began with an examination of the working of the first Congress ministry in 1937-9. The ministry phase was particularly interesting for, in the course of it, for the first time the Congress party changed from a being a movement of opposition to being a party in power. Mass aspirations fuelled by the Congress acceptance of office resulted in a large-scale bakasht agitation in several districts in rural Bihar. In the study I have specifically examined the nature of this movement and have highlighted the critical ecological context of the peasant agitation as well as the more straightforward political setting. The very considerable explanatory importance of ecological transitions and erosion of common property rights has, I argue, never been sufficiently understood nor alluded to in earlier literature. It is becoming increasingly clear, however, that environmental and common property resource factors fuelled local protest movements from which the Congress drew its strength. These movements aimed to force the Congress to abide by its promises to the people and to restore peasant rights eroded by landlord incursions and the resource crisis. The implications of this large-scale rural protest for Gandhian strategies of action and the overall nationalist programme are important. Once in power the party was forced to spell out its policies in the agrarian and labour fields and, in so doing, revealed a growing tendency to support the landed and the urban elites and maintain the institutions of the colonial state. This put severe strains on the Congress organization, threatening the party with the dangers of incorporation into the imperialist machinery. More specifically, I have examined the policies and strategies of what came to be known as the Left in the Congress (I have continued to use this term for lack of a better alternative) as it challenged the entrenched Congress leadership and carried out an intensive campaign to pressurize the Congress ministry to implement far-reaching agrarian legislation. Beyond this, I have examined the agrarian legislation of the first Congress ministry in detail (a task not undertaken by historians hitherto) to arrive at the nature and impact of agrarian legislation. One of my main findings here has consisted in exposing the remarkable extent to which the Congress in power became a conservative organization, adopting many of the policies of the colonial state to diffuse agrarian agitation.

The Second World War ended the Congress experiment in office and pushed it back on to a course of confrontation with the colonial state. The Congress organization, which in power had revealed severe strains, was now reinvigorated by a fresh burst of mass-based action culminating in the Quit India struggle of 1942. The attack on the colonial state was renewed and the party became a movement of opposition with widespread mass support once again. In this connection, I have concentrated on studying the impact of the Second World War on Indian nationalism, its effect on the tripartite struggle between the British, the Congress and the Muslim League and, more importantly, the nature of the mass-based actions during this period. While general studies of the constitutional problem between 1935 and 1942 have been undertaken by other scholars, notably by R. Moore in his study of *Churchill, Cripps and India*, I have specifically examined the impact of the constitutional impasse at the regional level and the crisis of the colonial state in 1942 in Bihar. The consequent erosion of colonial hegemony gave rise to rumours of the imminent collapse of state power and culminated in the Quit India agitation in 1942. The political attack on the colonial state and the mass actions associated with it did not end in 1942 but continued till 1945. Few historians recognize this fact, believing the Quit India agitation to have ended within a few months in 1942. Instead, as my evidence for Bihar shows, the agitation retreated underground, manifesting itself in terrorist attacks and an increase in crime and in dacoity from 1943-5. The underground terrorist organization in Bihar was actively assisted by villagers and my research on the organization of the Azad Dastas (an underground nationalist organization) and the dacoit gangs of Parasuram Singh has revealed the emergence of a complex network of social relationships between these gangs and the local people in the villages in south Bihar. This underground movement subsided only by 1945 when the Congress leadership came out of jail and made attempts to re-enter the political arena for a negotiated transfer of power.

In 1946 the second Congress ministry was formed in Bihar and other provinces as a stepping stone to the final transfer of power at the Centre in 1947. This time the situation was different from that which prevailed in 1937, for it was clear that there would be no turning back and that the phase of mass struggle was finally over. A new nationalist state was being formed in Bihar.

What were the constitutive elements of this nationalist state? The experience of the first Congress ministry had shown that the right wing in the Congress was unsympathetic to mass aspirations and had little hesitation in using the powers of its colonial predecessors to crush the agrarian agitation. The working of the second Congress ministry revealed that these attitudes had further hardened. The Congress had held out hopes of a popular government. This ostensibly implied a democratization of institutions, more representative government and fewer authoritarian constraints on the people. However, beyond some cosmetic changes, the institutions of the new nationalist state almost entirely betrayed popular expectations. The state continued to operate within the paradigm of authoritarian constraints and played an ever more crucial role in shoring up urban and rural elites. Instead of protecting the poorest sectors of the society, it reinforced their insecurity in the accelerated erosion of land rights, common rights and the rural ecological balance. This was most effectively revealed during the agrarian agitation in 1946-7 and during the legislative debates over zamindari abolition, when the Congress consistently failed to address the outstanding problems of the rural majority. Instead, the repressive forces of the state were also brought to bear time and again to crush agrarian agitation in districts throughout Bihar during 1946. Many incidents of police firing on agrarian rioters were reported, while in other cases the full apparatus of colonial 'emergency' legislation was imposed with impunity and with scant regard for tenant rights.

Often the institutions of the police and the forces of law and order worked exclusively on the side of the landed elites. The police in rural Bihar continued to be the feared and corrupt institution that it had been under the colonial state. Moreover, officials at many levels in local government continued to argue against tenant rights on disputed lands. The Congress also failed to tackle institutional corruption at a higher level.[1] This was particularly apparent in the context of the food

[1] These were trends that had manifested themselves under the first Congress ministry. In 1939, deeply disillusioned by the misuse of office and corruption in Congress ranks, Gandhi said, 'I would go to the length of giving the whole

crisis and the failure of the paddy levy order in 1946 imposed by the ministry to tackle the food problem. Very few of the agrarian elites, including many Congressmen, delivered their share of the levy and virtually no grain was collected in many districts. Within the Congress party itself, corrupt malpractices were revealed as the new breed of politicians developed an unhealthy readiness to support the landed elites. In fact, the early years of the nationalist state in Bihar had revealed an almost uninterrupted trend towards the development and consolidation of systems of exploitation by the new national institutions, in alliance with various dominant classes.

The party failed at another level as well. Congress policies and attitudes towards the minorities, and especially towards the Muslims, demonstrated an unhealthy association with Hindu communal forces. In consequence, Congress institutions failed to gain the confidence of Muslims of all sections of the community. Perhaps most serious of all, the police were seen to be clearly biased in favour of the majority community. Thus both during the period in office from 1937 to 1939 and during the critical year of 1946 a considerable number of District Congress Committees actively encouraged communal rioting. In this vital respect, the Congress in power revealed itself to be a God with feet of clay. While the economic betrayal of the poorest classes may have been serious enough, the failure to protect the religious minority constituted, in the long term, a far more serious failure on the part of Congress. Indeed, it was a failure which helped to set the scene for a sad history of communal friction in post-independence India.

It would be interesting to ask whether Bihar was a special case or whether the Congress record in office was similar in other provinces. A partial reading of the sources shows that the interests of the Bihar Congress leadership and those of the landed elites coalesced most effectively in Bihar. The relative backwardness of the province as compared to the western United Provinces or Punjab had constrained the emergence of a strong middle-level peasantry and bolstered the domination of the landed elites. Thus the specific configuration of forces in Bihar, of economic backwardness, landlord power and a conservative leadership served to structure exploitation. For the poor agricultural labourer in Bihar, his deprivation was institutionalized by law and made tolerable by tradition. However, the attitude of the

---

Congress organisation a decent burial, rather than put up with the corruption that is rampant.' See CWMG, vol. lxix, p. 210.

Congress leadership to popular agitation was not very different from that in other ministries, such as in U.P., Bombay, Madras or Orissa.[2] In Madras, the Congress Premier, C. Rajagopalachari, and the Home Minister of Bombay, K.M. Munshi, did not hesitate to use the repressive apparatus of their states to check popular agitation in 1937-9.[3] Indeed, in this period, all militant protests in Congress-ruled provinces, including struggles sanctioned by trade unions, were treated as 'law and order' problems. As such, the ministries made extensive recourse to section 144 of the Criminal Code against agitating workers and had no compunction in arresting peasant and trade union leaders. This was particularly the case in the United Provinces.[4] In a few other states, such as Kerala, the political course was distinctively different after 1934. Here the Congress leadership in the post civil-disobedience period had moved dramatically towards the Left while the socialist and communist forces rode the crest of a popular wave which aimed at ensuring a more equitable society. In contrast to the Left in other provinces, they were largely successful in pressurizing the Madras Congress ministry to accede to their demands for legislation on debt relief and tenurial problems.[5]

In Bihar, on the other hand, the reform efforts of the new state were diluted by the continued dependence of the Congress on the petty zamindars and rich peasants in the countryside and on small traders and industrial classes in the city. This was a result of the continued dominance of the right wing in the Bihar Provincial Congress Committee. To make a real dent in rural poverty the Congress party would have needed to launch a forthright attack on the inequities of rural land-holding distribution and to effectively implement minimum-wage legislation for the millions of often semi-bonded agricultural labourers. Instead, when the Congress came to power it introduced only very conservative legislation, which failed to have any radical

[2] Our understanding of the working of Congress ministries in other provinces is constrained by the lack of good secondary literature which critically examines the record of the Congress in office in provinces such as UP, Bombay, Madras or Orissa.

[3] Bipan Chandra, *India's Struggle for Independence*, p. 336.

[4] Ibid. In Kanpur, the Congress ministry ordered the arrest of many workers and trade union leaders taking part in the textile mill strikes in 1939.

[5] The Kerala Karshaka Sanghams under the CSP organized an intensive campaign, demanding the amendment of the Malabar Tenancy Act. Ibid, pp. 346-7.

impact in rural areas, in terms of ameliorating poverty. Future studies of the post-independence history of Bihar will, I fear, only tend to confirm the continuation of the kinds of negative trends which I have already partially uncovered in the province.

Glossary

Abpashi	Irrigation.
Abwab	Unauthorized cesses levied in addition to rent.
Aghani	Crop harvested in the month of *aghan*.
Ahar	A reservoir of water for agricultural purposes.
Alang	The main embankment of an ahar.
Amla	An employee of the zamindar for rent collection.
Basti	Village site.
Bakasht	Lands on which though tenant rights may accrue, landlords have direct control over cultivation.
Bandh	Embankments over a river bed that lead the water into subsidiary channels.
Bania	Merchant.
Bari	Cultivated land attached to homestead.
Batai	Produce-rent system under which the actual crop is divided between landlord and tenant.
Bataidar	Sharecropping tenant.
Beopari	Trader.
Bhadoi	Crops harvested in the month of *bhado* (rainy season).
Bhaoli	Produce-rent.
Bhit	Land cropped with any bhadoi or rabi crop.
Bhojpuri	A colloquial dialect spoken in the districts of Shahabad and Saran.
Bigha	An area of 20 kathas roughly equal to ¾ of an acre.
Chakran	Service tenures of all kinds.
Chowkidar	Village watchman.

Daffadar	Rural policeman.
Danabandhi	The system of produce under which the crop is appraised before harvest.
Fard Abpashi	Village irrigation record.
Fasli	The Hindi era prevalent in Bihar.
Gairmazura	Uncultivated land.
Gola	Granary.
Gomastha	Village officials who look after the interests of the zamindar.
Gunge	Market.
Guzastha	A tenancy on fixed rent or rate of rent found in northern Shahabad.
Hat	Village market place.
Hathiya	Rainfall in the beginning of October.
Istemrari	Permanent tenure.
Jungle	Forest.
Kamia	A bonded labourer.
Khanapuri	A preliminary record of rights.
Khalihan	Threshing floor.
Khas	Lands under direct management by the landlord.
Khaitan	A detailed record of rights for each holding.
Khatbey	Agricultural labourer.
Khudkasht	Owner cultivator.
Latha kunri	A lever with a bucket.
Lathi	Stick.
Lathials	Landlord's henchmen.
Mahajan	Moneylender.
Mahanth	Temple functionary.
Mahua	A fruit from which wine is distilled.
Malguzari	Rent.
Malik	The traditional proprietor of the village.
Mela	Fair.

Mohana	The mouth of a pain or a channel.
Mukarari	Tenure held in perpetuity at a fixed rent.
Nagdi	Cash-rent.
Nazrana	Fees; premium paid on settlement of land.
Nigar	Letting of surplus water when irrigation is over.
Paikar	Trader.
Parabandi	An allotment of water from a pain or reservoir to different villages by terms agreed upon or regulated by custom.
Patta	A kind of lease executed by the landlord in favour of the tenant.
Rabi	Crops harvested in spring.
Raiyat	Peasant.
Sairat	Miscellaneous exactions by zamindars.
Salami	Landlord's fees.
Sarkar	Government.
Satyagraha	Non-violent non-cooperation.
Tal	Long stretch of denuded plains.
Terai	Valley.
Tola	Subdivision of a village.
Zamindar	Proprietor.
Zirat	Proprietor's private land.

Bibliography

PRIMARY SOURCES

A. *Government Records*

Cabinet office papers, microfilm, Seeley library, Cambridge.

Proceedings of the Secretary of State for India in the Public and Judicial Department Series, India Office Library, London.

Proceedings of the Government of India in the Home department (Political Branch) Finance Department and Reforms office and Fortnightly Reports. National Archives of India, New Delhi.

Proceedings of the Government of Bihar and Orissa (later Government of Bihar) in the political department (special series) and in Bihar Pradesh History of the Freedom Movement Series, State Central Record Office, Patna.

Proceedings of the Criminal Investigation Department, Department of Intelligence, Patna.

B. *Official publications and published reports of organizations*

Bengal District Gazetteers, Patna, Calcutta, 1907.

Bengal Presidency, Report on the Administration of Bengal, 1901-1902, Calcutta, 1902.

Bengal Presidency, Report on the Administration of Bengal, 1908-1909, Calcutta, 1909.

Bihar Acts 1937, Government of Bihar, 1937.

Bihar Acts 1938, Government of Bihar, 1938.

Bihar Tenancy Act 1885, (as modified up to 1st August 1940), Government of Bihar 1941.

Bihar Administration Reports, 1931-1932, Patna.

Bihar and Orissa District Gazetteer, Monghyr, Patna, 1926.

Bihar and Orissa District Gazetteer, Shahabad, Patna, 1924.

Bihar District Gazetteer, Purnea, 1963.

Census of India 1911 for the province of Bihar and Orissa.

Census of India 1931 for the province of Bihar and Orissa.

Census of India 1941 for the province of Bihar.

Government of India Act 1935, Delhi, 1937.

Indian Annual Register, 1937-1939.

Indian National Congress Resolutions, 1934-1939.

Annual Report of the Agricultural Department for Bihar, 1937-40, Patna, 1938-43.

Report on the Administration of the Registration Department 1936-40, Patna, 1937-41.

Report on the Land Revenue Administration of the province of Bihar 1937-41, Patna, 1938-41.

Report on the Administration of the Police in the province of Bihar 1936-44, Patna, 1937-46.

Annual progress report on Forest Administration in the province of Bihar 1936-41, Patna, 1938-42.

Report of the Commission on Agricultural Wages in India, Delhi, 1952.

Report of the Committee appointed by the Council of the All-India Muslim League to inquire into Muslim grievances in Congress provinces, Lucknow, 1938.

Report of the Committee on Agrarian Conditions in Champaran, 2 vols., Patna 1911.

Report and Evidence of the Royal Commission on Agriculture in India, 1928.

Report of the Indian National Congress 1934-36.

Report and Memoranda of the Indian Franchise Committee, vols. 1-5, Calcutta, 1932.

Report of the Violence Enquiry Committee in Bihar, Patna, 1939.

Report of the Working of Municipalities in Bihar 1937-39, Patna, 1940.

Prantiya Kisan Sabha ki Report, November 1929-November 1935, Patna, 1935.

Report on food grain supply and statistical review of the relief operations in the distressed districts of Bihar and Bengal during the famine of 1874, Calcutta, 1876.

Report of the Commission appointed to enquire into the famine in Bengal and Orissa in 1866, 3 vols., Calcutta, 1867.

Minute by the Hon. Sir R. Temple, Lieutenant-Governor of Bengal, 31st October 1874.

Selection of papers relating to the famine of 1896-7 in Bengal, Calcutta, 1897.

Report of the Famine Enquiry Commission for Bengal, Delhi, 1945.

Report of the Agricultural Department, Bihar 1936/7 to 1939/40.

Report of the United Provinces Zamindari Abolition Committee, vols. I and II, Allahabad, 1948.

Final Report on the survey and settlement operations in the district of Gaya 1911-18, Patna 1919.

Final Report on the survey and settlement operations in the district of Patna, 1907-1912, Patna, 1914.

Final Report on the survey and settlement operations in the Kosi diara area in the district of Purnea 1923-1926, Patna, 1927.

Final Report on the survey and settlement operations in the district of Monghyr (North) 1905-1907, Patna, 1908.

Final Report on the survey and settlement operations in the district of Darbhanga 1896-1903, Patna, 1904.

Final Report on the survey and settlement operations in Saran district 1915-1921, Patna, 1923.

Final Report on the survey and settlement operations in the district of Shahabad 1907-1916, Patna, 1918.

Statement published by the government of India on the Congress party's responsibility for the disturbances in India, 1924-43, CMD 6430.

Triennial Review of Irrigation in India, 1918-1921, Calcutta, 1923.

C. Newspapers

Amrita Bazar Patrika, 1939-45.

Congress Socialist, 1936-9.

Indian Nation, 1937-9.

Janata, 1937-9.

Kisan Bulletin, 1937-9.

Nav Shakti, 1937-9.

People's Age, 1939-46.

Searchlight, 1936-46.

Statesman, 1939.

D. Papers and unpublished reports of Organizations

Papers of the All India Congress Committee, Nehru Memorial Museum and Library.

Report of the Kisan Enquiry Committee, 1936.

E. Private Papers

H. Dow MSS, India Office Library, London.

Flack MSS, Centre for South Asian Studies, Cambridge.

P.J. Grigg MSS, Churchill College Archives, Cambridge.

Stanton-Ife MSS, Centre for South Asian Studies, Cambridge.

H. Kemp MSS, Centre for South Asian Studies, Cambridge.

Linlithgow MSS, India Office Library, London.

H.B. Martin MSS, Centre for South Asian Studies, Cambridge.

G. Macdonald MSS, India Office Library, London.

W. Slim MSS, Churchill College Archives, Cambridge.

Zetland MSS, India Office Library, London.

Ramnandan Mishra MSS, Nehru Memorial Museum, New Delhi.

Jayaprakash Narayan MSS, Nehru Memorial Museum, New Delhi.

Rajendra Prasad MSS, Nehru Memorial Museum, New Delhi.

Sahajanand Saraswati MSS, Nehru Memorial Museum, New Delhi.

Indulal Yajnik MSS, Nehru Memorial Museum, New Delhi.

SECONDARY SOURCES

A. Unpublished Dissertations

V. Damodaran, 'Official Acceptance and Some Aspects of the Congress Ministry in Bihar 1937-39', M. Phil. thesis, J.N.U., 1985.

R. Grove, 'Conservation and Colonial Expansion: The Origins of Colonial Forest and Conservation Policy on St Helena, Mauritius and India, 1660-1860', Ph.D. thesis, Cambridge, 1988.

W. Hauser, 'The Bihar Provincial Kisan Sabha 1929-42: A study of an Indian Peasant Movement', Ph.D. thesis, Chicago, 1961.

I. Kamtekar, 'The End of the Colonial State in India 1942-47', Ph.D. thesis, Cambridge, 1988.

C.S. Mitra, 'Political Mobilisation and the National Movement in Eastern U.P. and Bihar 1937-42', D. Phil. thesis, Oxford, 1983.

G. Prakash, 'Production and Reproduction of Bondage: Kamias and Maliks in South Bihar, c. 1300 to 1930's', Ph.D. thesis, University of Pennsylvania, 1984.

R. Roy, 'Study of the Bihar Pradesh Congress Committee', Ph.D. thesis, Berkeley, 1966.

T.A. Rusch, 'The Role of the Congress Socialist Party in the Indian National Congress 1931-42', Ph.D. thesis, Chicago, 1955.

B. Journals

H. Alavi, 'The State in Post Colonial Societies, Pakistan and Bangladesh' in *New Left Review*, no. 74, July/August.

K.K. Banerjee, 'Grain Traders and the East India Company' in *Indian Economic and Social History Review*, 23, 4, 1986.

C.A. Bayly, 'Rallying around the Subaltern', *Journal of Peasant Studies*, vol. 16, no. 1, October 1988.

R. Brenner, 'Agrarian Class Structure and Economic Development in Pre-industrial Europe', *Past and Present*, 70, February 1976.

N. Charlesworth, 'The Middle Peasant Thesis and the Roots of Rural

Agitation in India, 1914-47', *Journal of Peasant Studies*, 7, 3, 1980.

A. Chakravarti, 'The Unfinished Struggle of the Santhal *Bataidars* in Purnea district 1938-1942', *Economic and Political Weekly*, vol. xxi, no. 43, 25 October 1986.

B.B. Chaudhuri, 'The Process of Depeasantisation in Bengal and Bihar 1885-1947', *Indian Historical Review*, vol. 11, no. 1, July 1975.

A.N. Das, 'Changel: Three Centuries of an Indian Village', *Journal of Peasant Studies*, vol. 15, no. 1, October 1987.

R. Guha and Madhav Gadgil, 'State Forestry and Social Conflict in British India', *Past and Present*, May, 1989.

Christopher Hill, 'Santhal *Bataidars* in Purnea, the Ecological Evolution of a Share-cropping system', *Economic and Political Weekly*, 22 August 1987.

Tajul Islam Hashmi, 'Communalisation of Class Struggle, the East Bengal Peasantry 1923-29', *Indian Economic and Social History Review*, 25, 2, 1988.

B. Jha, 'Birth and Growth of the Communists in Bihar', *Indian Left Review*, February 1976.

Hetakur Jha, 'Lower Caste Peasants and Upper Caste Zamindars in Bihar 1921-25', *Indian Economic and Social History Review*, XIV, 4, 1977.

S. Kulkarni, 'Forest Legislation and Tribals, Comments on Forest Policy Resolution', *Economic and Political Weekly*, vol. xxii, no. 50, 12 December 1987.

P. Mohapatra, 'Coolies and Colliers: A Study of the Agrarian Context of Labour Migration from Chotanagpur, 1880-1920', *Studies in History*, new series, vol. 1, no. 2, July-December 1985.

R. O'Hanlon, 'Recovering the Subject, Subaltern Studies and Histories of Resistance in South Asia', *Modern Asian Studies*, 22, 1, 1988.

V.K. Okhri, 'How Custom bound is the Indian Farmer', *Indian Economic and Social History Review*, 18, 1981.

G. Omvedt, 'Migration in Colonial India: The Articulation of

Feudalism and Capitalism by the Colonial State', *Journal of Peasant Studies*, vol. 7, no. 2, 1980.

J. Pouchepadass, 'Local Leaders and the Intelligentsia in the Champaran Struggle', *Contributions to Indian Sociology*, 8, 1974.

J.F. Richards, *et al.*, 'Changing Land Use in Bihar, Punjab and Haryana 1850-1970', *Modern Asian Studies*, 19, 3, 1985.

S. Sarkar, 'The Logic of Gandhian Nationalism: Civil Disobedience and the Gandhi-Irwin pact, 1930-1931', *Indian Historical Review*, vol. iii, no.1, July 1976.

M.H. Siddiqui, 'History and Popular Rebellion Mewat 1920-33', *Comparative Studies in History and Society*, vol. 28, no. 3, July 1986.

E.P. Thompson, 'Moral Economy of the English Crowd in the Eighteenth Century', *Past and Present*, no. 50, 1971.

A. Yang, 'A Conversation of Rumours, the Language of Popular Mentalities in Late Nineteenth-century Colonial India', *Journal of Social History*, no. 20, 1987.

C. Books

L. Amery, *The Empire at Bay : The Leo Amery Diaries*. London, 1988.

B. Anderson, *Imagined Communities : Reflections on the Origin and Spread of Nationalism*. London, 1983.

A. Ashraf, *The Muslim Elite*. New Delhi, 1982.

S. Amin, *Sugarcane and Sugar in Gorakhpur : An Inquiry into Peasant Production for Capitalist Enterprise in Colonial India*. Delhi, 1984.

D. Arnold, *Famine, Social Crisis and Historical Change*. London, 1989.

S. Bose, *Agrarian Bengal : Economy, Social Structure and Politics, 1919-47*. Cambridge, 1986.

P.R. Brass and M.F. Franda (eds.), *Radical Politics in South Asia*. Cambridge, Mass., 1973.

P. Brass (ed.), *Indian National Congress and Indian Society, 1885-1985, Ideology, Social Structure and Political Dominance*. Delhi, 1987.

C. Bridge, *Holding India to the Empire, the British Conservative Party and the 1935 Constitution.* Delhi, 1986.

F. Buchanan, *Eastern India.* London, 1838.

J. Brown, *Gandhi's Rise to Power in Indian Politics 1915-1922.* London, 1972.

P. Chatterjee, *The Land Question in Bengal.* Calcutta, 1984.

——, *Nationalist Thought and the Colonial World: A Derivative Discourse?* Delhi, 1986.

A.K. Chaudhuri, *Socialist Movement in India, the Congress Socialist Party 1934-47.* Calcutta, 1980.

B. Chandra, *Nationalism and Colonialism in Modern India.* Delhi, 1979.

B. Chandra, *et al.*, *India's Struggle for Independence 1857-1947.* Delhi, 1989.

J. Christian, *Bihar Proverbs.* London, 1891.

R. Coupland, *Indian Politics 1936-42.* India, 1943.

J.A. Curran, *Militant Hinduism in Indian Politics.* New York, 1951.

P. Custers, *Women in the Telengana Uprising, Rural Poor Women and Revolutionary Leadership 1946-47.* Calcutta, 1987.

A.N. Das, *Agrarian Unrest and Socio-economic Change in Bihar 1900-1980.* New Delhi, 1983.

—— (ed.), *Agrarian Movements in India.* London, 1982.

C. Dewy and Hopkins (eds.), *The Imperial Impact : Studies in the Economic History of Africa and India.* London, 1978.

A. Desai (ed.), *Peasant Struggles in India.* Bombay, 1979.

R.R. Diwakar, *Bihar through the Ages.* Calcutta, 1958.

R.P. Dutt, *India Today.* London, 1940.

K.K. Dutta, *Freedom Movement in Bihar.* 3 vols, Patna, 1957.

M.K. Durrani, *The Sixth Column.* London, 1955.

J.M. Everett, *Women and Social Change in India.* New Delhi, 1981.

F.R. Frankel and M.S.D. Rao, *Dominance and State Power in Modern India: Decline of a Social Order*, vol. I. Delhi, 1989.

R. Frykenberg (ed.), *Land Control and Social Structure in Indian History*. Madison, 1969.

M.K. Gandhi, *The Collected Works of Mahatma Gandhi*. Delhi, 1958-84.

——, *The Story of my Experiments with Truth*. London, 1982.

J. Gallagher, *et al.*, (eds.), *Locality, Province and Nation: Essays in Indian Politics 1870-1940*. London, 1973.

M. Gilbert (ed.), *Winston S. Churchill*, vols. 1-5. London, 1966.

S. Gopal (ed.), *Selected Works of Jawaharlal Nehru*. New Delhi.

G. Grierson, *Bihar Peasant Life*. Calcutta, 1885.

P. Greenough, *Prosperity and Misery in Modern Bengal: The Famine of 1943-44*. Oxford, 1982.

R. Guha, *Elementary Aspects of Peasant Insurgency in Colonial India*. Delhi, 1983.

—— (ed.), *Subaltern Studies: Writings on South Asian History and Society*, 5 vols. Delhi, 1982.

——, *The Unquiet Woods, Ecological Change and Peasant Resistance in the Himalaya*. Delhi, 1989.

A. Gramsci, *Selections from Prison Notebooks of Antonio Gramsci*. London, 1971.

A.K. Gupta (ed.), *Myth and Reality: The Struggle for Freedom in India, 1945-47*. Delhi, 1987.

R. Gupta, *The Bihar Peasantry and the Kisan Sabha*. Delhi, 1982.

M. Gwyer and A. Appadorai (eds.), *Speeches and Documents on the Indian Constitution 1921-47*. Oxford, 1957.

I. Habib, *The Agrarian System of Mughal India*. London, 1963.

M. Hauner, *India in Axis Strategy: Germany, Japan and Indian Nationalists in the Second World War*. Stuttgart, 1987.

M. Hasan, *Nationalism and Communal Politics in India*. Delhi, 1979.

D. Held (ed.), *States and Societies*. Oxford, 1983.

S. Henningham, *Peasant Movements in Colonial India, North Bihar 1917-42*. Canberra, 1984.

P. Hill, *Development Economics on Trial: The Anthropological Case for the Prosecution*. Cambridge, 1987.

R. Hilton, *Bond Men Made Free : Medieval Peasant Movements and the English Rising of 1381*. London, 1971.

Hindi sahitya ka itihas. Kashi, 1964.

E.J. Hobsbawm (ed.), *Bandits*. Harmondworth, 1985.

—— (ed.), *Peasants in History: Essays in Honour of Daniel Thorner*. Calcutta, 1980.

W.W. Hunter, *Annals of Rural Bengal*. New York, 1970.

F. Hutchins, *Spontaneous Revolution: The Quit India Movement*. Delhi, 1971.

A. Jalal, *The Sole Spokesman : Jinnah, the Muslim League and the Demand for Pakistan*. Cambridge, 1985.

S. Jha, *Political Elites in Bihar*. Delhi, 1972.

G.S Jones, *Languages of Class*. Cambridge, 1983.

M. Kishwar and R. Vanita (eds.), *In Search of Answers, Women's Voices*. London, 1984.

F. Krantz (ed.), *History from Below, Studies in Popular Protest and Popular Ideology in Honour of George Rude*. Montreal, 1985.

R. Kumar (ed.), *Essays in Gandhian Politics: The Rowlatt Satyagraha of 1919*. Oxford, 1971.

C. Levi-Strauss, *The Savage Mind*. London, 1966.

Lord Linlithgow, *Speeches and Statements 1936-43*.

D.A. Low (ed.), *Soundings in Modern South Asian History*. London, 1968.

——, *Lion Rampant: Essays in the Study of British Imperialism*. London, 1973.

—— (ed.), *The Congress and the Raj: Facets of the Indian Struggle, 1917-47*. London, 1977.

—— (ed.), *The Indian National Congress: Centenary Hindsights*. Delhi, 1988.

R.C. Majumdar (ed.), *Struggle for Freedom*, vol. xi, Bombay, 1969.

N. Mansergh, *et al.*, (ed.), *Transfer of Power 1942-47, Constitutional Relations between Britain and India*. London, 1971.

Y.B. Mathur, *Growth of Muslim Politics in India*. Delhi, 1979.

K. Mukherjee and R.S. Yadav (eds.), *Bhojpur Naxalism on the Plains of Bihar*. New Delhi, 1980.

B.R. Nanda and V.C. Joshi (eds.), *Studies in Modern Indian History*. Delhi, 1976.

J. Narayan, *Selected Writings of Jayaprakash Narayan*. Delhi, 1980.

S. Orwell and I. Angus (eds.), *The Collected Essays, Journalism, and Letters of George Orwell*, vol. 1. New York, 1968.

Gyanendra Pandey, *The Ascendancy of the Congress in Uttar Pradesh 1926-34, A Study in Imperfect Mobilisation*. Delhi, 1978.

——, *The Construction of Communalism in Colonial North India*. Delhi, 1990.

—— (ed.), *The Indian Nation in 1942*. Calcutta, 1989.

F. Piven and R. Cloward (eds.), *Poor Peoples' Movements: Why they Succeed, How they Fail*. New York, 1977.

K. Prasad, *The Economics of a Backward Region in a Backward Economy, a Case Study of Bihar in Relation to other States*, vols. 1 and 2. Calcutta, 1967.

R. Prasad, *Autobiography*. Bombay, 1957.

——, *Mahatma Gandhi in Bihar*.

K.N. Raj, *et al.*, (eds.), *Essays on the Commercialisation of Indian Agriculture*. Delhi, 1985.

T. Ranger and E.J. Hobsbawm (eds.), *The Invention of Tradition*. Cambridge, 1983.

M.A. Rasul, *A History of the All India Kisan Sabha*. Calcutta, 1974.

P. Robb (ed.), *Rural India: Land, Power and Society under British Rule*. London, 1983.

G. Rude, *The Crowd in History: A Study of Popular Disturbances in France and England, 1730-1848*. London, 1981.

S. Saraswati, *Mera Jeevan Sangharsh*. Patna, 1952.

——, *Rent Reduction in Bihar, How it Works*. Patna, 1939.

——, *Other Side of the Shield*. Bombay, 1938.

B.N. Sarkar, *Some Landmarks in Wartime Food Control in Eastern India*. Patna, 1944.

S. Sarkar, *'Popular Movements' and 'Middle Class' Leadership in Late Colonial India, Perspectives and Problems of a 'History from Below'*. Calcutta, 1983.

J.C. Scott, *The Moral Economy of the Peasant, Rebellion and Resistance in South East Asia*. Yale, 1979.

——, *Weapons of the Weak: Everyday Forms of Peasant Resistance*. Yale, 1985.

T. Shanin, *Peasants and Peasant Societies*. Penguin, 1971.

R. Shankardass, *The Congress Raj : Provincial Autonomy in Bombay*. Delhi, 1982.

M. Shepperdson and C. Simmons (eds.), *The Indian National Congress and the Political Economy of India 1885-1985*. Aldershot, 1988.

A.N. Sinha, *Mere Sansmaran*. Patna, 1942.

R. Sisson and S. Wolpert (eds.), *Congress and Indian Nationalism: The Pre-independence Phase*. London, 1988.

P. Sitarammayya, *History of the Indian National Congress, vol. 1, 1885-1935*. Madras, 1935.

E. Stokes, *The Peasant and the Raj : Studies in Agrarian Society and Peasant Rebellion in Colonial India*. Cambridge, 1978.

M. Taussig, *Devil and Commodity Fetishism in South America*. Chapel Hill, 1980.

M.L. Thomas and J. Grimmett, *Women in Protest 1800-1850*, London, 1982.

G.R. Thursby, *Hindu-Muslim Relations in British India*, Leiden, 1975.

B.R. Tomlinson, *The Indian National Congress and the Raj*. London, 1976.

Havildar Tripathi (ed.), *Ramanand Tiwari smriti granth*. Patna, 1985.

F. Tuker, *While Memory Serves*. London, 1950.

Lord Wavell, *The Viceroy's Journal*. Oxford, 1973.

E. Whitcombe, *Agrarian Conditions in Northern India. vol. 1. The United Provinces under British rule, 1860-1900*. Berkeley, 1972.

A.A. Yang (ed.), *Crime and Criminality in British India*. Tucson, 1985.

——, *Limited Raj: Agrarian Relations in Colonial India, Saran District 1793-1920*. Berkeley, 1989.

A.M. Zaidi and S.G. Zaidi (eds.), *The Encyclopaedia of the Indian National Congress*. New Delhi, 1981.

Oral Interviews

Martin Farrow, Cambridge, June 1988.

Stanton Ife, Cambridge, June 1986.

Maheswar Gope, Patna, March 1987.

Kuldip Singh Kula Rai, Patna, January 1988.

Munirka Singh, Suryapur village, Shahabad, January 1988.

Ramayan Singh, Patna, March 1987.

Uday Raj Sinha, Patna, January 1988.

Havildar Tripathi, Patna, March 1987.

Oral Transcripts

Interview with Ram Manohar Lohia, Centre for South Asian Studies, Cambridge.

Interview with Usha Mehta, Centre for South Asian Studies, Cambridge.

Interview with Sunil Mukherjee, courtesy Professor Bipan Chandra.

Interview with Jayaprakash Narayan, Centre for South Asian Studies, Cambridge.

Index